T3-BVC-575

CASED-HOLE LOG ANALYSIS AND RESERVOIR PERFORMANCE MONITORING

G.

CASED-HOLE LOG ANALYSIS AND RESERVOIR PERFORMANCE MONITORING

Richard M. Bateman

PRESIDENT, VIZILOG, INC.

International Human Resources Development Corporation

BOSTON

Library of Congress Cataloging in Publication Data
Bateman, Richard M., 1940–
Cased-hole log analysis and reservoir performance monitoring.

Bibliography: p.
Includes index.
1. Oil well logging. 2. Oil reservoir engineering.
I. Title
TN871.35.B25 1984 622'.3382 84–15777
ISBN 0–934634–92–0

Printed in the United States of America.

This book is dedicated to Rex Cantrell and John Aitken, who awakened my interest in the subject matter, and to the memory of Rex Curtis, who instructed me so wisely in its practice.

CONTENTS

Acknowledgments *xii*

1. Introduction *1*

OBJECTIVES OF PRODUCTION LOGGING

Reservoir Performance; Completion Problems; Production and Injection Profiles; Gauging Treatment Effectiveness

PETROLEUM RESERVOIRS

MODERN PRODUCTION LOGGING TOOLS

Tools for Formation Properties; Tools for Fluid Typing and Monitoring; Tools for Completion Inspection

APPENDIX: QUICK REFERENCE TO PRODUCTION LOGGING TOOLS

BIBLIOGRAPHY

2. Cased-Hole Logging Environment *25*

PLANNING A PRODUCTION LOGGING JOB

PRESSURE CONTROL EQUIPMENT

THE BOREHOLE ENVIRONMENT

CHOOSING PRODUCTION LOGS

ANSWERS TO TEXT QUESTIONS

3. Reservoir Fluid Properties *31*

PVT REFRESHER COURSE

Single-Component Hydrocarbon System; Multicomponent Hydrocarbon System; Oil Reservoirs; Condensate Reservoirs; Dry-Gas Reservoir; Composition of Natural Oils and Gases

FLUID PROPERTIES

Water; Gas; Oil; Practical Applications

APPENDIX: STANDARD PRESSURES AND TEMPERATURES

BIBLIOGRAPHY

ANSWERS TO TEXT QUESTIONS

4. Flow Regimes *65*

LAMINAR AND TURBULENT FLOW

UNIT CONVERSIONS

Velocity; Flow Rate

FLOW REGIMES
HOLDUP
BIBLIOGRAPHY
ANSWERS TO TEXT QUESTIONS

5. Flowmeters *75*

APPLICATIONS
PACKER AND BASKET FLOWMETERS
CONTINUOUS FLOWMETERS
FULL-BORE FLOWMETERS
COMBINATION TOOLS
INTERPRETATION OF FLOWMETER SURVEYS
RADIOACTIVE TRACERS AND THERMOMETERS
MEASUREMENTS IN DEVIATED HOLES
APPENDIX: AVAILABLE PROFILING TOOLS
BIBLIOGRAPHY
ANSWERS TO TEXT QUESTION

6. Radioactive Tracer Logs *91*

APPLICATIONS
WELL TREATMENT
TRACER EJECTOR TOOL
Velocity Shot; Timed-Run Analysis
CHOICE OF RADIOACTIVE TRACER MATERIALS
MONITORING NATURAL RADIOACTIVE DEPOSITS
BIBLIOGRAPHY
ANSWERS TO TEXT QUESTIONS

7. Fluid Identification *99*

TOOLS AVAILABLE
GRADIOMANOMETER
FLUID DENSITY TOOL (GAMMA RAY ABSORPTION)
CAPACITANCE (DIELECTRIC) TOOLS
RESONATOR (VIBRATOR)
FLUID SAMPLER
MANOMETER
OTHER MEASUREMENTS
BIBLIOGRAPHY
ANSWERS TO TEST QUESTION

8. Temperature Logging *113*

FUNDAMENTALS
CEMENT-TOP EVALUATION

LOST-CIRCULATION ZONES

TEMPERATURE PROFILES IN PRODUCTION AND INJECTION WELLS

General; Liquid Production; Gas Production; Water Injection; Gas Injection

LOGGING TECHNIQUES

Shut-in Temperature Surveys; Differential-Temperature Surveys; Radial Differential-Temperature Tool

BIBLIOGRAPHY

ANSWER TO TEXT QUESTION

9. Noise Logging *129*

TOOLS AVAILABLE

OPERATING PRINCIPLE

INTERPRETATION

BIBLIOGRAPHY

ANSWERS TO TEXT QUESTION

10. Interpretation *137*

HOLDUP EQUATIONS

PRACTICAL APPLICATIONS

FLOWMETER AND GRADIOMANOMETER COMBINATIONS

BIBLIOGRAPHY

ANSWERS TO TEXT QUESTIONS

11. The Gamma Ray Log *147*

ORIGIN OF NATURAL GAMMA RAYS

ABUNDANCE OF NATURALLY OCCURRING RADIOACTIVE MINERALS

OPERATING PRINCIPLE OF GAMMA RAY TOOLS

CALIBRATION OF GAMMA RAY DETECTORS AND LOGS

TIME CONSTANTS

PERTURBING EFFECTS ON GAMMA RAY LOGS

ESTIMATING SHALE CONTENT FROM GAMMA RAY LOGS

GAMMA RAY SPECTROSCOPY

INTERPRETATION OF NATURAL GAMMA RAY SPECTRA LOGS

APPENDIX: RADIOACTIVE ELEMENTS, MINERALS, AND ROCKS

BIBLIOGRAPHY

ANSWERS TO TEXT QUESTIONS

12. Pulsed Neutron Logging *173*

TOOLS AVAILABLE

PRINCIPLE OF MEASUREMENT

TDT-K; DNLL; TDT-M; TMD

LOG PRESENTATIONS

Sigma Curve; Tau Curve; Ratio Curve; Near and Far Count-Rate Display; Background and Quality Curves; Summary

CAPTURE CROSS SECTIONS

BASIC INTERPRETATION

Clean Formations; Shaly Formations; Finding Interpretation Parameters; Sigma-Ratio Crossplot

PRACTICAL LOG ANALYSIS

RESERVOIR MONITORING—TIME-LAPSE TECHNIQUE

LOG-INJECT-LOG

DEPARTURE CURVES

DEPTH OF INVESTIGATION

APPENDIX: INTERPRETATION OF PULSED NEUTRON LOGS USING THE DUAL-WATER METHOD

Finding Parameters; Finding ϕ_T *and* V_{dc}*; Solving for* ϕ_e *and* S_{we}*; Practical Applications*

BIBLIOGRAPHY

ANSWERS TO TEXT QUESTIONS

13. Cement Bond Logging *223*

PRINCIPLES OF OILWELL CEMENTING

PRINCIPLES OF CEMENT BOND LOGGING

TOOLS AVAILABLE

OPERATING PRINCIPLES

Amplitude Measurement; Travel-Time Measurement; Wave-Train Display; Δ_t *Stretching; Cycle Skipping; Gating Systems; Deviated Holes and Eccentered Tools*

INTERPRETATION

Cement Compressive Strength; Partial Cementation; Wave-Train Signatures; Free Pipe; Free Pipe in Deviated Hole; Well-Cemented Pipe; Microannulus/Channeling

LEG QUALITY CONTROL

CEMENT EVALUATION TOOL (CET)

BIBLIOGRAPHY

ANSWERS TO TEXT QUESTIONS

14. Casing Inspection *259*

CALIPER LOGS

Tubing Profiles; Casing Profiles

ELECTRICAL-POTENTIAL LOGS

ELECTROMAGNETIC DEVICES

Electromagnetic Thickness Tool (ETT); Pipe Analysis Log (PAL)

BOREHOLE TELEVIEWERS

BIBLIOGRAPHY

Appendixes: Production Logging Charts and Tables

A. CONVERSION FACTORS BETWEEN METRIC, API, AND U.S. MEASURES *284*

B. AVERAGE FLUID VELOCITY VS. TUBING SIZE *288*

C. AVERAGE FLUID VELOCITY VS. CASING SIZE *290*

D. AVERAGE FLUID VELOCITY *293*

E. QUICK GUIDE TO BIPHASIC FLOW INTERPRETATION *296*

F. HOLDUP AND FLOW RATE CHARTS *297*

BIBLIOGRAPHY FOR APPENDIXES *311*

General Bibliography *313*

Index *315*

ACKNOWLEDGMENTS

This book is largely based on the courses I have taught during the past ten years to engineers and geologists whose most urgent need was for a basic understanding of logging tools and an effective, practical method for analyzing their measurements. In collecting materials for such courses, I have received invaluable help from many individuals, service companies, oil companies, and authors who have published materials in technical journals such as the *Log Analyst,* the *Journal of Petroleum Technology,* and others. In many cases it has been difficult, if not impossible, to trace the source of the materials used in this book. The logging fraternity is a closely knit one and, in the interests of getting the job done, tends to share publication materials rather freely. Wherever possible I have given credit to my sources, and where not possible I thank the many dedicated friends who have provided me with figures, charts, and log examples. I also owe a debt of gratitude to the hundreds of students who have given me valuable feedback on both the content and style of my presentations.

Richard M. Bateman

HOUSTON
JUNE 1984

CHAPTER ONE

INTRODUCTION

OBJECTIVES OF PRODUCTION LOGGING

The objectives of production logging can be categorized as follows:

1. monitoring of reservoir performance
2. diagnosis of completion problems
3. delineating production and injection profiles
4. gauging treatment effectiveness

Reservoir Performance

Monitoring reservoir performance is an important aspect of production logging. It allows an operator to establish the overall behavior of a reservoir and hence make intelligent decisions regarding field-wide production or injection strategies. In particular, the information sought includes details concerning:

Water breakthrough
Water coning
Gas breakthrough
Abnormal formation pressures
Thief zones
Pressure maintenance

In this book, methods of making measurements in a completed well to gather the needed data will be discussed.

Completion Problems

Completion problems may arise due to the mechanical state of the completion string. Relatively simple measurements can be made to pinpoint such problems as:

Casing leaks
Packer leaks
Bad cement jobs
Plugged perforations
Corrosion

Many different tools and techniques may be used in the search for mechanical problems and each will be discussed later in the text.

Production and Injection Profiles

An important diagnostic tool for both producing and injecting wells is the ability to derive flow profiles from production logging devices.

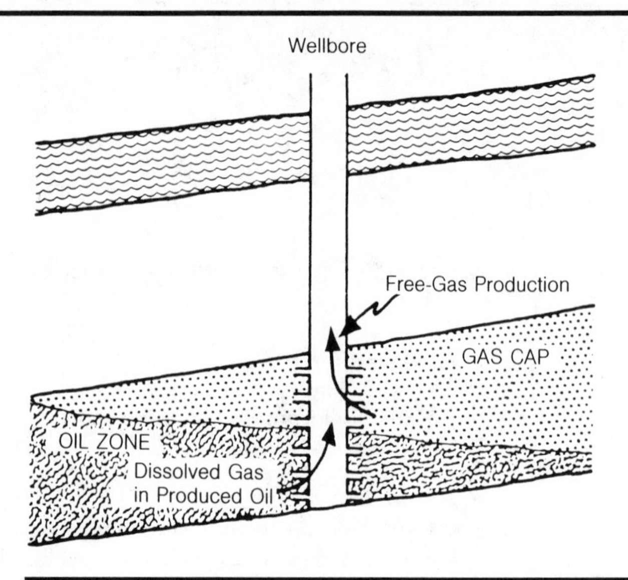

FIGURE 1.1 *Gas Produced with Oil from Associated Gas Cap and Solution in the Oil. Reprinted by permission of the* **Petroleum Engineer International** *from Cesak and Schultz 1956.*

These devices seek to answer the questions, (1) How much of what comes from where? and (2) How much goes where? In most wells, a production or injection profile will be a surprise to the operator. In few cases does the flow come from, or go to, the expected reservoir unit.

Gauging Treatment Effectiveness

Finally, productive logging can be used to establish the effectiveness of well treatments such as frac jobs and/or secondary and tertiary floods.

PETROLEUM RESERVOIRS

In order to place these production/injection problems in perspective and illustrate the conditions that may arise during the life of a well or field, it is worthwhile to recall the four principal types of petroleum reservoirs:

1. solution-gas drive
2. gas-cap expansion
3. water drive
4. gravity drainage (segregation)

Gas problems may be expected with solution-gas drive and gas-cap expansion reservoirs as illustrated in figures 1.1 and 1.2. Gas problems may be caused by:

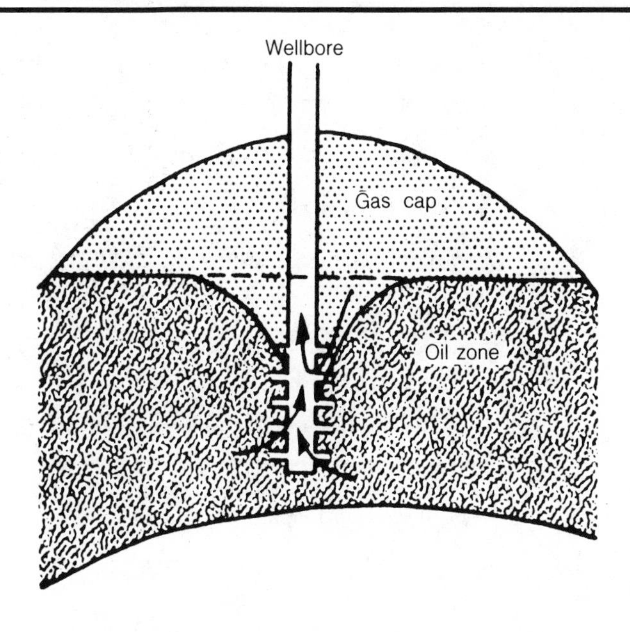

FIGURE 1.2 *Coning of Free Gas into a Well Across Bedding Planes. Reprinted by permission of the* **Petroleum Engineer International** *from Cesak and Schultz 1956.*

Gas breakthrough
Inadequate pressure maintenance
Production rate too high
Formation pressure below bubble point
Incorrect completion

Water problems may be expected with water drive and gravity drainage reservoirs—as illustrated in figures 1.3, 1.4, and 1.5—and can be traced to the following causes:

Coning of water
Permeability problems
Completion problems

MODERN PRODUCTION LOGGING TOOLS

Modern production logging tools may be categorized in three general groups as making measurements of:

Formation properties through casing and/or tubing;
Fluid type, flow rate, and movement within the casing/tubing vicinity; and
The status of the completion string.

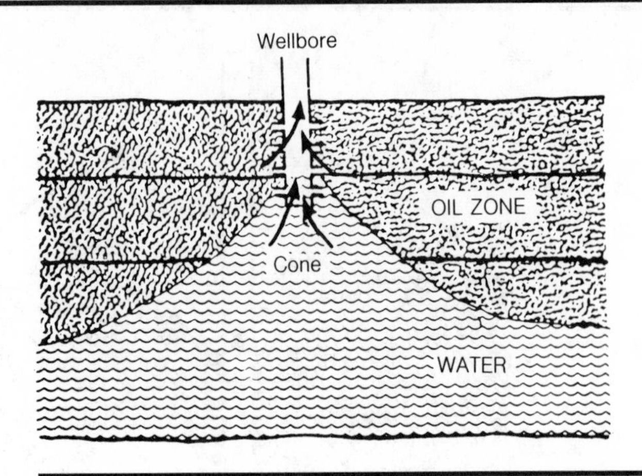

FIGURE 1.3 *Water Coning Across Bedding Planes. Reprinted by permission of the* **Petroleum Engineer International** *from Cesak and Schultz 1956.*

Tools for Formation Properties

Tools that measure formation properties from within a completed well include:

pulsed neutron logs	(TDT, NLL, TMD)
gamma ray logs	(GR)
natural gamma spectra logs	(NGT)
inelastic gamma logs	(IGT)
carbon/oxygen logs	(C/O, GST)

The first three will be dealt with extensively in this book. All have a common characteristic; they depend on interactions of nuclear particles (neutrons) and/or radiation (gamma rays) that have the ability to pass through steel casing.

Tools for Fluid Typing and Monitoring

Tools that distinguish oil from gas and water and monitor their flow rates include:

flowmeters	packer, basket, and continuous
temperature	absolute, differential, and radial
fluid density	gradiomanometer and gamma-gamma
radioactive tracers	
noise logs	

Each of these devices will be discussed and their uses illustrated with field examples.

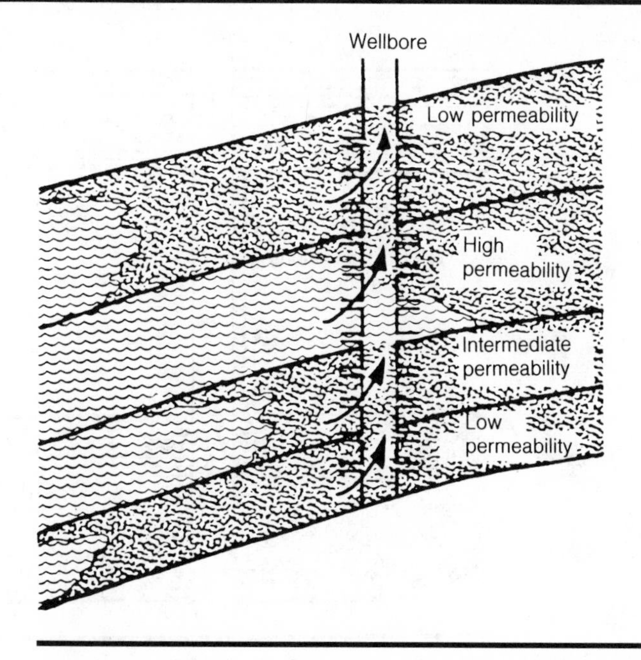

FIGURE 1.4 *Irregular Water Encroachment and Early Breakthrough in High-Permeability Layers. Reprinted by permission of the* **Petroleum Engineer International** *from Cesak and Schultz 1956.*

Tools for Completion Inspection

Tools for monitoring the mechanical status of the completion string include:

cement bond logs	(CBL, CET)
casing-collar logs	(CCL)
casing inspection logs	(ETT, PAL, calipers)
casing potential logs	

In addition to these direct measurements of the status of the completion string, all the fluid typing and monitoring tools may be used to infer completion problems. For example, a temperature log may indicate a tubing leak.

APPENDIX: QUICK REFERENCE TO PRODUCTION LOGGING TOOLS

Table 1A.1 summarizes the majority of the common production logging tools and their spheres of measurement. Examples of each type of log are shown in figures 1A.1 through 1A.16.

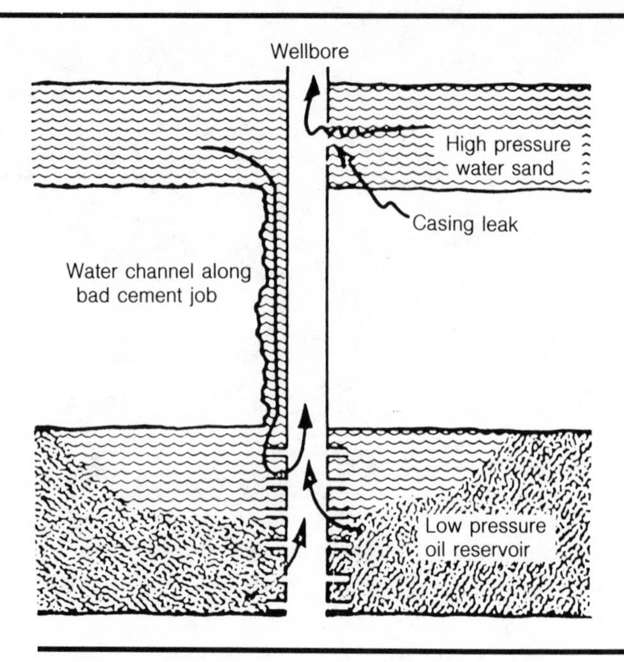

FIGURE 1.5 *Water Production Through Casing Leak and Channel Leak. Reprinted by permission of the* Petroleum Engineer International *from Cesak and Schultz 1956.*

TABLE 1A.1. *Common PL Devices*

	Sphere of Measurement		
Tool Name	Formation Properties	Fluid Type or Flow	Status of Completion String
Pulsed neutron	X		
Gamma ray	X		
GR spectra	X		
Inelastic gamma	X		
Carbon/oxygen	X		
Flowmeters		X	X
Temperature		X	X
Fluid density		X	
Gadiomanometer		X	
Radioactive tracers		X	X
Noise logs		X	X
Cement bond log			X
Casing-collar locator			X
Electromagnetic thickness tool			X
Pipe-analysis log			X
Calipers			X

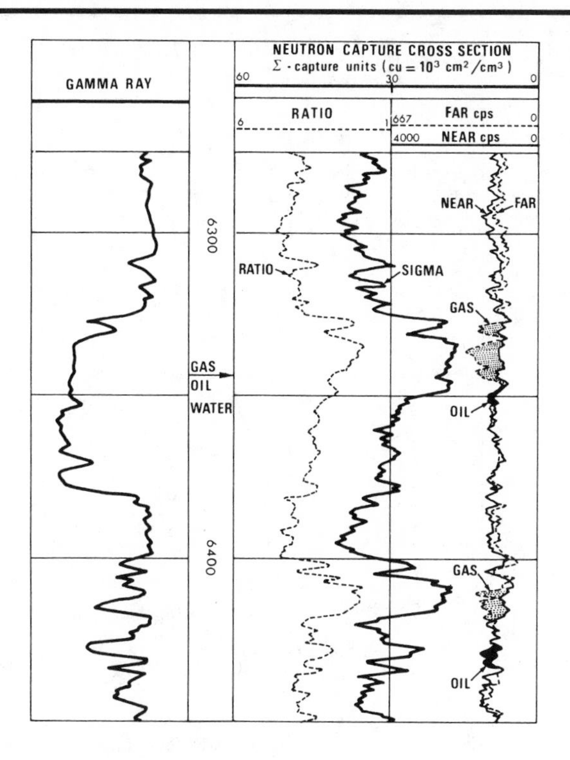

Abbreviation:	TDT, NLL, TMD
Use:	Water saturation through tubing/casing
Rating:	300°F
	17,000 psi
Size:	1¹¹⁄₁₆-in. OD
Limitations:	Needs salty formation water
Comments:	Measures formation capture cross section, Σ, and porosity-related curve (ratio)

FIGURE 1A.1 *Pulsed Neutron Log. Courtesy Schlumberger Well Services.*

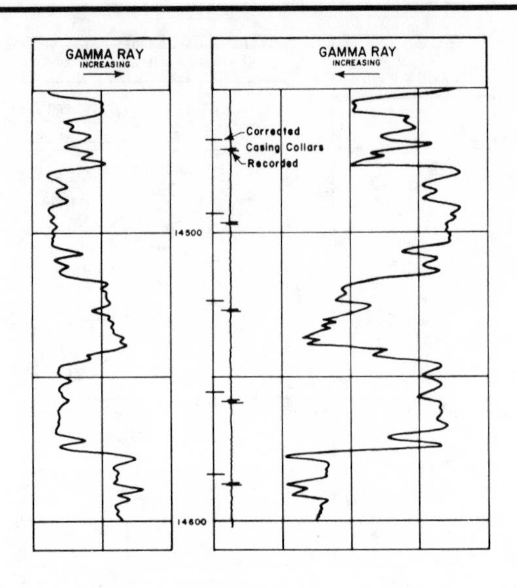

Abbreviation: GR
Use: Correlation
Shale content
Radioactive tracers
Rating: 350°F
20,000 psi
Size: 1 11/16-in. OD and others (see tabulation)
Limitations: None
Comments: Loses resolution if run too fast or in multiple tubing strings
Measures total natural gamma ray activity
Usually run with casing collar locator in cased hole

OD (inches)	Max. Press. (psi)	Max. Temp. (°F)	Min. Pipe Size
1 11/16	20,000	350	2-in. tubing
2	20,000	300	2½-in. tubing
2⅝	25,000	500	3½-in. ID pipe
3⅜	20,000	400	4½-in. casing

FIGURE 1A.2 *Gamma Ray Log. Courtesy Schlumberger Well Services.*

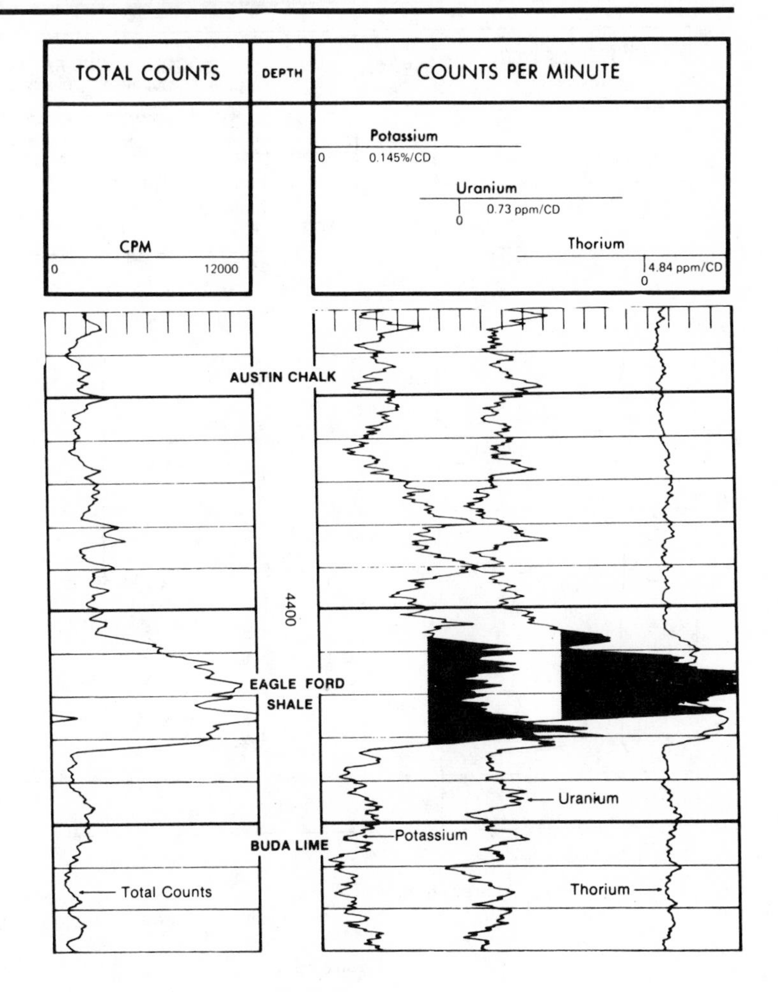

Abbreviation:	NGT, Spectralog, CSNG
Use:	Mineral identification Fracture finding Clay typing
Rating:	350°F 20,000 psi
Size:	3⅝-in. OD
Limitations:	Needs to be run slowly
Comments:	Measures relative concentrations of K, U, and T

FIGURE 1A.3 *Spectral Gamma Ray Log. Courtesy Dresser Atlas.*

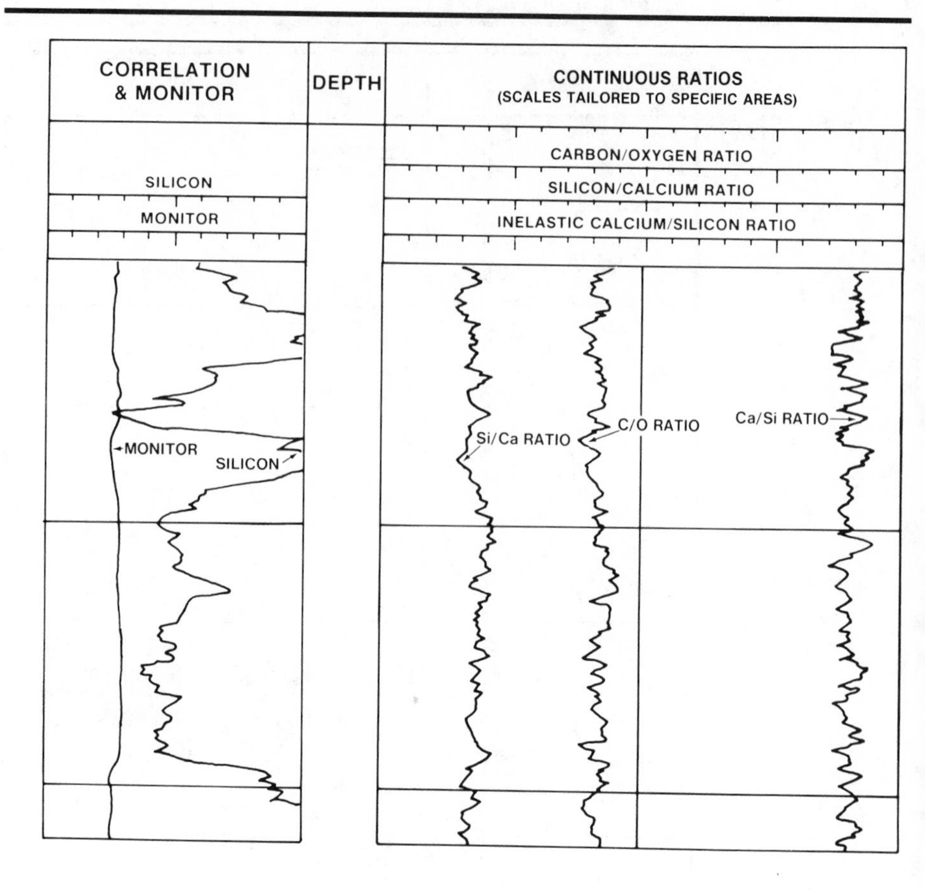

Abbreviation:	C/O
Use:	Formation evaluation through casing/tubing
Rating:	275°F
	15,000 psi
Size:	3⅞ in.
Limitations:	Station measurements or exceedingly slow logging speed
Comments:	Measures relative concentrations of atoms in formation
	Common ratio found is carbon/oxygen or C/O

FIGURE 1A.4 *Carbon/Oxygen Log. Courtesy Dresser Atlas.*

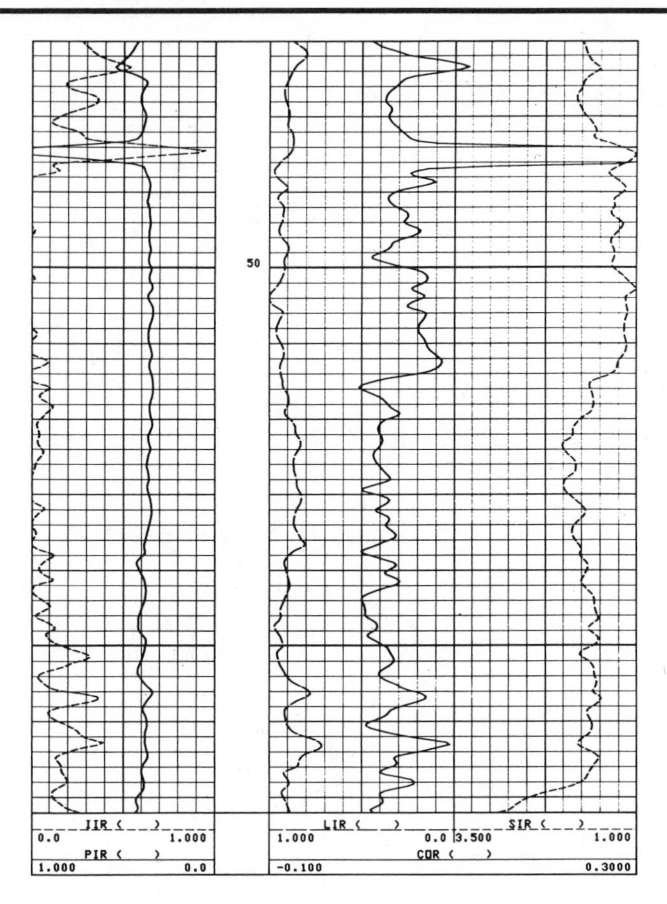

Yield Ratio	*Interaction*	*Name*	*Label*
C/O	Inelastic	Carbon-Oxygen Ratio	COR
Cl/H	Capture	Salinity-Indicator Ratio	SIR
H/(Si + Ca)	Capture	Porosity-Indicator Ratio	PIR
Fe/(Si + Ca)	Capture	Iron-Indicator Ratio	IIR
Si/(Si + Ca)	Capture and Inelastic	Lithology-Indicator Ratio	LIR
S/(Si + Ca)	Capture	Anhydrite-Indicator Ratio	AIR

Abbreviation: IGT
Use: Formation evaluation through casing/tubing
Rating: ______ °F
______ psi
Size: ______
Limitations: Station measurements or exceedingly slow logging speed
Comments: Produces curves of porosity, salinity, water saturation, etc. Operates in both Inelastic and Capture mode

FIGURE 1A.5 *Inelastic Gamma Log. Courtesy Schlumberger Well Services.*

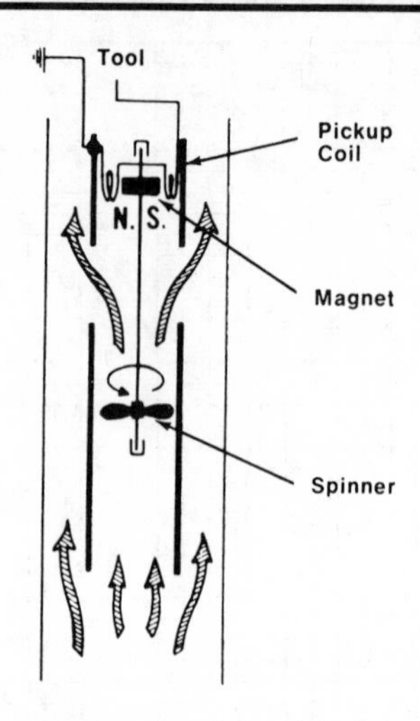

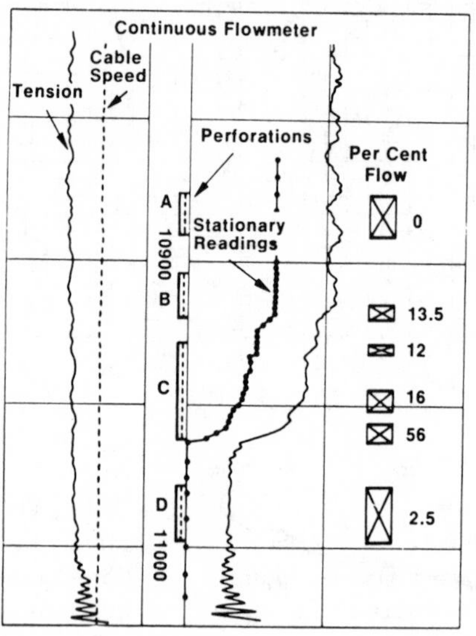

Use:	Production or injection profiles
Rating:	350°F 15,000 psi
Size:	1$^{11}/_{16}$- and 2$^{1}/_{8}$-in. OD
Limitations:	Only useful in mid- to high flow rates
Comments:	Measures rate of spin of a propeller Need to know flow area and cable speed to find volumetric flow rate

FIGURE 1A.6 *Continuous Flowmeter. Logged while injecting water. Courtesy Schlumberger Well Services.*

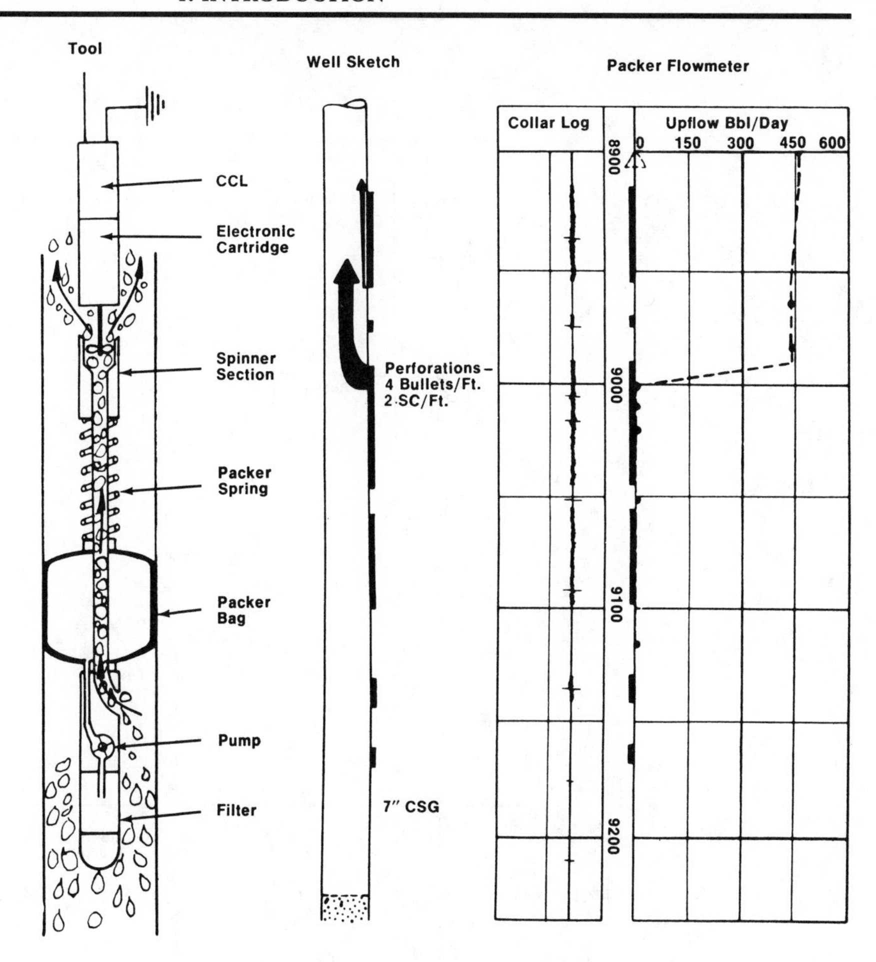

Use:	Low-flow measurement device
Rating:	285°F 10,000 psi
Size:	1¹¹⁄₁₆- and 2⅛-in. OD
Limitations:	Not for high flow rate, tends to choke flow stream. Handles flow rates from 10 to 1900 B/D
Comments:	Essential for inclined flow conditions

FIGURE 1A.7 *Packer/Diverter Flowmeter. Production often comes from only a fraction of the total intervals perforated. This flow profile shows 97% of the production entering the casing in one ten-foot zone. Courtesy Schlumberger Well Services.*

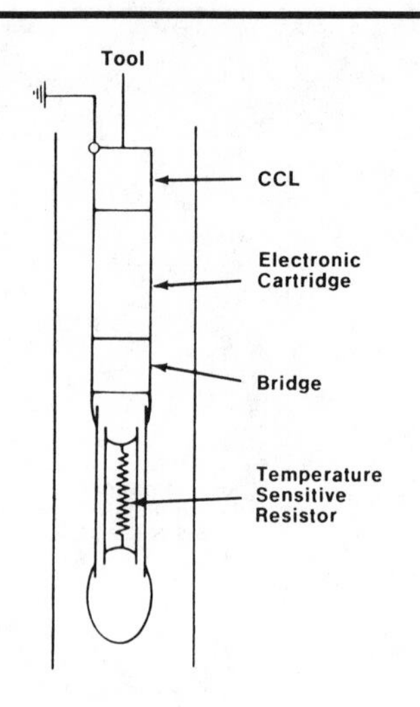

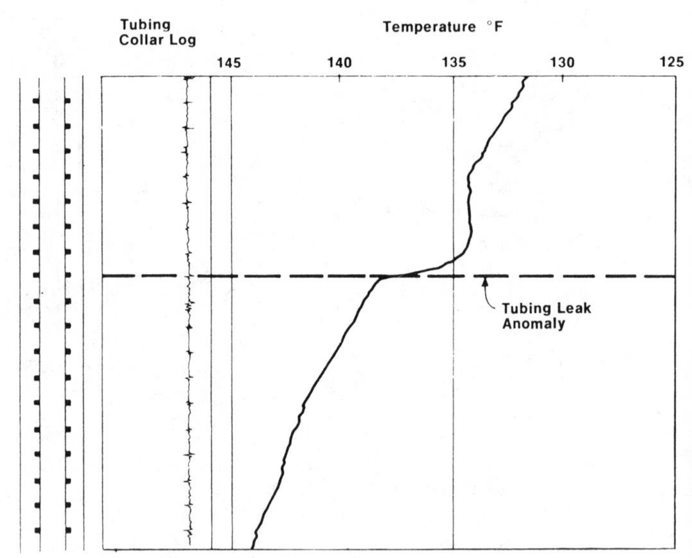

Abbreviation:	HRT, RDT
Use:	Flow indicator Channel finder Qualitative production/injection profiling
Rating:	350°F 15,000 psi
Size:	$1^{11}/_{16}$-in. OD
Limitations:	Maximum recording speed 10,000 ft/hr
Comments:	Response dictated by mass flow rate, geothermal gradient, and thermal conductivity of formation

FIGURE 1A.8 *Temperature Logging. Courtesy Schlumberger Well Services.*

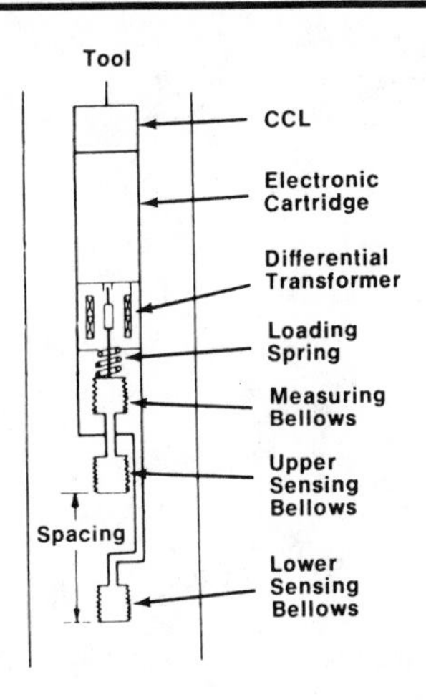

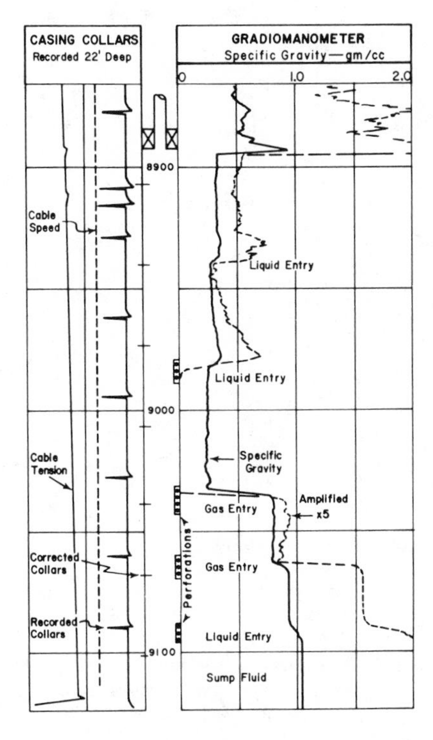

Abbreviation:	Gradio	Limitations:	Range 0.0 to 1.6 g/cc
Use:	Fluid density Holdup	Comments:	Tools available include: Gradiomanometer Gamma-gamma Vibrators
Rating:	350°F 15,000 psi		
Size:	1$^{11}/_{16}$-in. OD		

FIGURE 1A.9 *Fluid Density Logging. Courtesy Schlumberger Well Services.*

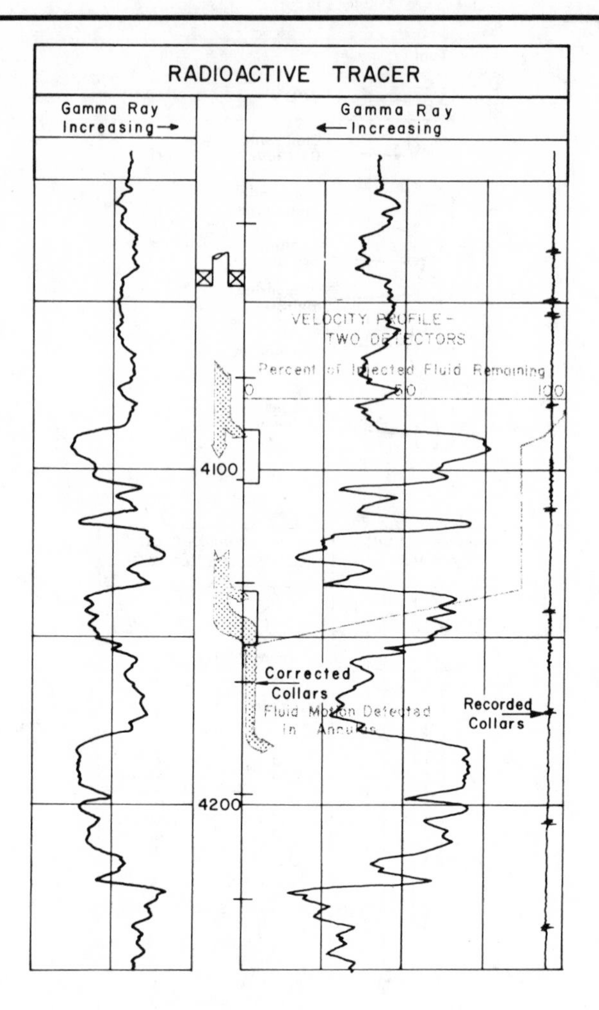

Use:	Flow rates
	Injection profiles
	Treatment effectiveness
Rating:	275°F
	20,000 psi
Size:	$1^{11}/_{16}$-in. OD
Limitations:	For injection wells only
Comments:	Can handle a very wide range of injection rates
	Follows flow behind pipe

FIGURE 1A.10 *Radioactive Tracers. The recording shows a base gamma ray log and a profile in percentage of total injection, plotted from velocity measurements of tracer material. By proper positioning of the selective ejector and sensitive scintillation detectors, the path and velocity of the injected fluid is determined. The radioactive tagging of the fluid permits identification of fluid motion within the annulus. Courtesy Schlumberger Well Services.*

Abbreviation:	Sibilation, Sonan, BATS
Use:	Distinguish oil, gas, and water flow
Rating:	350°F
	17,000 psi
Size:	$1\frac{11}{16}$-in. OD
Limitations:	Minimum ID 2 in.
Comments:	Splits noise spectrum into three of four bands

FIGURE 1A.11 *Noise Logs. Courtesy Dresser Atlas.*

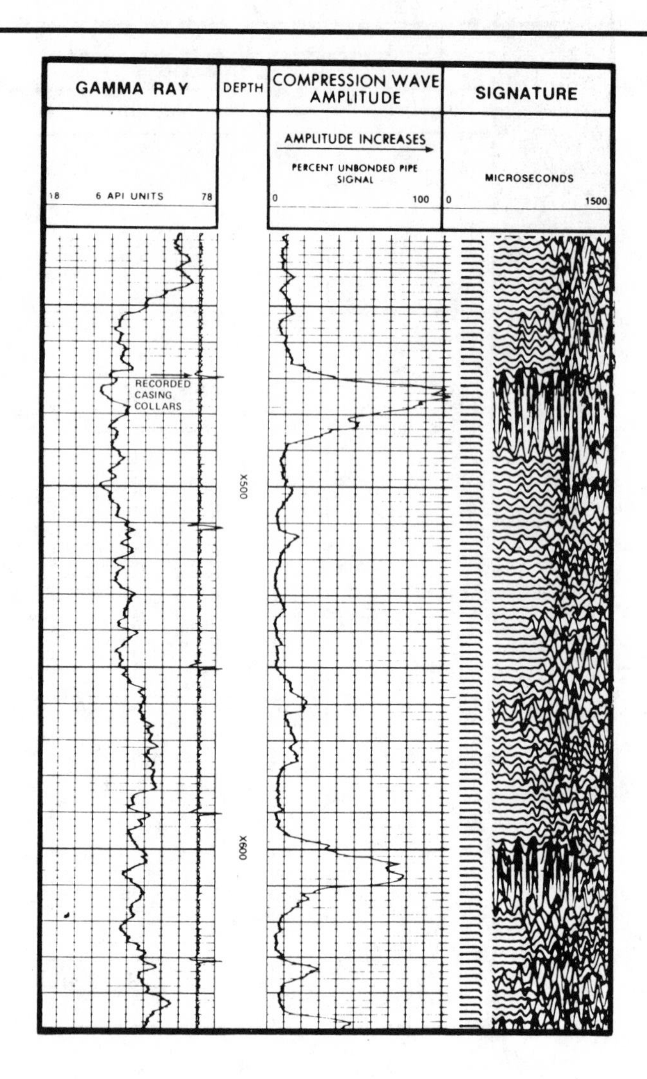

Abbreviation: CBL, CBL-VDL, CET
Use: Monitoring presence of cement, strength of bond, etc.
Rating: 350°F
20,000 psi
Size: 3⅝- and 1¹¹⁄₁₆-in. OD
Limitations: Requires fluid in well
Comments: Needs good centralization

FIGURE 1A.12 *Cement Bond Log. Courtesy Dresser Atlas.*

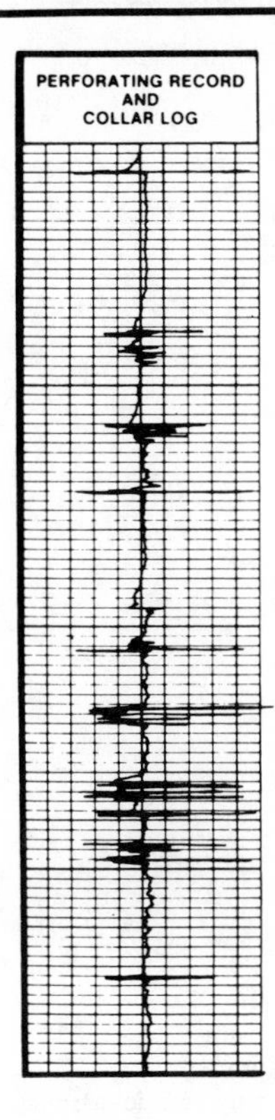

Abbreviation:	CCL
Use:	Pinpointing casing/tubing collars, perforations, packers, etc.
Rating:	350°F 20,000 psi
Size:	$1\frac{11}{16}$-in. OD
Limitations:	None
Comments:	If not memorized, may be recorded off-depth to main measurement

FIGURE 1A.13 *Casing-Collar Log. Courtesy Dresser Atlas.*

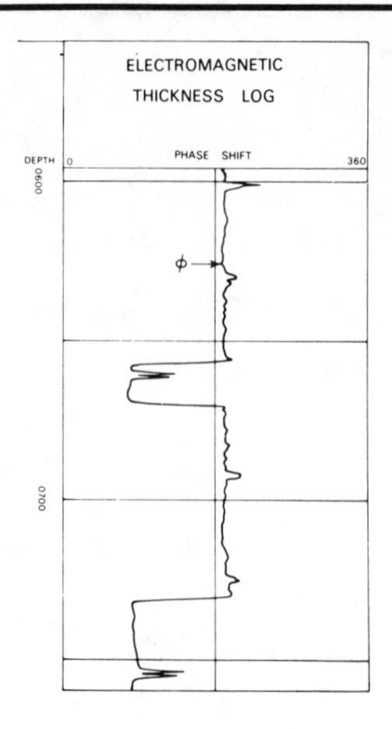

Abbreviation:	ETT, Magnelog
Use:	Pipe inspection
Rating:	285°F
	10,000 psi
Size:	2⅝-, 4.5-, and 5-in. OD
Limitations:	Min. ID 3½-in. casing
	Max. ID 9-in. casing
Comments:	Two-coil electromagnetic device that measures phase shift of signal

FIGURE 1A.14 *Electromagnetic Thickness Tool Log. Courtesy Schlumberger Well Services.*

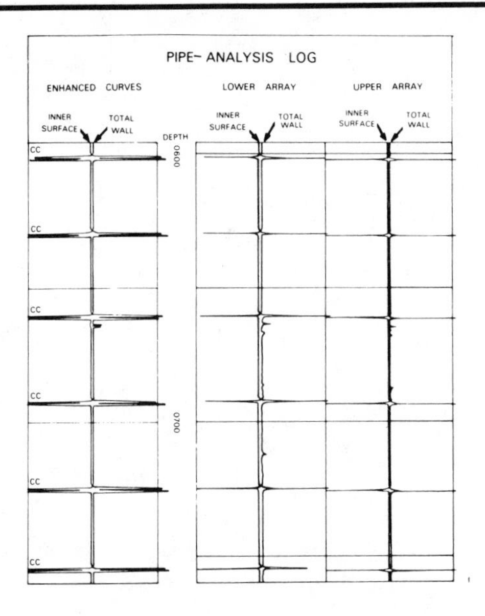

Abbreviation:	PAL, Vertilog (CAT)
Use:	Pipe inspection
Rating:	255°F
	10,000 psi
Size:	4.5-in. OD
Limitations:	Blind to vertical splits
Comments:	Differentiates between inner and outer casing wall anomalies

FIGURE 1A.15 *Pipe-Analysis Log. Courtesy Schlumberger Well Services.*

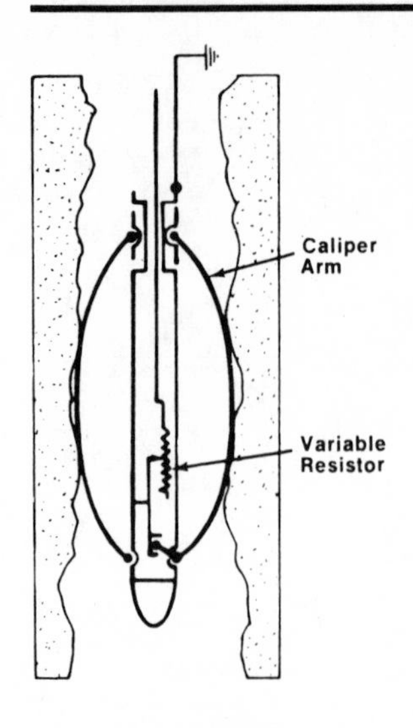

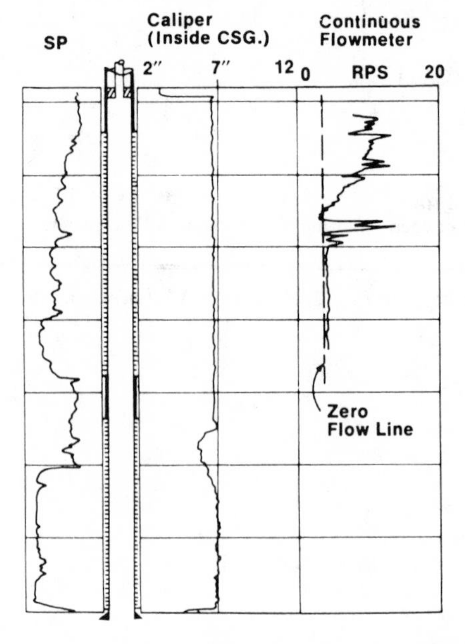

Use:	Pipe inspection Hole diameter measurement
Rating:	350°F 15,000 psi
Size:	$1^{11}/_{16}$-in. OD and up
Limitations:	Measures from 2 to 18 in.
Comments:	Essential for flow analysis in barefoot completions Useful for pipe inspection.

FIGURE 1A.16 *Caliper and Caliper Log Inside Casing.*

BIBLIOGRAPHY

Cesak, N. J. and Schultz, W. P.: "Time Analysis of Problem Wells," *Pet. Eng. Intl.* (September 1956) 13–30.

Fertl, W. H.: "Well Logging and Its Application in Cased Hole," paper SPE 10034 presented at the 1982 Intl. Petroleum Exhibition and Technical Symposium held in Beijing, China.

Raymer, L. L. and Burgess, K. A.: "The Role of Well Logs in Reservoir Modeling," paper SPE 9342 presented at the SPE 55th Annual Technical Conference and Exhibition, Dallas, Sept. 21–24, 1980.

Timur, A.: "Open Hole Well Logging," paper SPE 10037 presented at the 1982 Intl. Petroleum Exhibition and Technical Symposium held in Beijing, China.

CHAPTER TWO

CASED-HOLE LOGGING ENVIRONMENT

PLANNING A PRODUCTION LOGGING JOB

Planning is an important part of a production logging job. Frequently these jobs can only be done in safety during daylight. Thus, the correct type of equipment must be available for the expected well conditions. Before attempting any production logging job the following check list should be consulted:

1. full well-completion details
2. full production history
3. all openhole logs
4. PVT data

Lack of any part of this data will result in delays that may jeopardize the entire job.

PRESSURE CONTROL EQUIPMENT

Practical details should not be forgotten. In particular, considerable care and attention should be given to the matter of working on a well that has pressure at the wellhead. It is a good idea to plan well in advance with the logging-service company using the following check list:

1. wellhead connection
2. riser requirements
3. tubing restrictions (minimum ID)
4. tubing-head pressure
5. safety (H_2S? pressure/temperature ratings)

In general, when working against wellhead pressure the logging cable will be a single-conductor armored cable about ¼ inch in diameter. To seal the wellhead assembly against well fluids, a stuffing box or hydraulic packing gland will be used. For high pressures, a "grease-seal" assembly will be used. In order to get logging tools into and out of the well in a safe and efficient manner, a section of riser will be needed. A typical setup is illustrated in figure 2.1. Note that, above the wireline blowout preventer, this pressure-control assembly has (1) a tool trap, (2) multiple sections of riser, and (3) the pressure sealing equipment.

When retrieving a tool from the well, it is sometimes difficult to gauge exactly where the cable head is in the riser. If it is pulled

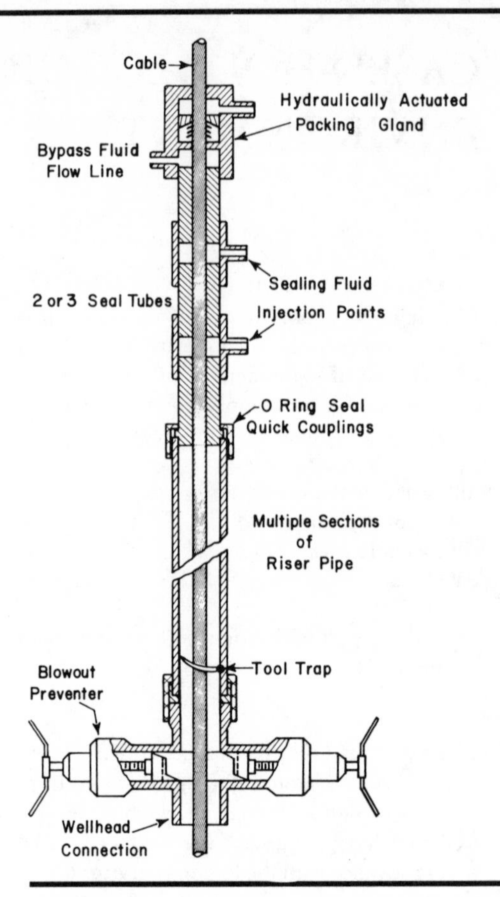

FIGURE 2.1 *Production Logging Wellhead Pressure-Control Assembly. Courtesy Schlumberger Well Services.*

up against the pressure sealing assembly too briskly, the tool may shear off the end of the cable and drop back into the well. To prevent this undesirable event, the tool trap catches the tool at the base of the riser.

The cable itself is at all times subject to an extrusion force, since the portion inside the riser experiences wellhead pressure, while the portion outside the riser experiences atmospheric pressure. The upward force is thus the difference in pressure multiplied by the cross-sectional area of the cable itself. Sometimes this upward force can be surprisingly large and tools will not go down the well unless "ballasted" with additional weights.

QUESTION #2.1

Tubing head pressure is 4986 psi. The logging cable OD is $\frac{7}{32}$ in. The tool weighs 20 lb and is 16 ft long.

a. Calculate the upward force on the cable.
b. If weights are available, each 4 ft long and weighing 26 lb, how many are needed to make the tool go down the well?
c. In that case, how long a riser is required?
d. If the top of the BOP is 10 ft above ground level, the grease-seal equipment measures 10 ft, and the sheave assembly requires 6 ft of clearance, how tall must the workover rig be in order to log this well?

It is also important to plan the arrangement of the Christmas tree—the objective being to be able to log the well without disturbing the dynamic behavior of the production or injection process. Sometimes this consideration is forgotten in the planning, with the result that the only way to get production logging tools into and out of the well is by shutting in the well. This is undesirable, since a well may take hours or days to reach equilibrium again after being shut in. Figure 2.2 illustrates an ideal Christmas-tree setup. Note the numbered items in the figure:

#1: valve on the riser side of the production line
#2: valve on the production line
#3: valve on the well side of the production line
#4: pressure gauge on the riser
#5: bleed-off valve on the riser

QUESTION #2.2

a. What happens if item #1 is missing?
b. What happens if item #3 is missing?
c. Why are items #4 and #5 needed?

THE BOREHOLE ENVIRONMENT

In many of the problems that arise in completed wells, quantitative analysis will require detailed knowledge of flow rates, casing and tubing sizes and weights, as well as the types of flow that are occurring. For example, in the analysis of flowmeter data, fluid speed needs to be related to volumetric flow rate. Conversion from one set of units to another can be facilitated by using Appendix A: Conversion factors between metric, API, and customary (U.S.) measures; Appendix B: Average fluid velocity vs. tubing size; and Appendix C: Average fluid velocity vs. casing size.

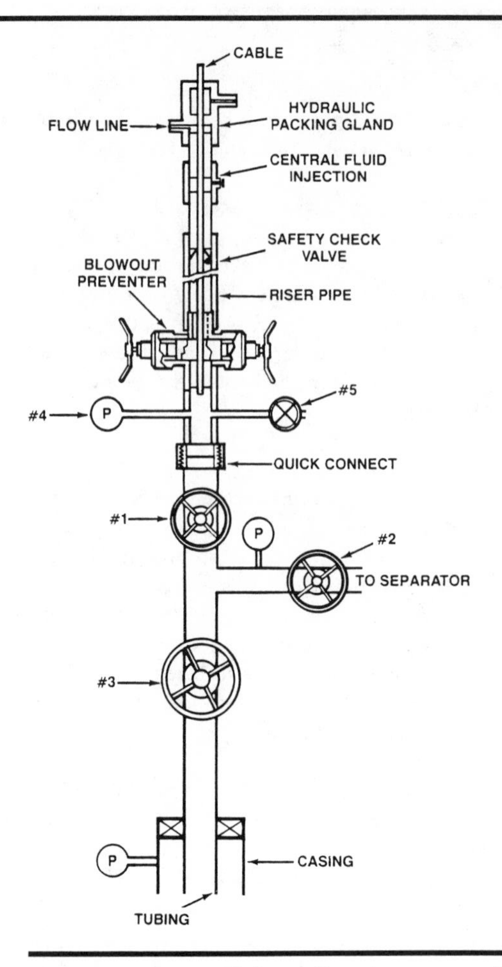

FIGURE 2.2 *Correct Christmas-Tree Arrangement for Production Logging.*

QUESTION #2.3
Use Appendix C to find the flow rate in 7-in., 26-lb casing if the fluid speed is 9.1 ft/min.

CHOOSING PRODUCTION LOGS

Depending on the type of problem encountered, a choice will exist regarding the correct tool or logging technique to be used. The following suggestions are offered as a quick guide. A more informed choice can be made after studying the individual tools in the chapters that follow.

Flow rates	
Low (0 and up)	Radioactive tracer log
Low to medium (10 to 1900 B/D)	Packer or diverter flowmeter
Medium to high (50 to 5000 B/D)	Full-bore flowmeter
High (3000 B/D and up)	Continuous flowmeter
Fluid type	
Oil/water	Gradiomanometer
Oil/gas	Gradiometer or densimeter/vibrator
Gas/water	Densimeter/vibrator or gamma-gamma log
Formation content	
High-salinity water	Pulsed neutron log
Low-salinity water	Carbon/oxygen log
Casing/tubing inspection	Electromagnetic thickness tool Flux-leakage type tool Multifingered calipers
Cement/channeling	Cement bond log Radial differential thermometer Temperature log Noise log Radioactive tracer log

Answers to Text Questions

QUESTION #2.1

Cable area = $\pi/4 \times (7/32)^2 = 0.03758$ sq in.

Differential pressure = 4986 psi.

a. Upward force = $4986 \times 0.03758 = 187.4$ lb.
   Weights required = $(187.4 - 20)/26 = 6.44$.
b. So use 7 weights.
c. Riser requirements = 16-ft tool + $(7 \times 4$ ft$)$ = 44 ft.
d. Rig height = $10 + 44 + 10 + 6 = 70$ ft.

QUESTION #2.2

a. The well must be shut in before tools can be run in or out of the well.
b. No effect on ability to log well.
c. To bleed off pressure in the riser before undoing the quick-connect riser connection.

QUESTION #2.3

500 B/D

CHAPTER THREE

RESERVOIR FLUID PROPERTIES

PVT REFRESHER COURSE

For a complete understanding of the behavior of producing wells, it is necessary to keep in mind the fundamental principles that govern the properties of hydrocarbon liquids and gases. Only then can downhole flow rates be accurately found from surface flow rates. The correct choice of a flowmeter tool depends on the expected flow rate; and whether or not free gas is present will influence the choice of a fluid-typing tool.

In order to keep clear the different sets of conditions of pressure, volume, and temperature, subscripts will be used in this text as follows:

- o oil
- w water
- g gas
- sc standard conditions
- wf well flowing conditions (at depth)
- s solution
- b bubble point
- bc brine concentration

Additionally, the following symbols will be used:

- ρ density
- μ viscosity
- R gas/oil ratio or solubility
- p pressure
- q flow rate
- B formation volume factor
- V volume
- C concentration
- γ specific gravity
- T temperature

and the following abbreviations:

- BFPD barrels of fluid per day
- BGPD barrels of gas per day
- BOPD barrels of oil per day
- BWPD barrels of water per day
- cf/B cubic feet per barrel
- scf/D standard cubic feet per day

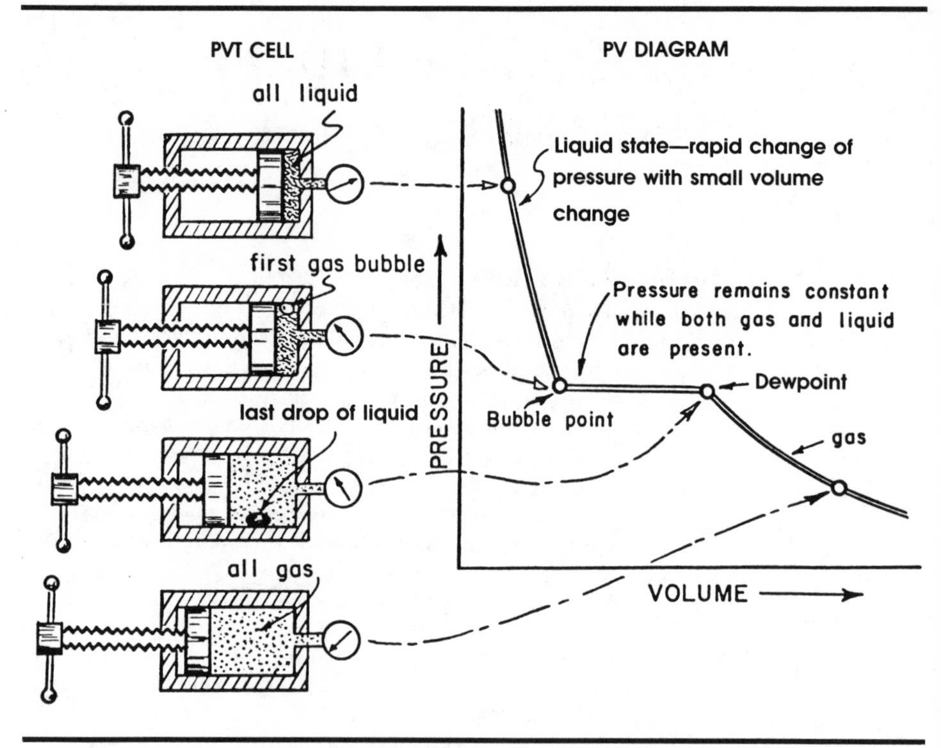

FIGURE 3.1 *Phase Behavior of a Single-Component Hydrocarbon. Courtesy Schlumberger Well Services.*

Single-Component Hydrocarbon System

The behavior of a single-component hydrocarbon system is illustrated in figure 3.1. On the graph, pressure is plotted against volume, and the resulting curve is called the *vapor-pressure curve.* As the pressure on a fixed mass of hydrocarbon liquid is reduced, its volume increases slightly until the *bubble point* is reached. Further increasing the volume available leaves the pressure constant and more and more of the liquid hydrocarbon converts to the gaseous phase. As the volume continues to increase, eventually the *dew point* is reached and no further liquid hydrocarbon is present; from then on, further increases in volume result in reductions in pressure. The PV diagram shown in figure 3.1 is for a fixed temperature. The effects of temperature can be understood by reference to figure 3.2. Note that at high temperature the dewpoint line and the bubble-point line coincide at the *critical point.* In summary, a single-component hydrocarbon (methane, ethane, etc.) can exist as a gas, a liquid, or a gas–liquid mixture depending on the pressure and temperature to which it is subjected.

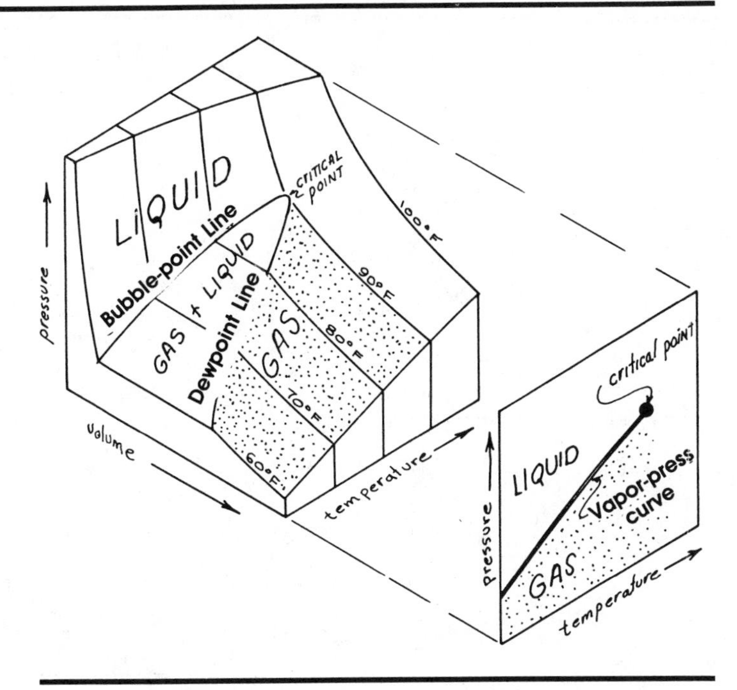

FIGURE 3.2 *Three-Dimensional Diagram of a Single-Component System. Courtesy Schlumberger Well Services.*

Multicomponent Hydrocarbon System

In actual reservoirs, the hydrocarbons found are never single-component systems. Rather, they are mixtures of several different hydrocarbons; and the behavior of the mix is different from that of any single component. In particular, there is no single vapor-pressure line. Rather, an envelope exists between the bubble-point line and the dewpoint line within which gas and liquid coexist. Figure 3.3 illustrates this concept.

Note that on the PV plane the bubble point and the dewpoint are found as discontinuities and no straight-line portion exists on the PV graph (see fig. 3.4). What, therefore, distinguishes one type of reservoir from another? What kind of production may be expected from a multicomponent hydrocarbon system? The answers lie in the starting and ending points on a pressure–temperature plot and their positions relative to the envelope between the bubble-point and dewpoint lines.

Oil Reservoirs

Figure 3.5 shows a plot of pressure vs. temperature for an oil-producing reservoir. At formation conditions, pressure and temperature are such as to place point A′, representing the original conditions,

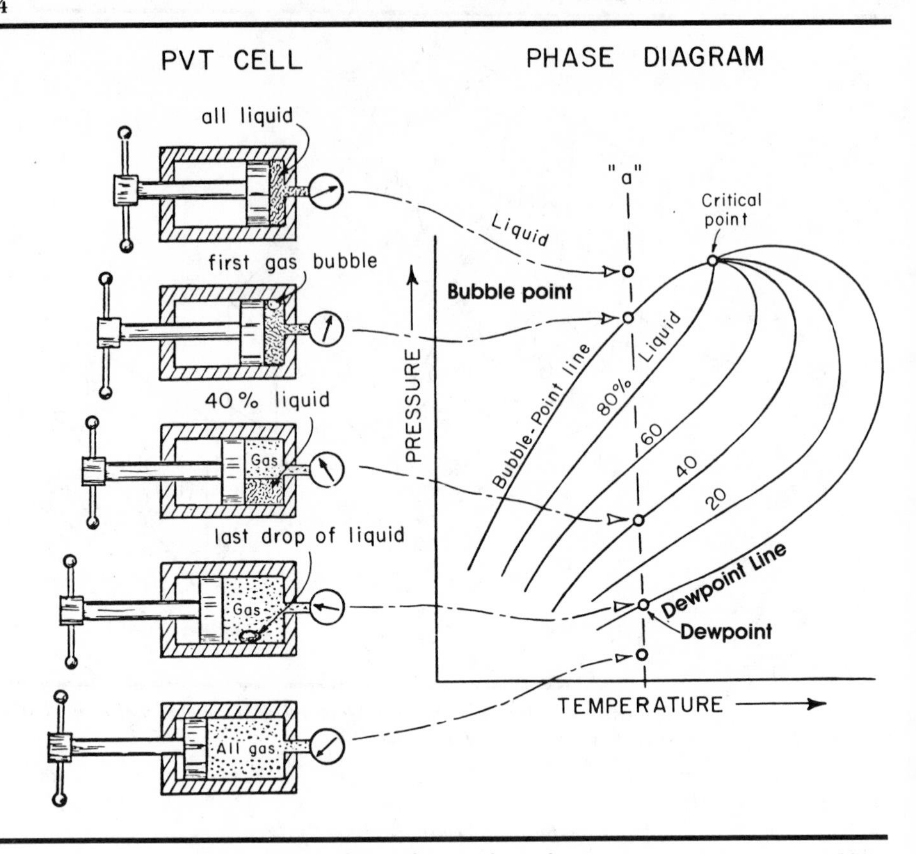

FIGURE 3.3 *Phase Behavior of a Multicomponent System. Courtesy Schlumberger Well Services.*

above the critical pressure. The multicomponent hydrocarbon therefore exists as an undersaturated liquid. As the pressure in the reservoir is drawn down by the production process, point A—the bubble point—is reached. Some gas can now start to come out of solution. The path from the producing horizon up the production string to the separator is illustrated by the dashed line. The final step to the stock tank leaves a point that falls on a line representing some percentage of oil and some percentage of gas. Typically, 80 to 90% of the original fluid is recovered in the form of liquid (oil).

Condensate Reservoirs

Figure 3.6 illustrates a reservoir where the starting point A′ is above the critical temperature. Thus, all of the multicomponent system exists as a gas. During the production process, however, the tempera-

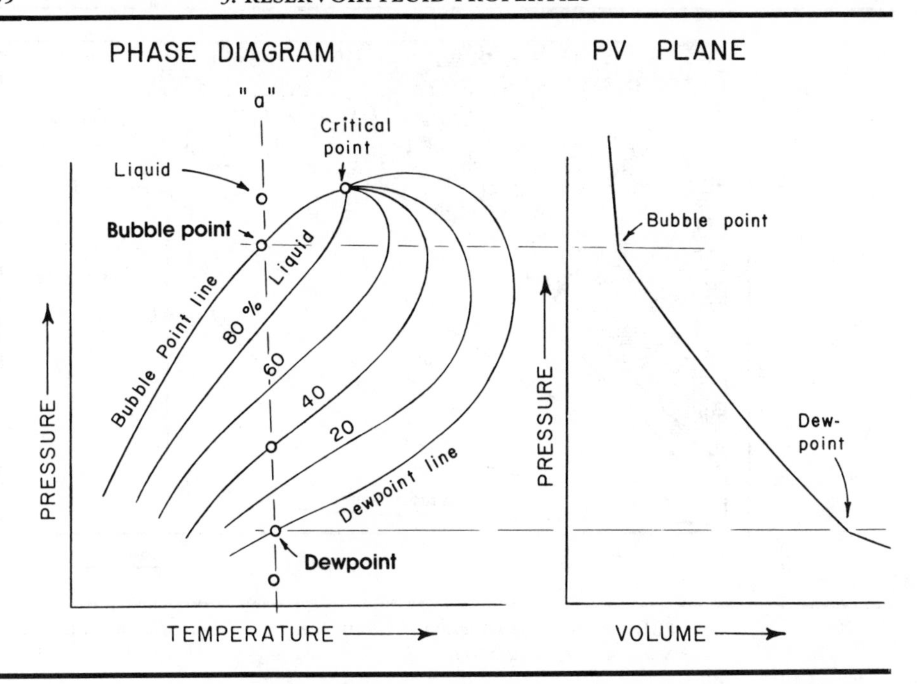

FIGURE 3.4 *Bubble-Point and Dewpoint Determination for a Multicomponent System. Courtesy Schlumberger Well Services.*

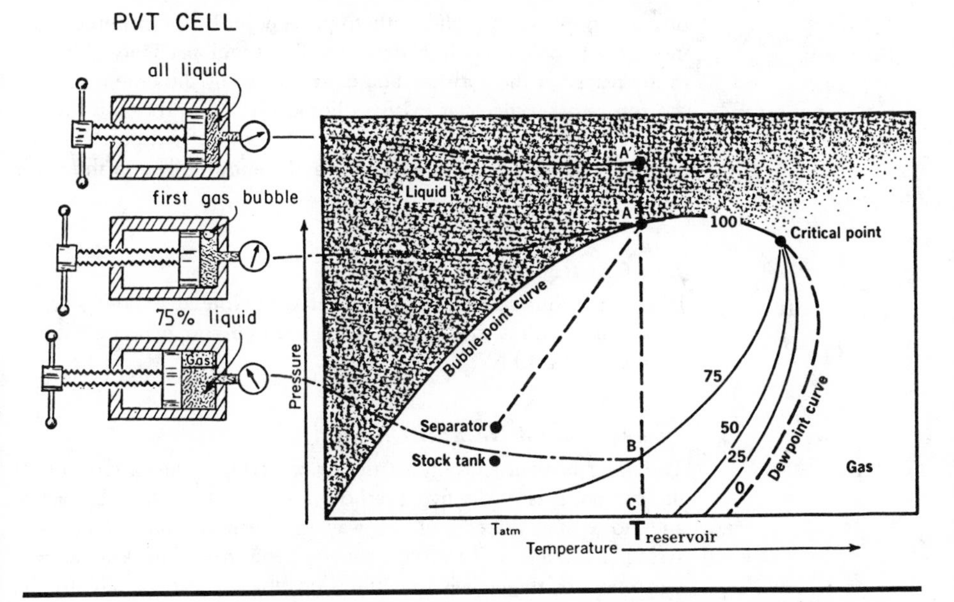

FIGURE 3.5 *Phase Diagram of a Normal GOR Well. Courtesy Schlumberger Well Services.*

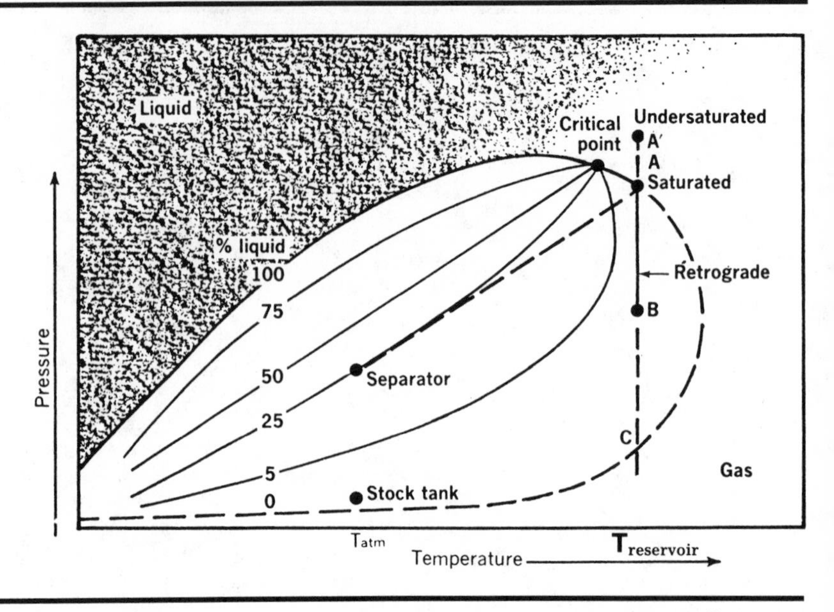

FIGURE 3.6 *Phase Diagram of a Retrograde Condensate-Gas Well. Courtesy Schlumberger Well Services.*

ture and pressure fall sufficiently to place point B, for example, back inside the bubble-point-line–dewpoint-line envelope. Thus, although in the reservoir the hydrocarbon exists as gas, by the time it reaches the separator, some of it exists as liquid oil. This process is known as retrograde condensation. Typically, 25% of the hydrocarbon may be recovered as oil at the separator and somewhat less in the stock tank.

Dry-Gas Reservoir

Figure 3.7 illustrates a dry-gas reservoir. Note that both starting and ending points are outside the envelope and hence no liquid recovery is possible.

Composition of Natural Oils and Gases

The exact behavior of any particular reservoir is thus a function of the components of the hydrocarbon mixture placed there by nature and, to a small extent, of the way in which the multicomponent system is produced. By varying the temperature and pressure at various separator stages it is sometimes possible to increase slightly the liquid (oil) recovery. The range of hydrocarbon types commonly found is given in tables 3.1 and 3.2.

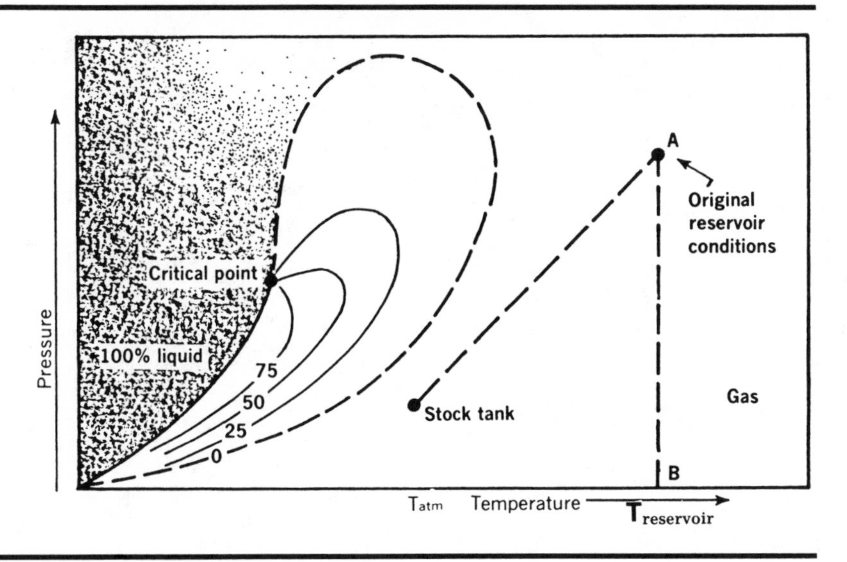

FIGURE 3.7 *Phase Diagram of a Dry-Gas Reservoir. Courtesy Schlumberger Well Services.*

FLUID PROPERTIES

Downhole flow rates can differ quite markedly from surface-recorded flow rates. For example, water is compressible, thus less water flows at downhole conditions than at surface. This can be expressed as

$$q_{wwf} = q_{wsc} \cdot B_w.$$

where B_w is the water formation volume factor (FVF).

By contrast, oil flow rates downhole are greater than oil flow rates at surface. Although oil too is compressible, it can also accept greater volumes of dissolved gas at reservoir conditions and therefore expands. This can be expressed as

$$q_{owf} = q_{osc} \cdot B_o.$$

where B_o is the oil formation volume factor.

Gas is highly compressible, hence downhole gas flow rates will be much smaller than those recorded on surface at standard conditions. Thus we have

$$q_{gwf} = q_{gsc} \cdot B_g.$$

However, the fact must also be taken into account that there are really three sources of the gas seen at the surface. Some will have come out of solution in the oil and/or water and some will be free

TABLE 3.1 *Composition of Natural Gases*

Field	Hugoton	Austin	Leduc Gas Cap D-3	Viking, Kinsella	West Cameron, Blk 149
State	Oklahoma, Texas	Michigan	Alberta	Alberta	Louisiana (Gulf)
Formation	Permian Dolomite	Stray Sand	Devonian	Cretaceous Sand	Miocene Sand
Depth (ft)	3000	1200	5000	—	7150
Mole percentage:					
Nitrogen, N_2	15.5	7.3	7.41	0.24	—
Carbon dioxide, CO_2	—	—	0.72	2.26	0.30
Helium, He	0.58	0.4			—
Methane, CH_4	71.51	79.74	72.88	88.76	96.65
Ethane, C_2H_4	7.0	9.1	9.97	4.76	2.05
Propane, C_3H_8	4.4	2.8	5.09	2.67	0.47
Isobutane, C_4H_{10}	0.29	0.1	0.72	0.42	0.08
n-Butane, C_4H_{10}	0.70	0.4	1.76	0.21	0.09
Isopentane, C_5H_{12}	0.02	0.1	0.99	0.38	0.03
n-Pentane, C_5H_{12}	—				0.02
Hexanes, C_6H_{14}	—	0.05	0.46	0.30	0.31
Heptane+	—	0.01			
	100.00	100.00	100.00	100.00	100.00

Note: Reprinted, by permission, from Donald Katz et al.: *Handbook of Natural Gas Engineering* (New York: McGraw-Hill, 1959).

TABLE 3.2 *Analysis of Reservoir Oils Containing Dissolved Gases*

Field	Leduc D-2	Leduc D-3	Paloma	Oklahoma City, Wilcox	Rodessa	Keokuk	Schuler (Jones Sand)
State or Province	Alberta	Alberta	California	Oklahoma	Louisiana	Oklahoma	Arkansas
Reservoir:							
Depth (ft)	5,000	5,300	10,600	6,200	5,950	4,026	7.600
Pressure (psia)	1,774	1,908	4,663	2,630	2,600	1,455	3,520
Temperature (°F)	149	153	255	132	192	130	198
Mole percentage:							
Nitrogen, N_2	—	—	—	—	—	—	1.00
Carbon dioxide, CO_2	—	—	—	—	—	—	0.80
Methane, CH_4	28.6	30.3	55.8	37.7	40.88	25.60	42.85
Ethane, C_2H_6	10.9	13.1	5.81	8.7	4.53	8.88	6.60
Propane, C_3H_8	9.4	9.4	6.42	6.3	2.60	12.41	4.10
Isobutane, C_4H_{10}	2.5	1.8	1.31	1.4	1.25	1.93	3.64
n-Butane, C_4H_{10}	4.4	4.9	3.97	3.0	1.82	7.56	
Pentanes, C_5H_{12}	4.8	4.5	3.67	3.3	3.48	5.53	3.10
Hexanes, C_4H_{14}	39.4	36.0	2.61	39.6	4.43	38.09	3.83
Heptane+	—	—	20.41	—	41.01		34.08
	100.0	100.0	100.00	100.0	100.00	100.00	100.00
Molecular weight, heptanes+	201	193	237	225	220	195	243
Specific gravity as liquid heptanes+	0.840	0.840	0.891	0.840	0.824	0.839	0.8759

Note: Reprinted, by permission, from Donald Katz et al.: *Handbook of Natural Gas Engineering* (New York: McGraw-Hill, 1959).

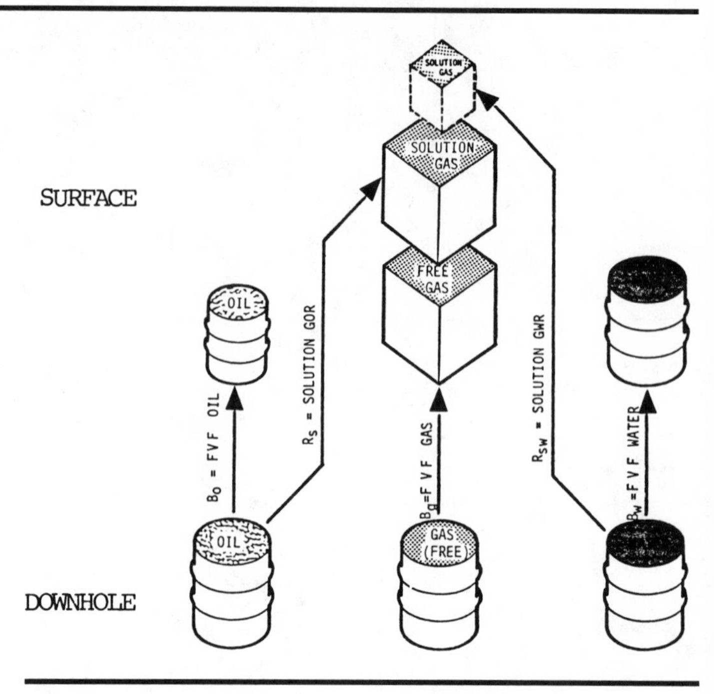

FIGURE 3.8 *Relation Between Surface and Downhole Volumes. Courtesy Schlumberger Well Services.*

gas at reservoir conditions (assuming reservoir pressure is below the bubble point). Hence,

gas at surface = free gas + solution gas.

Figure 3.8 illustrates these concepts. In order to be able to calculate downhole flow rates from the surface rates, a series of charts may be used or a computer or programmable calculator (e.g., the HP Petroleum Fluids Pac).

Water

FORMATION VOLUME FACTOR. The properties of water are determined by the amount of salt dissolved in it and by pressure and temperature. Figure 3.9 relates these three variables. Note that increasing temperature tends to expand a given volume of water, whereas increasing pressure tends to contract it. Thus, values of B_w tend to be close to 1.0.

B_w can be expressed in a number of equivalent forms, e.g.,

$$B_w = V_{wwf}/V_{wsc},$$
$$B_w = q_{wwf}/q_{wsc}, \text{ or}$$
$$B_w = \rho_{wsc}/\rho_{wwf}.$$

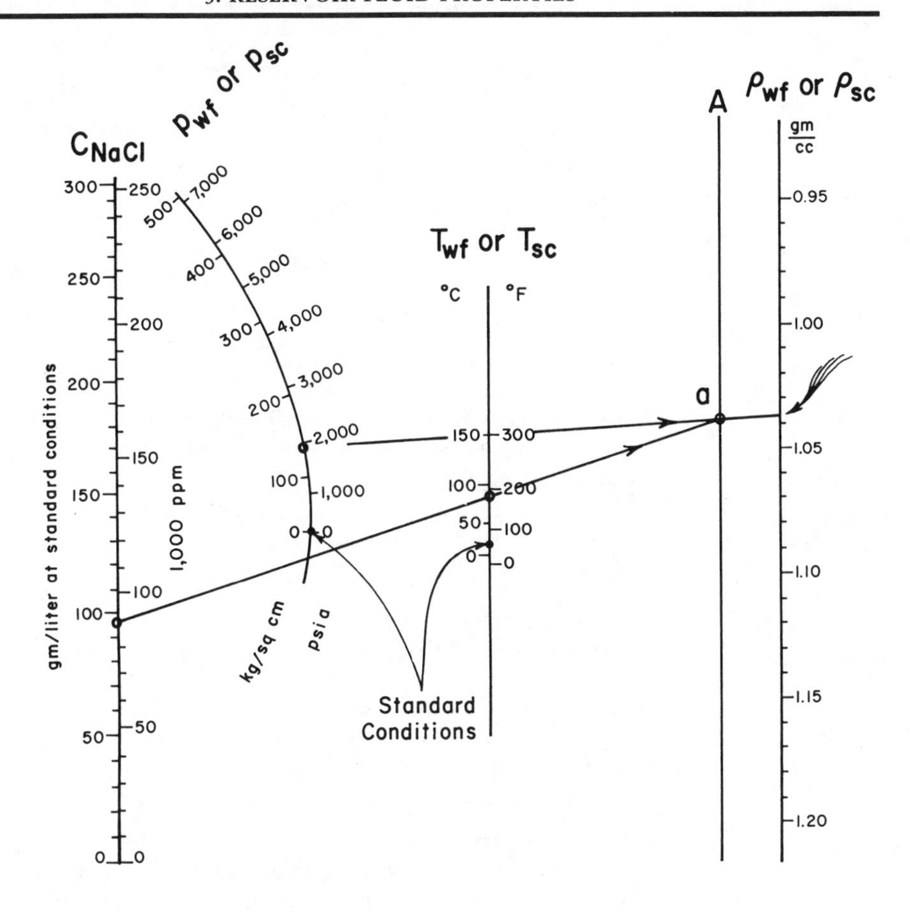

Find ρ_{wwf}.

Given

$C_{NaCl} = 90{,}000$ ppm.
$T_{wf} = 200°F$.
$p_{wf} = 2{,}000$ psia.

1. Locate Point a by a line from $C_{NaCl} = 90{,}000$ through $T_{wf} = 200$ to A.
2. Draw a line $p_{wf} = 2{,}000$ through point a.
3. Read: $\rho_{wwf} = 1.032$ gm/cc.

FIGURE 3.9 *Densities of NaCl Solutions. Courtesy Schlumberger Well Services.*

QUESTION #3.1 B_w AND q_{wwf}
Flowing pressure = 3,000 psi.
Flowing temperature = 200°F.
Water salinity = 150,000 ppm NaCl.
Surface flow rate = 300 BWPD (barrels of water per day).

a. Find ρ_{wsc} = _______ g/cc.
b. Find ρ_{wwf} = _______ g/cc.
c. Hence B_w = _______, and
d. q_{wwf} = _______ BWPD.

GAS SOLUBILITY IN WATER. Gas is soluble in water; the solubility is a function of temperature and water salinity. Figure 3.10 gives a means of finding R_{sw}.

QUESTION #3.2 R_{sw}
Flowing pressure = 2,000 psi.
Flowing temperature = 200 °F.
Water salinity = 25,000 ppm NaCl.

a. Find R'_{sw} from the upper chart.
b. Find F_{bc} from the lower chart.
c. Find $R_{sw} = R'_{sw} \cdot F_{bc}$ = ______ cf/B.

WATER VISCOSITY. The viscosity of water can be an important item of data when interpreting spinner (flowmeter) surveys and repeat formation tester permeability tests. Water viscosity can be determined from figure 3.11.

QUESTION #3.3 μ_{wwf}
Temperature = 190°F.
Water salinity = 100,000 ppm NaCl.

Find μ_{wwf} = _______ cp.

Gas

GAS FORMATION VOLUME FACTOR B_g. The behavior of natural gases is nonideal. Thus, the ideal gas law needs a "fudge factor" to make it truly reflect the behavior of natural hydrocarbon gases. This fudge factor is given the symbol Z and is known as the *supercompressibility factor:*

$$\frac{1}{B_g} = \frac{V_{sc}}{V_{wf}} = \frac{1}{Z} \cdot \frac{T_{sc}}{T_{wf}} \cdot \frac{p_{wf}}{p_{sc}}.$$

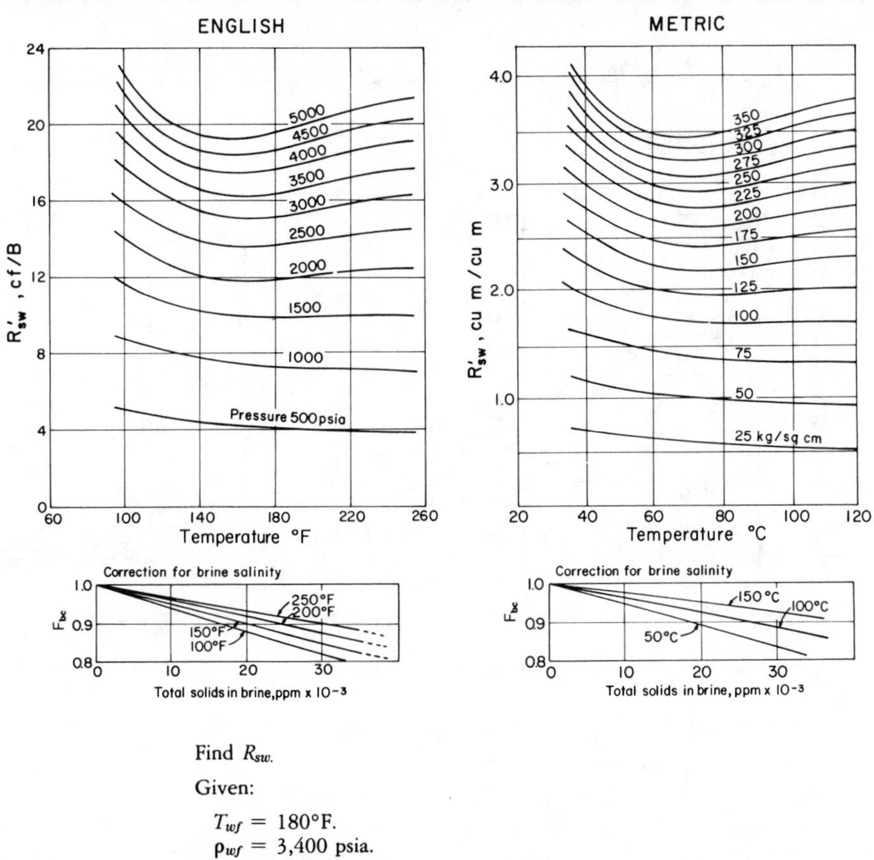

Find R_{sw}.

Given:

T_{wf} = 180°F.
p_{wf} = 3,400 psia.
C_{NaCl} = 20,000 ppm.

1. (Top) Enter the abscissa at T_{wf} = 180.
2. Go up to p_{wf} = 3,400.
3. Read R'_{sw} = 16 cf/B.
4. (Bottom) Enter the abscissa at 20 (20,000 ppm).
5. Go up to T = 180.
6. Read F_{bc} = 0.91.
7. $R_{sw} = R'_{sw} F_{bc} = 16 \times 0.91 = 15$ cf/B.

FIGURE 3.10 *Solubility of Gas in Water. Courtesy Schlumberger Well Services; after Dodson and Standing 1944.*

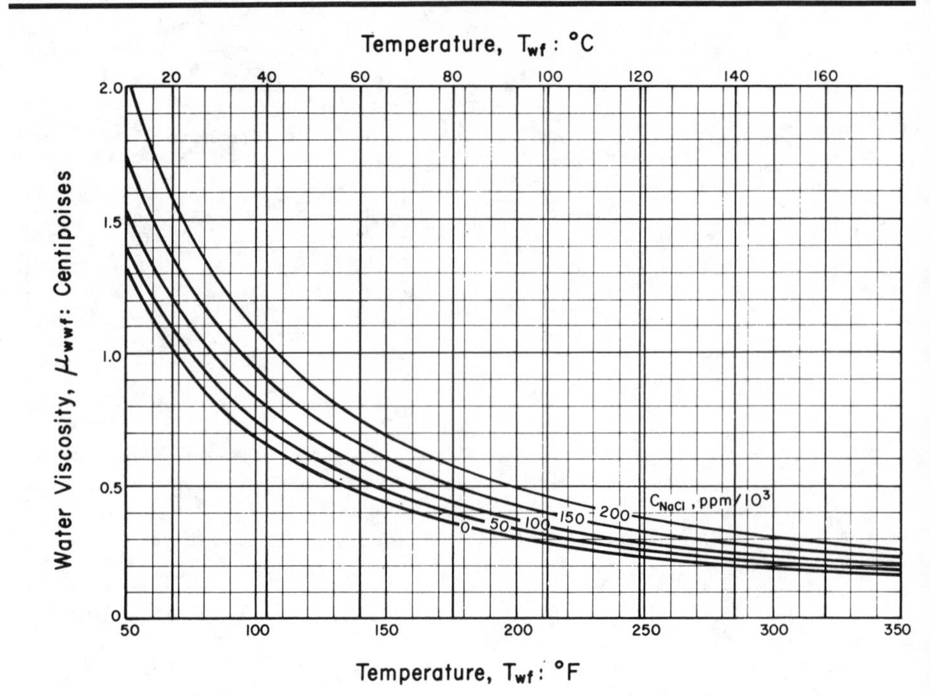

Find μ_{wwf}.

Given:

C_{NaCl} = 150,000 ppm.
T_{wf} = 200°F.

1. Enter abscissa at T_{wf} = 200.
2. Go up to C_{NaCl} = 150,000.
3. μ_{wwf} = 0.43 cp.

FIGURE 3.11 *Water Viscosity. Courtesy Schlumberger Well Services.*

If the Z factor is known for the conditions encountered and the gas in question, then B_g may be found by direct solution of this equation. However, Z is not normally known. A short-cut method uses figure 3.12, which requires only temperature, pressure, and gas gravity as inputs; $1/B_g$ may be read directly as output.

QUESTION #3.4 B_g
Pressure = 3000 psi.
Temperature = 200°F.
Gas gravity = 0.7.

Find $1/B_g$ = ____.

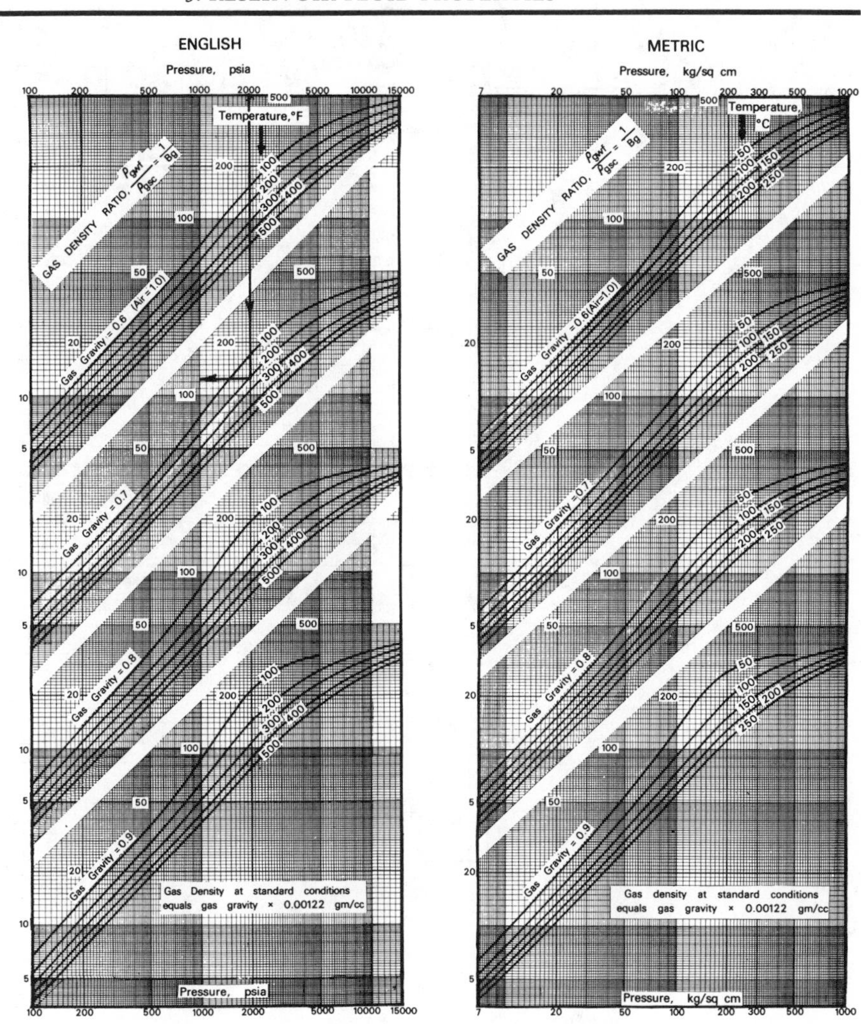

Find V_{gwf}.

Given:

$V_{gsc} = 400$ cu ft.
$\gamma_g = 0.70$.
$T_{wf} = 200°F$.
$p_{wf} = 2000$ psia.

1. Select $\gamma_g = 0.70$ section. Enter abscissa at 2000 psia.
2. Go vertically to 200°F.
3. Go left to $1/B_g = 125$.
4. $1/B_g = 125 = V_{gac}/V_{gwf} = 400/V_{gwf}$, so
5. $V_{gwf} = 3.2$ cu ft.

FIGURE 3.12 *Quick Solution for* **B_g**. *Courtesy Schlumberger Well Services; after Standing and Katz 1942.*

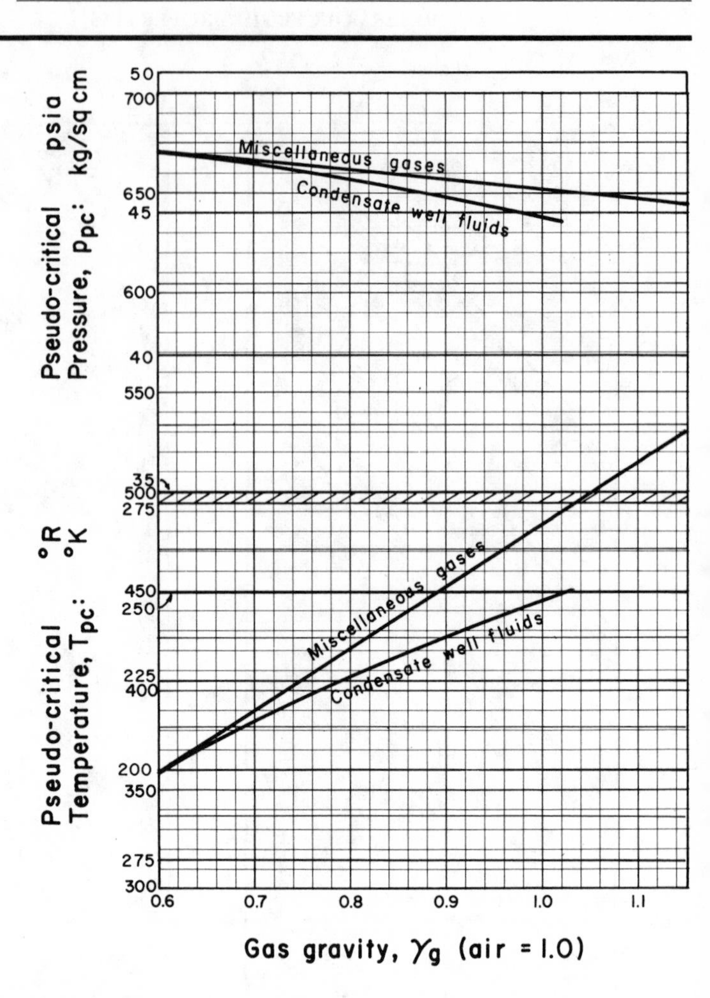

Find T_{pe} and P_{pc}.

Given: $\gamma_g = 0.75$, average gases.

1. Enter abscissa at 0.75.
2. Go up to $T_{pc} = 406°R$ and $P_{pc} = 664$ psia.

FIGURE 3.13 *Pseudo-Critical Natural-Gas Parameters. Courtesy Schlumberger Well Services; after Brown and Katz 1948.*

When more accurate results are required, the Z factor must be determined. This is a more laborious task and involves:

1. finding the pseudo-critical pressure (P_{pc}),
2. finding the pseudo-critical temperature (T_{pc}),
3. finding the pseudo-reduced pressure (P_{pr}),
4. finding the pseudo-reduced temperature (T_{pr}), and
5. using the latter two to find Z.

Figure 3.13 may be used to find P_{pc} and T_{pc} as a function of gas gravity to air and gas type.

QUESTION #3.5 T_{pc} AND p_{pc}

Gas gravity to air is 0.8. (Assume "miscellaneous gas.")

a. Find T_{pc} = ______ °F.
b. Find P_{pc} = ______ psi.

T_{pc} and P_{pc} may be converted to pseudo-reduced values depending on the flowing-well conditions:

$$P_{pr} = p_{wf}/P_{pc} \quad \text{and} \quad T_{pr} = T_{wf}/T_{pc}.$$

The values of P_{pr} and T_{pr} thus found can be entered on figure 3.14 to find the Z factor.

QUESTION #3.6 P_{pr}, T_{pr}, AND Z

Given:

T_{pc} = 420°F.
P_{pc} = 662 psi.
T_{wf} = 149°F (°R = °F + 460).
p_{wf} = 2913 psi.

a. Find P_{pr} = ______.
b. Find T_{pr} = ______.
c. Find Z = ______.

Once Z is established, B_g can be calculated either from the equation

$$\frac{1}{B_g} = \frac{1}{Z} \times \frac{T_{sc}}{T_{wf}} \times \frac{p_{wf}}{p_{sc}},$$

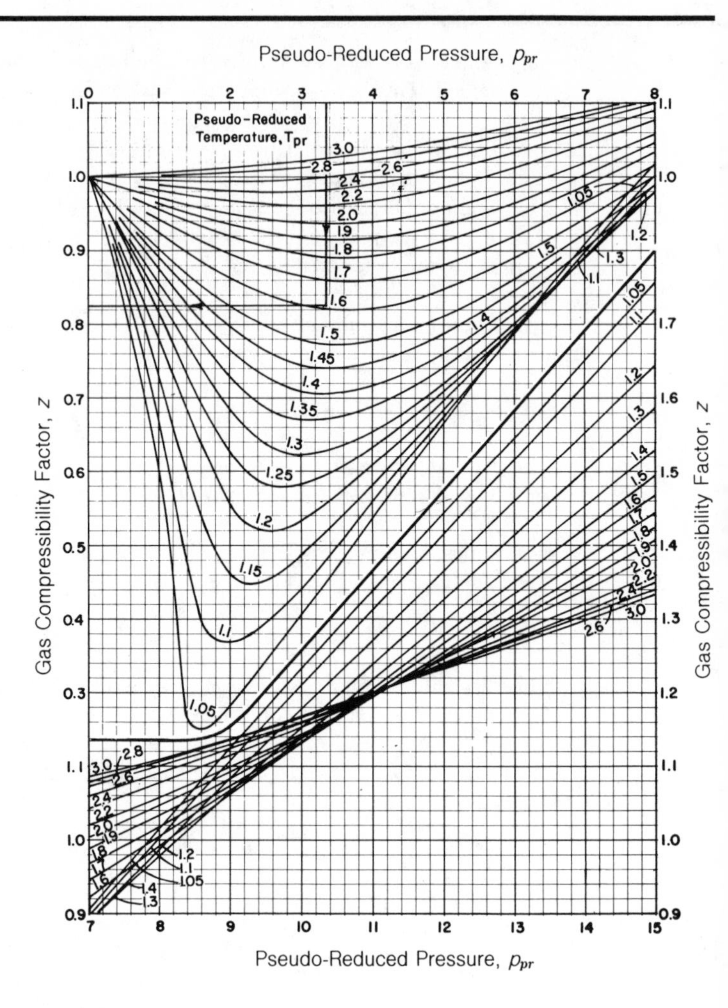

Find Z.

Given:

p_{wf} = 2000 psia.
P_{pc} = 650 psia.
T_{wf} = 200°F (600R).
T_{pc} = 410°R.

1. $P_{pr} = p_{wf}/P_{pe} = 2000/650 = 3.07$.
2. $T_{pr} = T_{wf}/T_{pc} = 660/410 = 1.61$.
3. Enter abscissa (top) at 3.07 (p_{pr}).
4. Go down to T_{pr} of 1.61, between 1.6 and 1.7 lines.
5. Read $Z = 0.828$.

FIGURE 3.14 ***Natural-Gas Deviation Factor. Courtesy Schlumberger Well Services; after Standing and Katz 1942.***

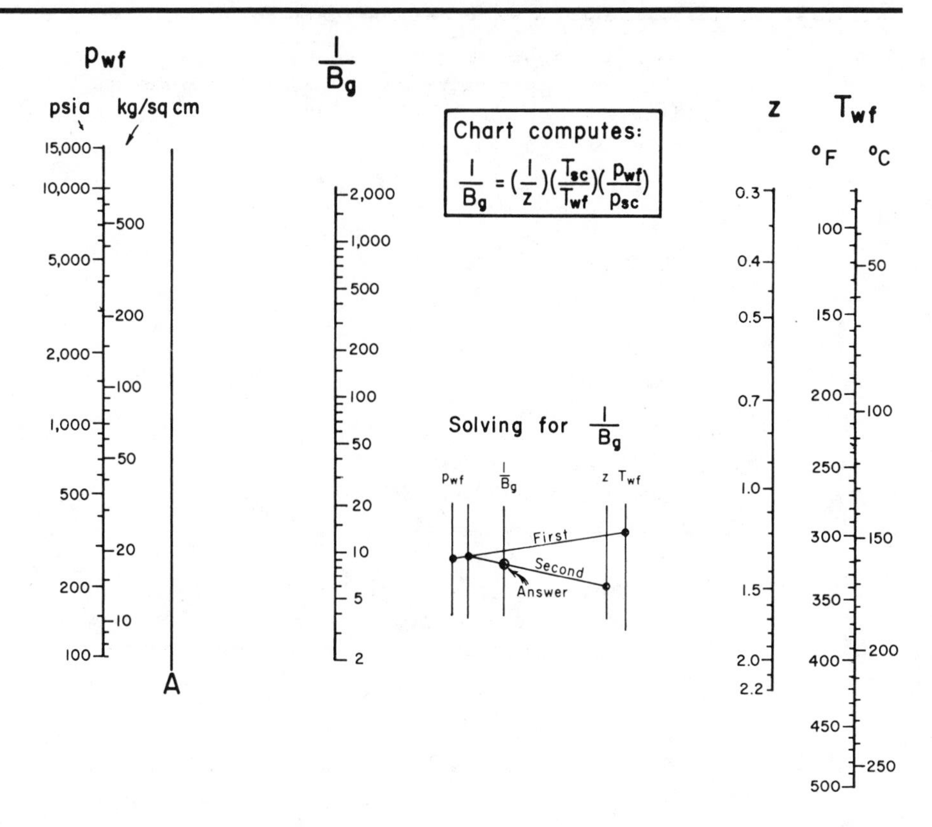

Find $1/B_g$.

Given:

p_{wf} = 140 kg/sq cm.
T_{wf} = 93°C.
Z = 0.828.

1. Enter p_{wf} scale at 140 kg/sq cm.
2. Follow lines as in small diagram.
2. Read $1/B_g$ = 135.

FIGURE 3.15 *Gas Formation Volume Factor. Courtesy Schlumberger Well Services.*

or by use of the nomogram given in figure 3.15. Note that values for T in the above equation must be in degrees Rankin (°R). To convert from Farenheit to Rankin, use:

$$°R = °F + 460.$$

QUESTION #3.7 $1/B_g$ AND q_{gwf}

Given:

$p_{wf} = 2913$ psi.
$T_{wf} = 149°F$.
$Z = 0.76$.

a. Find $1/B_g$.
b. If the well flows 1 MMscf/D, what is q_{gwf}?

BOTTOMHOLE GAS DENSITY. The bottomhole gas density is an important item of data for correct interpretation of gradiomanometer surveys. It may be calculated directly using the equation

$$\rho_{gwf} = \gamma_g \times 0.001223 \cdot 1/B_g \text{ g/cc.}$$

Figure 3.16 offers a nomogram that performs the same calculation.

QUESTION #3.8 ρ_{gwf}

Given:

$1/B_g = 100$.
$\gamma_g = 0.7$ (to air).
Find ρ_{gwf} = ___ g/cc.

GAS VISCOSITY. Gas viscosity is a function of gas gravity, temperature, and pressure. Figure 3.17 offers a means of finding μ_g.

QUESTION #3.9 μ_g

Given:

$\gamma_g = 0.6$.
$T_{wf} = 200°F$.
$p_{wf} = 3000$ psi.

Find μ_g = ___ cp.

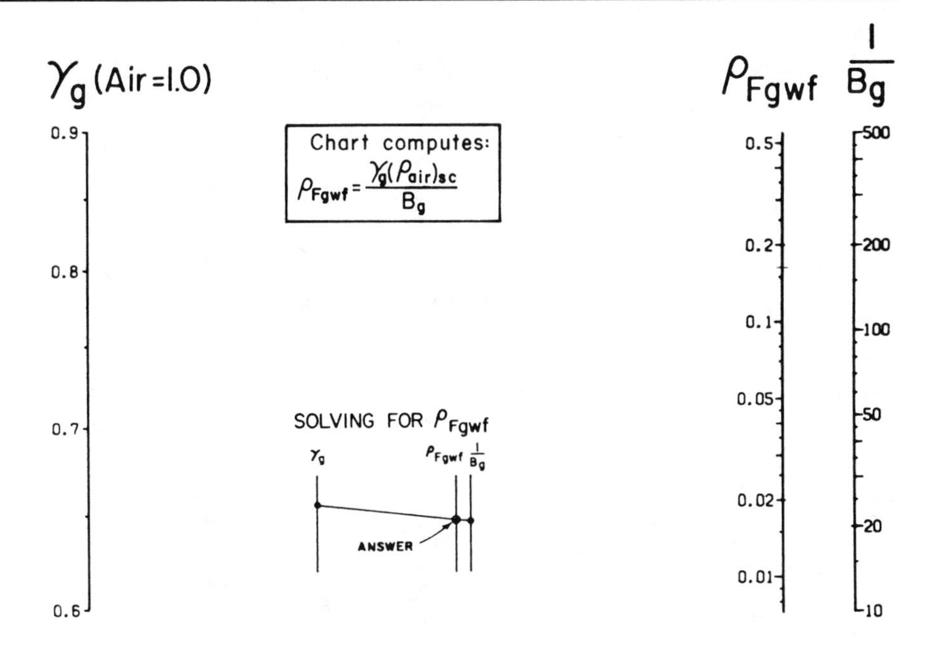

Find ρ_{Fgwf}.

Given:

$\gamma_g = 0.75.$
$1/B_g = 140.$

1. Connect $\gamma_g = 0.75$ and $1/B_g = 140$, as in small diagram.
2. Read $\rho_{Fgwf} = 0.13$ cm/cc.

FIGURE 3.16 *Gas Density. Courtesy Schlumberger Well Services.*

Oil

BUBBLE-POINT PRESSURE. The bubble-point pressure is a critical item of data. It is used to make many of the estimates necessary for correct prediction of downhole conditions during the planning of production logging and/or interpretation of the results. Bubble-point pressure (p_b) depends on T_{wf}, R_s, p_{wf}, γ_o, and γ_g. Figure 3.18 combines all these factors on one chart.

QUESTION #3.10 p_b

Given:

$\gamma_g = 0.7.$
$\gamma_o = 40$ API.
$T_{wf} = 200°$F.
$R_{sb} = 1000$ cf/B.

Find p_b = ____ psi.

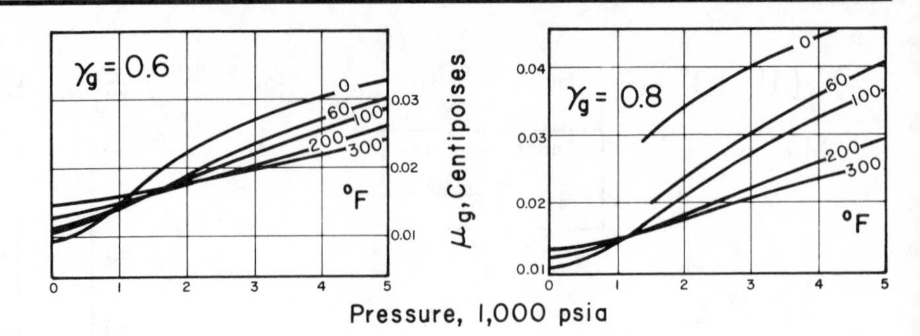

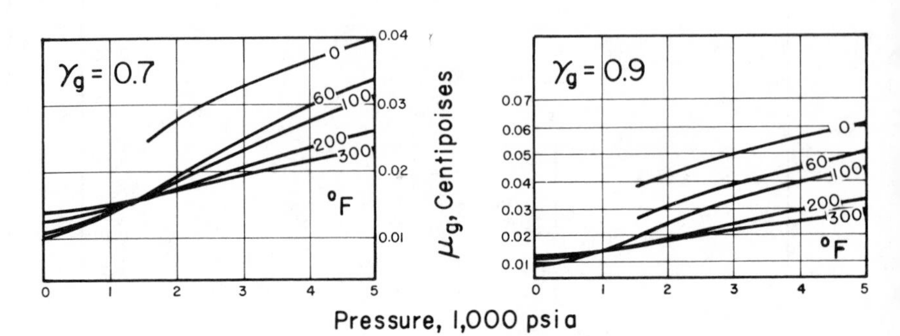

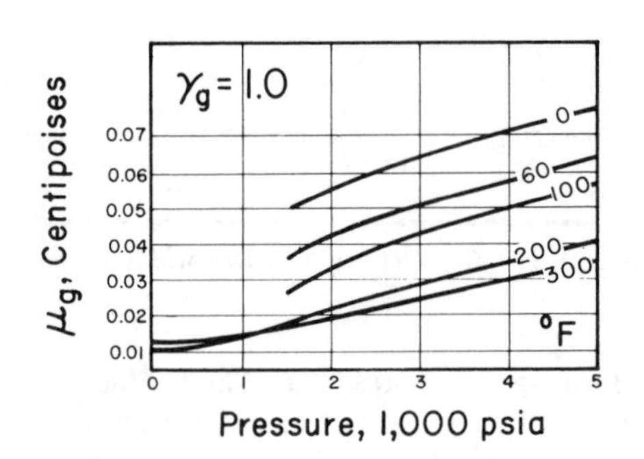

Find μ_{gwf}.

Given:

$\gamma_g = 0.70.$
$p_{wf} = 2000$ psia.
$T_{wf} = 200°F.$

1. Enter $\gamma_g = 0.70$ chart at $p_{wf} = 2000$ psia.
2. Go up to $T_{wf} = 200°F$.
3. $\mu_{gwf} = 0.018$ cp.

FIGURE 3.17 *Gas Viscosity. Reprinted by permission of the* Oil and Gas Journal *(from May 12, 1949).*

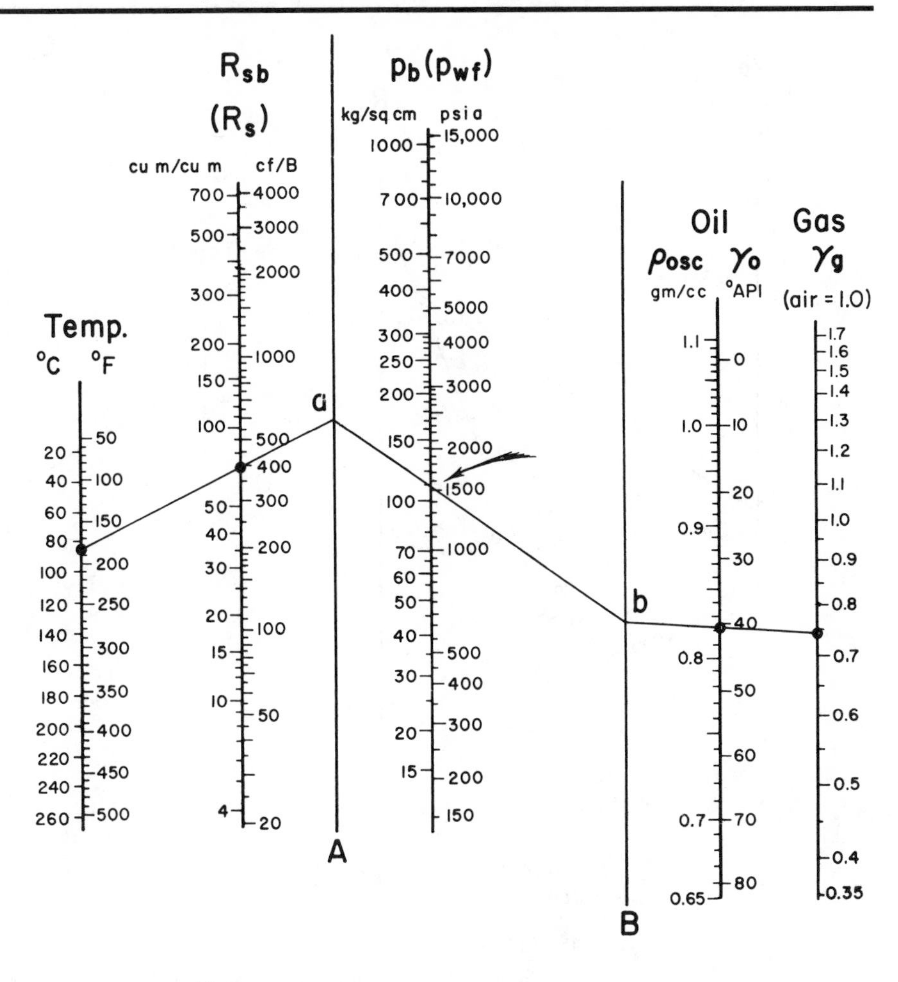

Find p_b.

Given:

T_{wf} = 180°F.
q_{osc} = 600 B/D.
q_{gac} = 240 Mcf/D.
γ_g = 0.75.
γ_o = 40° API.

1. $R = \dfrac{240{,}000\ \text{cf/D}}{600\ \text{B/D}} = 400\ \text{cf/B}.$
2. $R_{sb} = R$, since the field-usage definition of p_b stipulates given flow rates of oil and gas, taken here to be q_{osc} and q_{gsc} (above).
3. On the nomograph, located Point a by a line through T_{wf} = 180°F and R_{sb} = 400.
4. Locate Point b by a line through γ_g = 0.75 and γ_o = 40° API.
5. Connect a and b: p_b = 1560 psia.

FIGURE 3.18 *Bubble-Point Pressure. Courtesy Schlumberger Well Services.*

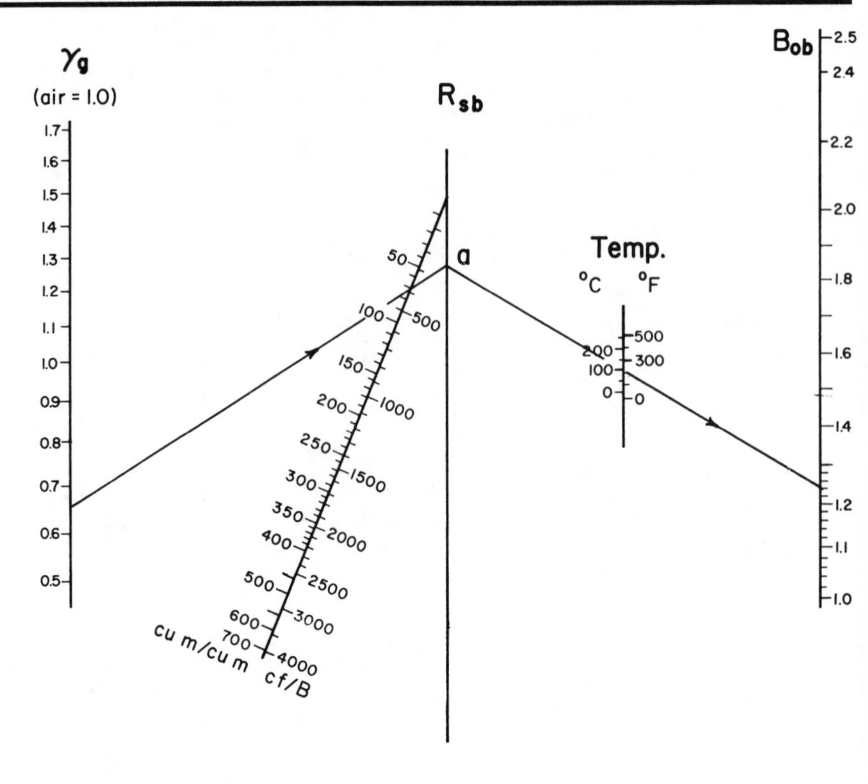

Find B_{ob}.

Given:

R_{sb} = 400 cf/B.
T_{wf} = 180°F.
γ_g = 0.65.
γ_o = 45° API.

1. Locate Point a by drawing a line through γ_g = 0.65 and R_{sb} = 400.
2. Draw a line from Point a through T_{wf} = 180°F, to B_{ob}.
3. B_{ob} = 1.24.

FIGURE 3.19 *Oil Formation Volume Factor at* p_b. *Courtesy Schlumberger Well Services.*

OIL FORMATION VOLUME FACTOR. By definition,

$B_o = V_{owf}/V_{osc}$ or $B_o = q_{owf}/q_{osc}$.

The oil formation volume factor is a function of γ_g, R_{sb}, and T. Its value at the bubble-point pressure can be estimated from figure 3.19.

QUESTION #3.11 B_{ob} AND q_{owf}

Given:

γ_g = 0.8.
R_{sb} = 1500 cf/B.
T = 200°F.

a. Find B_{ob} = ____.
b. If q_{osc} = 1000 BOPD,
find q_{owf} = ____ BOPD.

The nomogram in figure 3.19 holds only for conditions at bubble-point pressure. At well flowing pressures above and below the bubble point, it is necessary to apply the following algorithms: (1) for under-saturated oils (above bubble-point pressure),

$$B_o = B_{ob}\,[1 - C_o\,(p_{wf} - p_b)],$$

where C_o is the oil compressibility, a function of ρ_{ob}; and (2) for saturated oils (below bubble-point pressure),

$$B_o = 1 + k(B_{ob} - 1),$$

where k is a function of p_{wf}/p_b. Figure 3.20 serves to find the values of C_o (upper portion) and k (lower portion).

QUESTION #3.12 C_o AND k

a. If ρ_{ob} = 0.745,
find C_o = ____.
b. If p_{wf}/p_b = 0.375,
find k = ____.

QUESTION #3.13 B_o

Using the answers from question 3.12, find B_o for the following conditions:

a. undersaturated oil, where

B_{ob} = 1.44,
p_b = 3800, and
p_{wf} = 4800.

b. saturated oil, where

B_{ob} = 1.44, and
p_{wf}/p_b = 0.375.

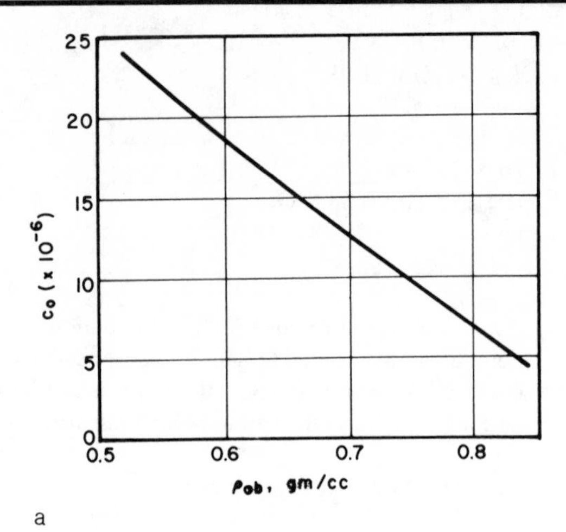

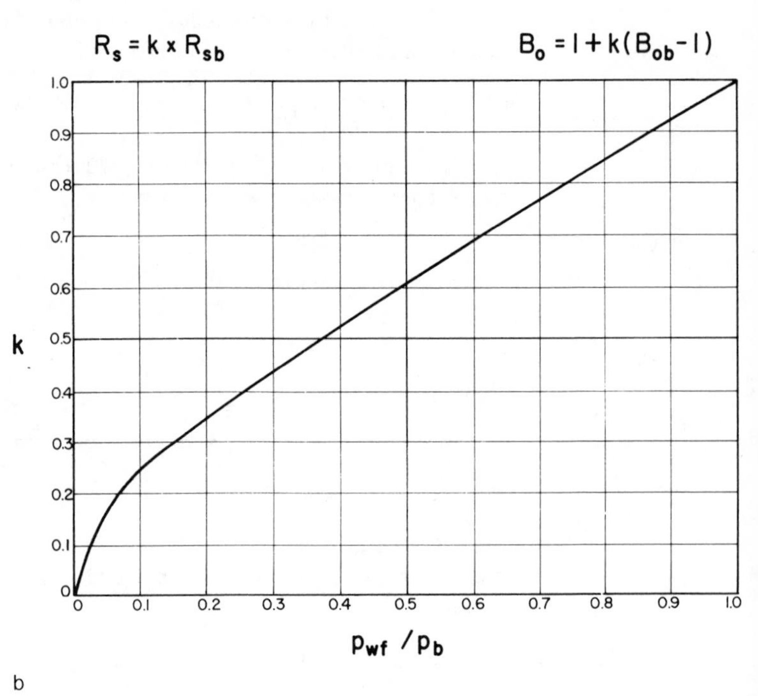

FIGURE 3.20 *Finding* C_o *and* k. *Courtesy Schlumberger Well Services.*

OIL DENSITY. Oil density at well flowing conditions is another vital piece of data for interpretation of gradiomanometer surveys. It is a function of γ_g, R_s, B_o and γ_o. Figure 3.21 combines all the appropriate data on one chart and gives ρ_{owf} as a result.

QUESTION #3.14 ρ_{owf}

Given:

$$\gamma_g = 0.7.$$
$$R_s = 1000 \text{ cf/B}.$$
$$B_o = 1.44.$$
$$\gamma_o = 40 \text{ API}.$$

Find ρ_{owf} = ____ g/cc.

ρ_{owf} may also be calculated directly using the equation

$$\rho_{owf} = \frac{(141.5)/(131.5 + \gamma_o) + 0.0002178 \cdot \gamma_g \cdot R_s}{B_o}.$$

OIL VISCOSITY. Oil viscosity is a function of γ_{osc}, T_{wf}, R_{sb} and Δp, the incremental pressure above the bubble-point pressure. It can be determined using figure 3.22, which is built in two parts. The first part gives μ_{ob}, the oil viscosity at the bubble-point pressure. For conditions above p_b, μ_o increases by a factor, $\Delta\mu$, given by the second part of the chart.

QUESTION #3.15 μ_{ob} AND μ_{owf}

Given:

$$\gamma_{ocs} = 40 \text{ API}.$$
$$T_{wf} = 190\ °\text{F}.$$
$$R_{sb} = 1000 \text{ cf/B}.$$
$$\Delta p = 1000 \text{ psi}.$$

a. Find μ_{ob} = ____ cp.
b. Find μ_{owf} = ____ cp.

Practical Applications

The whole purpose of this section has been to equip the analyst with a practical tool for actual field cases where decisions must be made regarding tool rating and interpretation techniques. A few examples should serve to guide the user in the analysis of day-to-day problems.

Before planning a production logging survey, the production his-

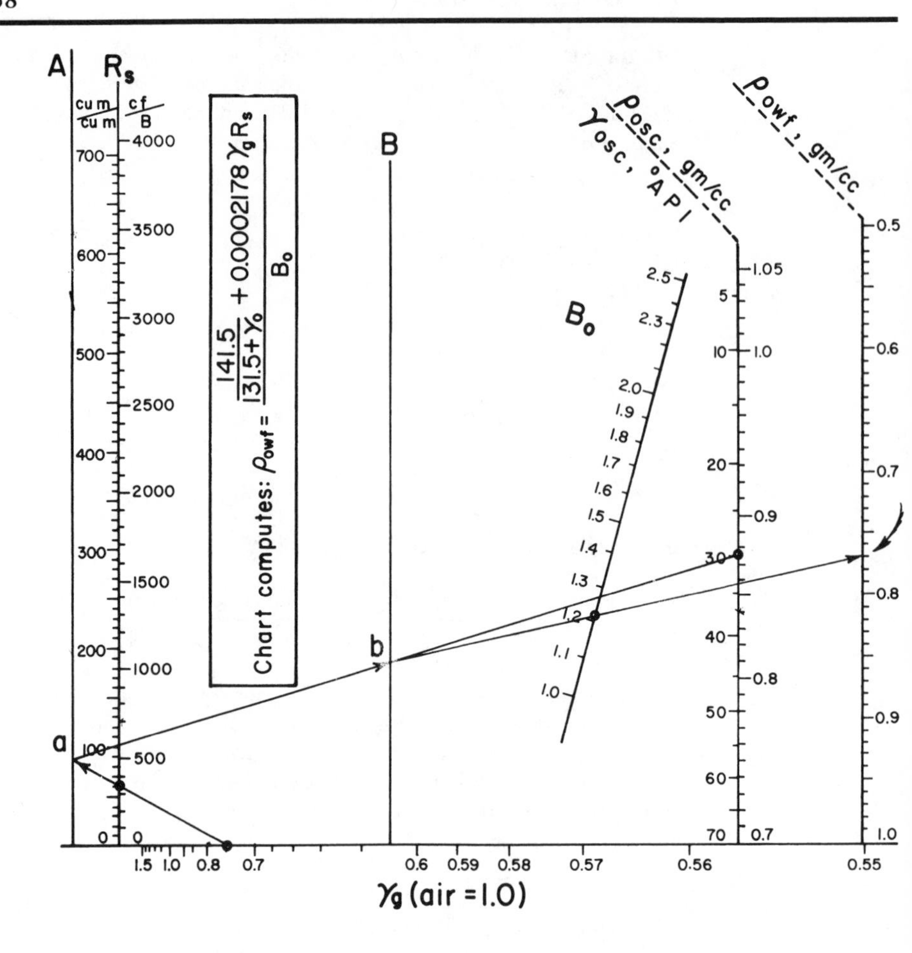

Find ρ_{owf}.

Given:

γ_o = 30° API.
γ_g = 0.75.
R_s = 350 cf/B.
B_o = 1.21.

1. Locate Point a by a line from γ_g = 0.75 through R_s = 350.
2. Locate Point b by a line from Point a to γ_o = 30° API.
3. Draw a line from Point b through B_o = 1.21.
   ρ_{owf} = 0.77 gm/cc.

FIGURE 3.21 *Oil Density at Well Conditions. Courtesy Schlumberger Well Services.*

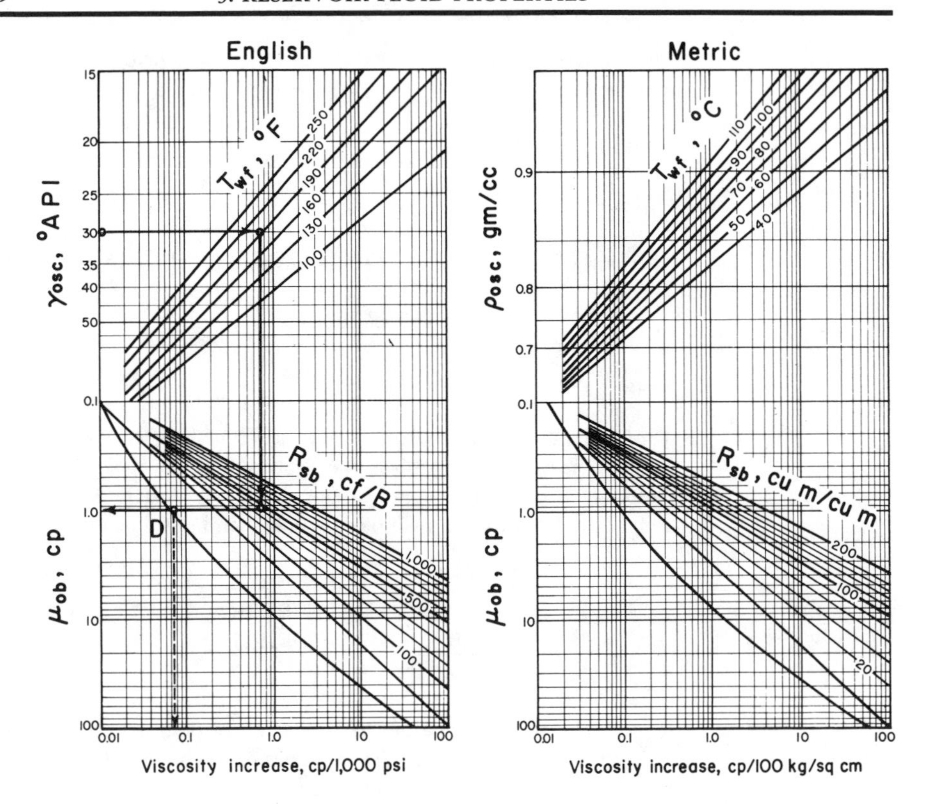

Find μ_{owf}.

Given:

γ_o = 30° API.
T_{wf} = 200°F.
p_b = 1700 psia.
p_{wf} = 2700 psia.
R_{sb} = 400 cf/B.

1. Enter ordinate at $\gamma_{osc} = 30°$ API.
2. Go right to $T_{wf} = 200$.
3. Go down to $R_{sb} = 400$.
4. Go left to answer, locating Point D on the way: $\mu_{ob} = 1.0$ cp.
5. Since $p_{wf} > p_b$, $\mu_{owf} > \mu_{ob}$. From Point D, go down to read: viscosity increase = 0.07 cp/1000 psi.
6. $\mu_{owf} = \mu_{ob} + \Delta\mu(p_{wf} - p_b)/1000 = 1.0 + 0.07(2700 - 1700)/1000 = 1.07$ cp.

FIGURE 3.22 *Oil Viscosity. Courtesy Schlumberger Well Services.*

tory of the well should be examined. Typically, production rates may be quoted as:

900 BFPD
GOR 500 cf/B
Water cut 15%

The first job is to translate those figures into the three components:

$q_{wsc} = 900 \cdot 0.15 = 135$ BWPD.
$q_{osc} = 900 - 135 = 765$ BOPD.
$q_{gsc} = 765 \cdot 500 = 382.5$ Mcf/D.

The next question is, "what will be the downhole flow rate?" In order to answer that question, the values of the formation volume factor must be found. For example, given $B_w = 1.1$, $B_o = 1.3$, and $1/B_g = 150$, the downhole flow rate can be estimated as:

$$q_{wwf} = 135 \cdot 1.1 = 148.5 \text{ BWPD}$$
$$q_{owf} = 765 \cdot 1.3 = 994.5 \text{ BWPD}$$
$$q_{gwf} = \frac{382.5}{150} \cdot \frac{1000}{5.615} = \underline{454.1} \text{ BGPD}$$

for a total downhole rate of 1597.1 barrels of fluid per day.

The next question will be "can free gas be expected to flow downhole at well flowing conditions?" If $p_{wf} < p_b$, then the answer is a definite "yes." However, there remains the question "how much free gas will be flowing downhole?" To answer this question, it is necessary to find the solubility of the gas in oil at pressures below the bubble point. This may be done by applying the equation

$$R_s = k \cdot R_{sb},$$

where k can be found as a function of p_{wf}/p_b from figure 3.20. For example, a well has the following characteristics:

$\gamma_g = 0.7$.
$\gamma_o = 40$ API.
$T_{wf} = 200°F$.
$p_b = 3800$ psi.
$p_{wf} = 2800$ psi.
$R_{sb} = 1000$ cf/B.

From figure 3.20, k is found to be 0.8. Therefore, $R_s = 800$ cf/B. This means that there are 200 cubic feet of gas flowing free (1000 − 800) for each barrel of oil. If the well flows at 600 BOPD, then the bottomhole flow rate of free gas is $600 \cdot 200 = 120{,}000$ scf/D. Note that this is expressed at standard conditions. If $1/B_g = 120$, the actual downhole flow rate is 1000 scf/D or 178 BGPD.

A quick graphic method of solving these problems is to return to figure 3.18, which is reproduced in figure 3.23. From the pivot-point **b** defined by γ_o and γ_g, two lines may be drawn, one through p_b and another through p_{wf}. These will define Points **a** and **a′**. Joining **a** to T_{wf} defines R_{sb}. Joining **a′** to T_{wf} defines R_s. The difference between R_{sb} and R_s then quantifies the amount of free gas flowing downhole:

$$q_{gwf} = q_{osc}(R_{sb} - R_s).$$

QUESTION #3.16 q_{owf}, q_{wwf}, AND q_{gwf}

Given:

q_{osc} = 50 B/D.
q_{gsc} = 80 Mcf/D.
q_{wsc} = 300 B/D.
γ_o = 450 API.
γ_g = 0.65.
NaCl = 80,000 ppm.
p_{wf} = 3200 psi.
T_{wf} = 170°F.

a. Find q_{owf} = ______ BOPD.
b. Find q_{wwf} = ______ BOPD.
c. Find q_{gwf} (free) = ______ cf/D.

APPENDIX: STANDARD PRESSURES AND TEMPERATURES

Place	Standard Pressure (psi)	Standard Temperature (°F)
Arkansas	14.65	60
California	14.73	60
Colorado	15.025	60
Illinois	14.65	60
Kansas	14.65	60
Louisiana	15.025	60
Michigan	14.73	60
Mississippi	15.025	60
New Mexico	15.025	60
Oklahoma	14.65	60
Texas	14.65	60
Utah	15.025	60
West Virginia	14.85	60
Wyoming	15.025	60
U. S. Federal Leases	14.73	60
Canada	14.696	59 (15°C)

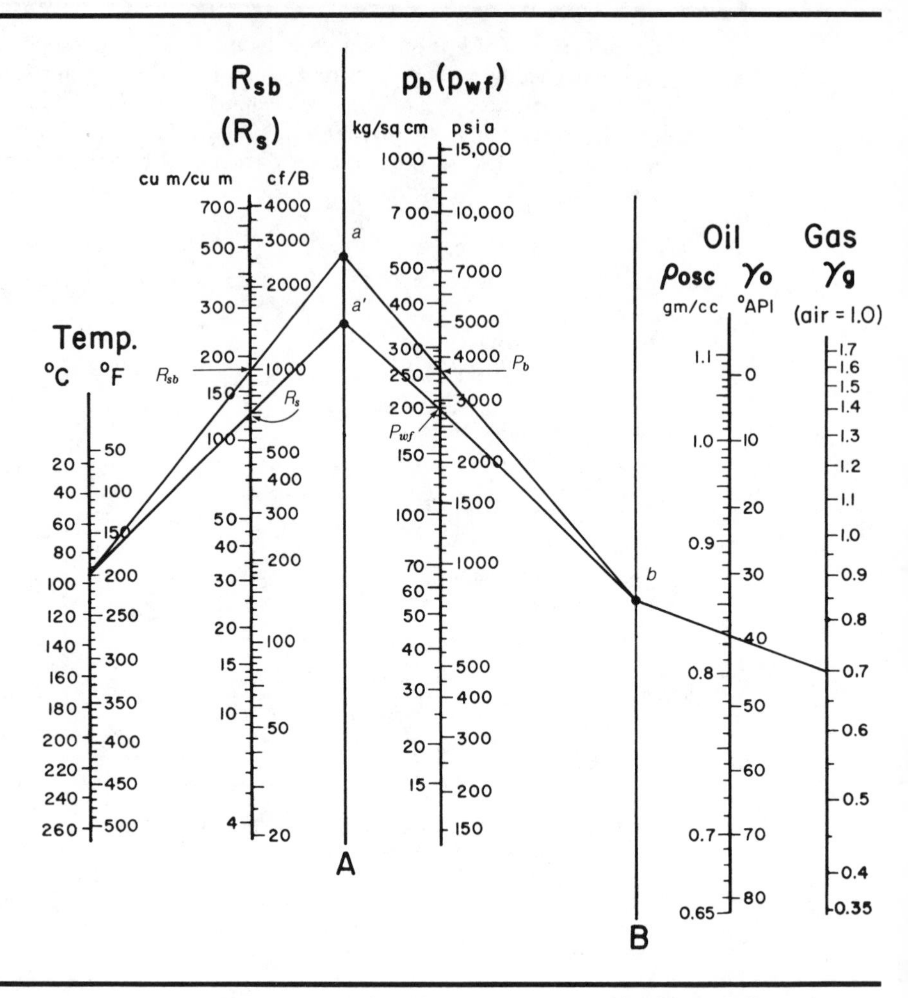

FIGURE 3.23 *Finding Gas Rate. Courtesy Schlumberger Well Services.*

BIBLIOGRAPHY

Brown, G. G., Katz, D. L., Oberfell, G. G., and Alden, R. C.: "Natural Gasoline and the Volatile Hydrocarbons," Natural Gas Association of America, Tulsa (1948).

Calhoun, John C.: "Fundamentals of Reservoir Engineering," University of Oklahoma Press, Norman, Okla. (1953).

Dodson, C. R. and Standing, M. B.: "Pressure-Volume-Temperature and Solubility Relations for Natural-Gas-Water Mixture," *Drill. and Prod. Prac.*, API (1944), 173.

"Fluid Conversions in Production Log Interpretation," Schlumberger Limited, New York (1974).

"Production Log Interpretation," Schlumberger (1970).

"Reservoir and Production Fundamentals," Schlumberger (1980).

Standing, M. B.: "A Pressure-Volume-Temperature Correlation for Mixtures of California Oils and Gases," *Drill. and Prod. Prac.* API (1947) 275.

Standing, M. B. and Katz, D. L.: "Density of Crude Oils Saturated with Gas," *Trans.*, AIME (1942) 146.

Answers to Text Questions

QUESTION #3.1

a. $\rho_{wsc} = 1.115$
b. $\rho_{wwf} = 1.08$
c. $B_w = 1.032$
d. $q_{wwf} = 310$ BWPD

QUESTION #3.2

a. $R'_{sw} = 12$ cf/B
b. $F_{bc} = 0.9$
c. $R_{sw} = 10.8$ cf/B

QUESTION #3.3

$\mu_{wwf} = 0.4$ cp

QUESTION #3.4

$1/B_g = 190$

QUESTION #3.5

a. $T_{pc} = 420°R$
b. $P_{pc} = 662$ psi

QUESTION #3.6

a. $P_{pr} = 4.4$
b. $T_{pr} = 1.45$
c. $Z = 0.76$

QUESTION 3.7

a. $1/B_g = 222.6$
b. $q_{gwf} = 4491.6$ scfg/D

QUESTION #3.8

$\rho_{gwf} = 0.086$ g/cc

QUESTION #3.9

$\mu_g = 0.02$ cp

QUESTION #3.10

$p_b = 3900$ psi

QUESTION #3.11

a. $B_{ob} = 1.9$
b. $q_{owf} = 1900$ BOPD

QUESTION #3.12

a. $C_o = 10 \times 10^{-6}$
b. $k = 0.5$

QUESTION 3.13

a. $B_o = 1.43$
b. $B_o = 1.22$

QUESTION #3.14

$\rho_{owf} = 0.68$ g/cc

QUESTION #3.15

a. $\mu_{ob} = 0.2$ cp
b. $\mu_{owf} = 0.22$ cp

QUESTION #3.16

a. $q_{owf} = 80$ BOPD
b. $q_{wwf} = 306$ BOPD
c. q_{gwf}(free) $= 98$ cf/D

CHAPTER FOUR

FLOW REGIMES

In this chapter, types of fluid flow will be discussed. In particular, three important concepts will be covered:

1. laminar and turbulent flow
2. superficial velocity
3. holdup

A proper understanding of these concepts is essential before any quantitative interpretation of production logs is possible.

LAMINAR AND TURBULENT FLOW

The flow of fluids in pipes can take place in a smooth, "streamlined" fashion or in a turbulent mode. The controlling factors are the fluid density, the superficial velocity, the pipe diameter, and the fluid density. The superficial velocity ($\overline{V}$) is defined as the volumetric flow rate divided by the area of the pipe available for flow; that is,

$$\overline{V} = 4q/\pi d^2.$$

The actual velocity of the fluid will be greater than $\overline{V}$ in the center of the pipe and less than $\overline{V}$ at the fluid/pipe interface, just as a river flows more rapidly at its middle than at its banks. A plot of fluid velocity across the pipe is shown in figure 4.1. Note that the velocity contrast between the center of the pipe and the superficial (or average) velocity is greater in the case of laminar flow than in the case of turbulent flow.

If a logging tool, such as a spinner flowmeter, takes a measurement of fluid velocity in the center of the pipe, then it holds that some corrections must be made before converting that velocity to a volumetric flow rate. The normal practice is to take 85% of the fluid velocity in the center of the pipe as being the superficial velocity. This is based on the assumption that flow is turbulent.

The kind of flow that will occur is predicted by use of the Reynolds number, N, which is defined as

$$N = \rho V d/\mu,$$

where:

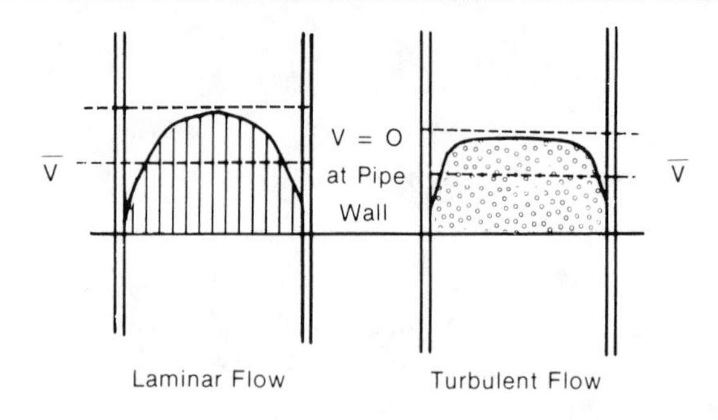

FIGURE 4.1 *Velocity Profiles for Laminar and Turbulent Flow. Courtesy Dresser Atlas.*

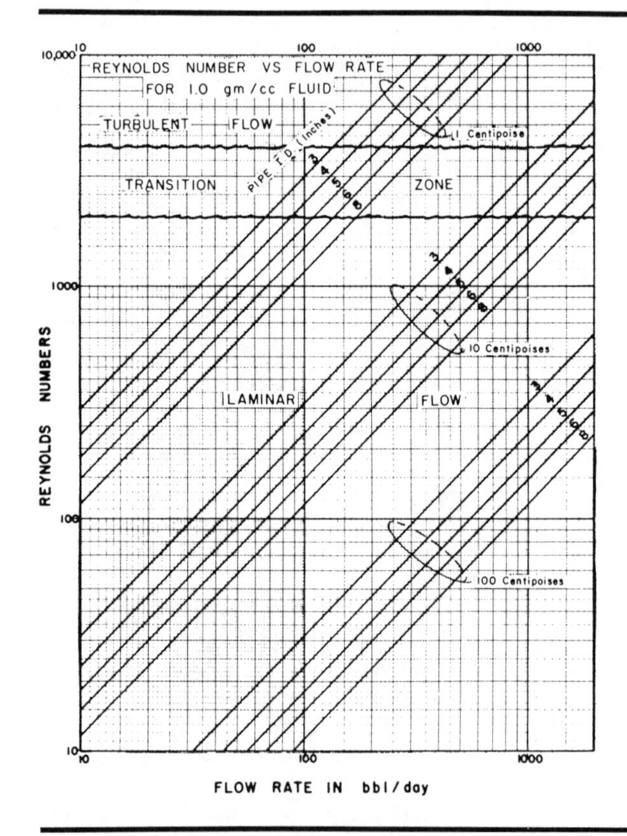

FIGURE 4.2 *Flow Rate vs. Reynolds Number. Courtesy Schlumberger Well Services.*

ρ = fluid density in g/cc,
V = superficial velocity in cm/s,
d = inside pipe diameter in cm, and
μ = fluid viscosity in poise.

Reynolds numbers in laminar flow are less than 2000. In turbulent flow, they are greater than 4000. Figure 4.2 shows a plot of flow rate vs. Reynolds number for a number of pipe IDs and viscosities.

QUESTION #4.1
Given:

ρ = 1.0 g/cc.
μ = 1 cp.
d = 5 inches.
q = 100 B/D.

a. Find N from figure 4.2.
b. What kind of flow would you expect?

Returning to the problem of correcting flowmeter-measured velocities to superficial velocities we can now quantify the correction factor as a function of the Reynolds number. Figure 4.3 plots the correction factor directly as a function of N.

UNIT CONVERSIONS

At this point, it is worthwhile to pause and face the problem of units. Until we are all comfortable with the SI system of units, there will be confusion when dealing with speeds, flow rates, etc., that are mixed. Flowmeters measure fluid velocities in feet/minute. Flow rates are quoted in barrels/day. Pipe IDs are in inches, etc. Thus, some handy conversion formulae are in order.

Velocity

$$V = 0.363(q/d^2),$$

where:

V = superficial velocity in cm/s,
q = flow rate in B/D, and
d = pipe ID in inches.

And:

$$V \text{ ft/min} = 1.969 V \text{ cm/s}.$$

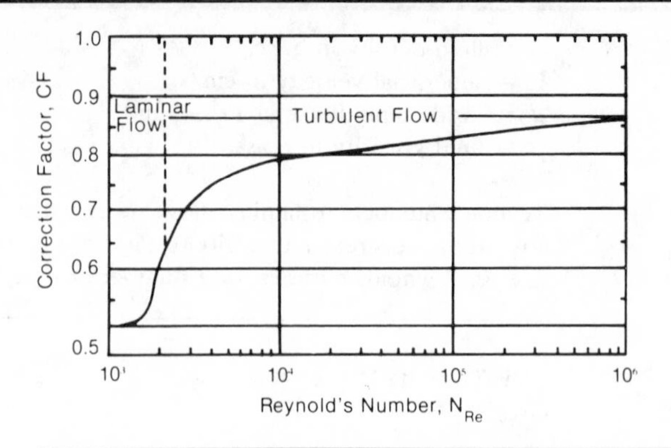

FIGURE 4.3 *Correction Factor vs.* N. *Courtesy Dresser Atlas.*

Flow Rate

$q = 2.75 V d^2$,

where:

q = flow rate in B/D,
V = superficial velocity in cm/s, and
d = pipe ID in inches.

A more complete set of conversion factors can be found in appendix A.

FLOW REGIMES

If more than one fluid is flowing in the pipe, the flow regime becomes more complex. For example, liquid and gas mixtures can form a number of flow regimes depending on the relative amounts of each in the total flow pattern. These can be categorized as:

bubble flow	bubbles of gas in liquid
slug flow	gas bubbles reach pipe ID
froth flow	gas carries liquid froth with it
mist flow	liquid carried as a mist in the gas

Figure 4.4 plots dimensionless liquid velocity against dimensionless gas velocity and exhibits these various flow regimes. Note that, in general, the lighter phase in the mixture travels fastest. Bubbles of gas in a glass of beer can be seen to rise through the static liquid. Experiments show that the difference in velocity (slip velocity) of

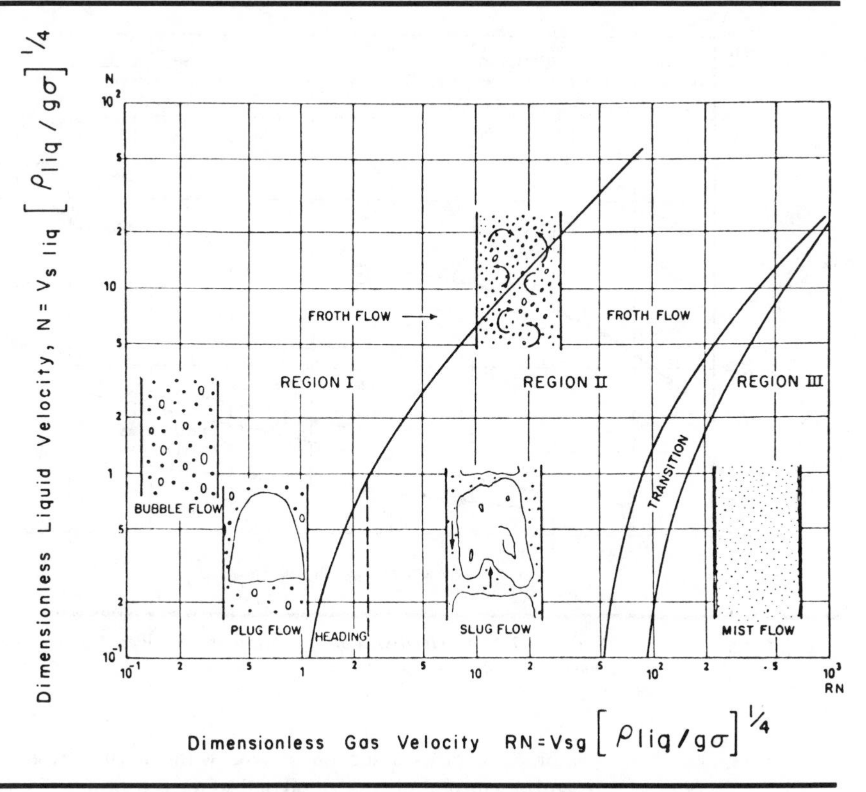

FIGURE 4.4 *Gas/Liquid Flow Regimes. Courtesy Schlumberger Well Services. Reprinted by permission of the SPE-AIME from Ros 1961, fig. 3, p. 1042. © 1961 SPE-AIME.*

two phases is generally related to their density difference. Typical values for slip velocities are:

oil–water mix	20 to 30 ft/min
oil–gas mix	5 to 10 ft/min
gas–water mix	40 to 50 ft/min

Slip velocity as a function of density difference is shown in figure 4.5. The subscripts *hp* and *lp* refer to the heavy and light phases respectively.

HOLDUP

The concept of two (or more) phases moving together up a pipe, but each with its own velocity, leads to the concept of *holdup.* To understand holdup, it is vital to bear in mind that in an oil–water mix, for example, the oil travels up the pipe faster than the water. This has a remarkable effect on production logging tools that measure fluid density.

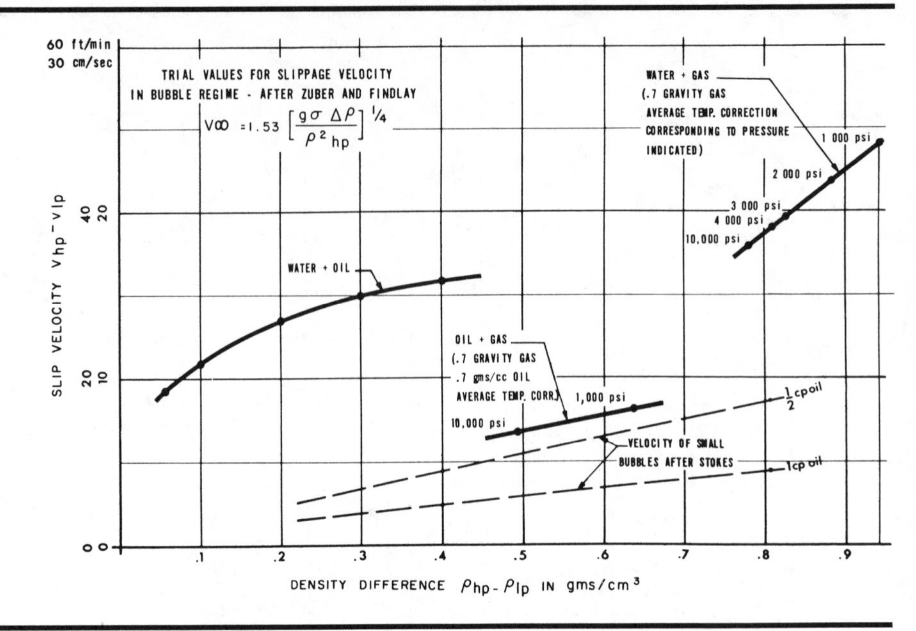

FIGURE 4.5 *Slip Velocities. Courtesy Schlumberger Well Services, with data from Zuber and Findlay 1965.*

Figure 4.6 shows a section of pipe with internal cross-sectional area A, in which a mixture of oil and water is flowing. The velocity of the water is V_w. The velocity of the oil is V_o. Since the oil is traveling faster by an amount V_s, the slip velocity, we may write

$$V_o = V_w + V_s.$$

The flow rates are q_w and q_o. If we define y_w as the fraction of the cross-sectional area taken up by the water, we may write:

$$q_w = y_w A V_w,$$
$$q_o = (1 - y_w) A V_o, \text{ and}$$
$$q_t = q_o + q_w (\text{total flow}).$$

Combining all these equations it is possible to derive an expression for y_w:

$$y_w = \frac{(AV_s - q_t) \pm \sqrt{(q_t - AV_s)^2 + 4q_w AV_s}}{2AV_s}.$$

That is, we find that y_w, the fraction of the pipe area actually occupied by water, is a quadratic function of the actual water cut. Figure 4.7 shows a plot of y_w, the water holdup, against total flow as a function of water cut for flow in a 6-in. ID pipe. Note that, for

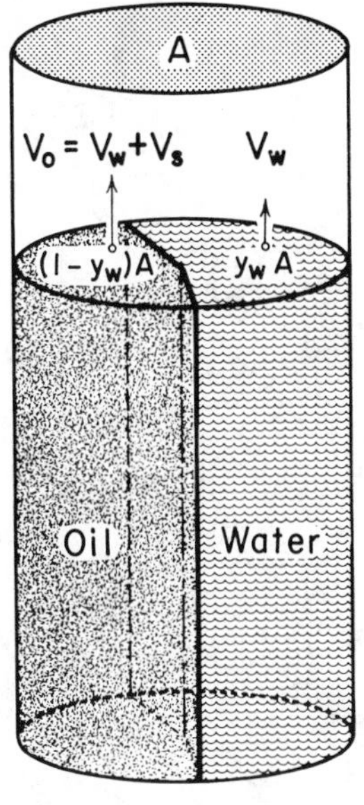

FIGURE 4.6 *Simplified Model Showing the Relationship Between Flow Rate* (q), *Holdup* (y), *and Velocity* (V). *Courtesy Schlumberger Well Services.*

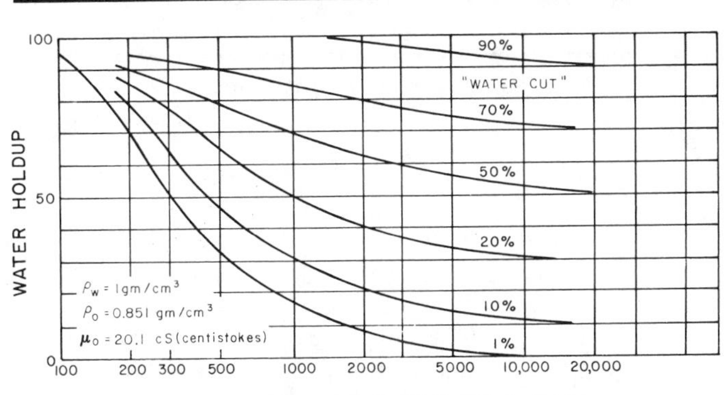

FIGURE 4.7 *Water Holdup vs. Flow Rate. Courtesy Schlumberger Well Services.*

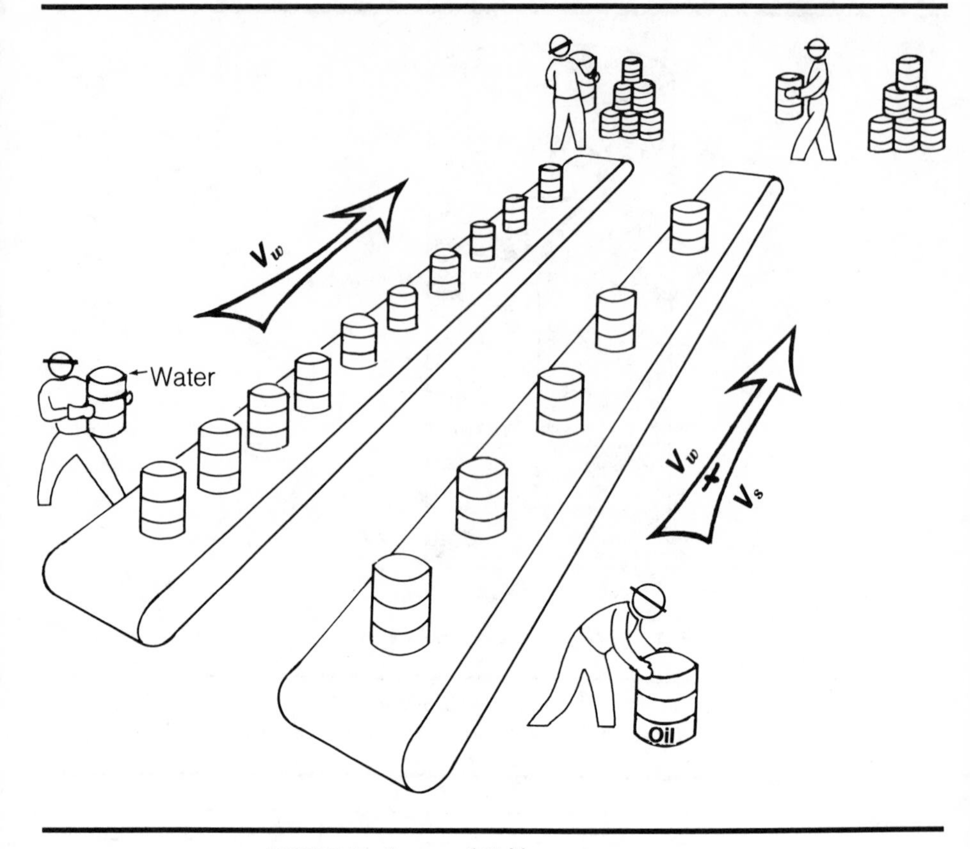

FIGURE 4.8 *Concept of Holdup.*

example, with a 50% water cut at a flow rate of 200 B/D some 90% of the pipe area is in fact occupied with water.

This surprising result may be grasped more easily by reference to figure 4.8. Two conveyerbelts are transporting barrels. One travels a constant speed of V_w transporting water and the other travels faster, at a speed of $V_w + V_s$, transporting oil. On both belts, one barrel is placed every minute. At the other end, one barrel of each is off-loaded each minute. So the "production" of oil and water is equal on the two belts. However, someone unaware of the difference in speed of the two belts could conclude that the system was "producing" more water than oil since there are more barrels of water on the top belt than barrels of oil on the bottom belt. The gradiomanometer tool, for example, just measures liquid density and has no knowledge of flow rates or slip velocities. Proper interpretation of the average density requires computation of the holdup.

QUESTION #4.2 y_w

Pipe area = 200 cm².
Slip velocity = 10 cm/s.
Total flow rate = 4800 cc/s.
Water cut = 50%.

a. Find y_w.
b. If ρ_o = 0.6 g/cc and ρ_w = 1.0 g/cc, what will be the density of the flowing mixture?

BIBLIOGRAPHY

"Interpretative Methods of Production Well Logs," second edition, Dresser Atlas (1982).

"Production Log Interpretation," Schlumberger (1973).

Ros, N. C. J.: "Simultaneous Flow of Gas and Liquid as Encountered in Well Tubing," *J. Pet. Tech.* (October 1961).

Zuber and Findlay: "Average Volumetric Concentration in Two-Phase Flow Systems," *J. of Heat Transfer* (1965) 87, 453.

Answers to Text Questions

QUESTION #4.1

a. $N = 1847.6$
b. Laminar

QUESTION #4.2

a. $y_w = 0.6$
b. $\rho_{mix} = (0.6 \cdot 1.0) + (0.4 \cdot 0.6) = 0.84$ g/cc

CHAPTER FIVE

FLOWMETERS

APPLICATIONS

Tools that measure flow rate or trace where fluids have gone have applications in both producing and injecting wells. Specifically, these types of tools may be used for:

1. production profiles
2. injection profiles
3. finding casing and tubing leaks
4. gauging the effectiveness of well treatments

The tools available for making the appropriate measurements are:

Packer flowmeters (including basket type)
Continuous flowmeters (spinners)
Full-bore flowmeters
Radioactive tracers (gamma ray)
Thermometers

Each tool has its particular merits for certain well conditions and ranges of flow rates. Table 5.1 summarizes each tool's applicability.

PACKER AND BASKET FLOWMETERS

Packer flowmeters and basket flowmeters both operate in a similar mode. The tool is lowered through the tubing and then held stationary in the casing. The total flow is then forced to pass through a central tube containing a spinner. This is achieved in the case of the packer flowmeters by means of an inflatable packer bag. The basket flowmeters open overlapping metal "petals." Figure 5.1 illustrates a packer flowmeter with its packer bag inflated. Stationary measurements are normally made in the casing between sets of perforations. The flow into or out of any set of perforations is deduced from the measured flow above and below that set of perforations. Raw tool readings are in terms of spinner revolutions per second. Depending on the diameter and pitch of the spinner used, the measured rps are converted into a flow rate in barrels per day. Presentation of the log may be in the form of a flow profile or by contribution to total flow by zone. Figure 5.2 illustrates a basket flowmeter survey for a producing well.

The advantages of packer and basket flowmeters are: (1) in biphasic flow, holdup becomes insignificant, because since flow is re-

TABLE 5.1 *Flow-Measuring Devices*

Tool	Flow Rate
Packer/basket	Low flow (10 to 1900 B/D)
Continuous flowmeters	High flow (3000 B/D and up)
Full-bore flowmeters	Medium flow (50 B/D and up)
Radioactive tracers	Any rate
(Thermometers)	(Qualitative only)

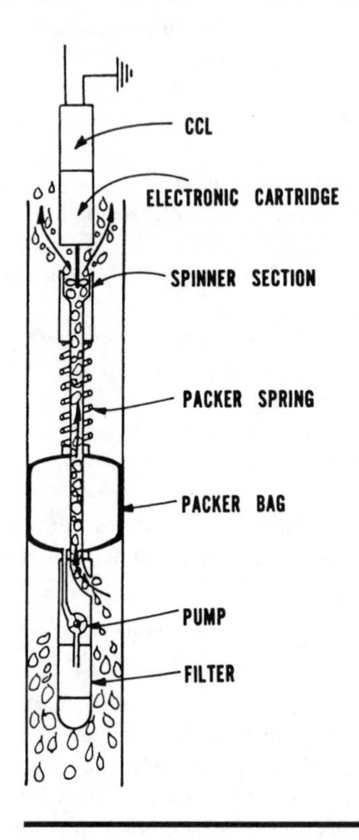

FIGURE 5.1 ***Inflatable Packer Flowmeter. Courtesy Schlumberger Well Services.***

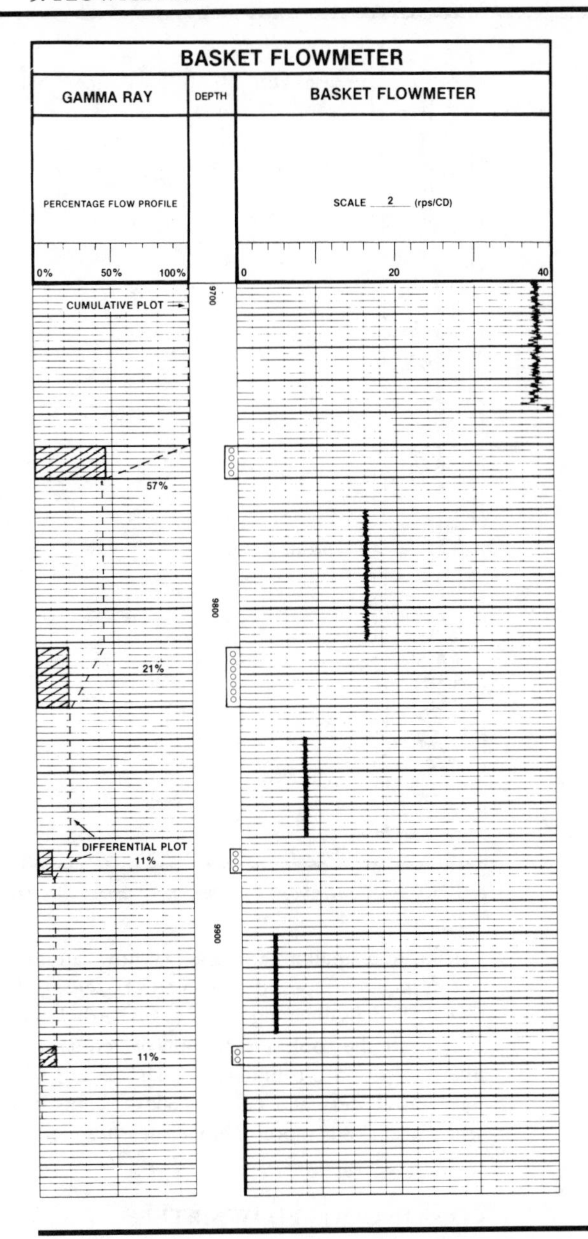

FIGURE 5.2 *Basket Flowmeter Survey. Courtesy Dresser Atlas.*

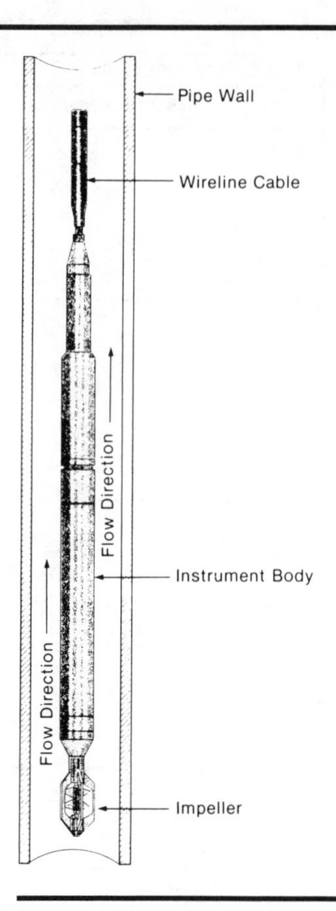

FIGURE 5.3 *Continuous Flowmeter. Courtesy Dresser Atlas.*

stricted to a small area, the actual flow rate in the tool is very high. (2) In inclined pipes, the problems of segregation of fluids are minimized. The disadvantages are: (1) A fairly large pressure drop develops across the tool, and this may eventually blow the tool uphole like a piston in a cylinder. (2) The packer bags are quite prone to mechanical wear and once a leak develops it is impossible to get a seal and make a representative measurement.

CONTINUOUS FLOWMETERS

The continuous flowmeter is so named because, unlike the packer flowmeter which makes station measurements, it makes a continuous log of flow against depth. Figure 5.3 shows the general form of these tools. Figure 5.4 shows details of the spinner section. Note that the cross-sectional area "seen" by the spinner is very small compared to the total area available to wellbore fluids. For this reason, these tools are less sensitive than the packer flowmeter, but can record much higher flow rates without danger of being shot out of the hole.

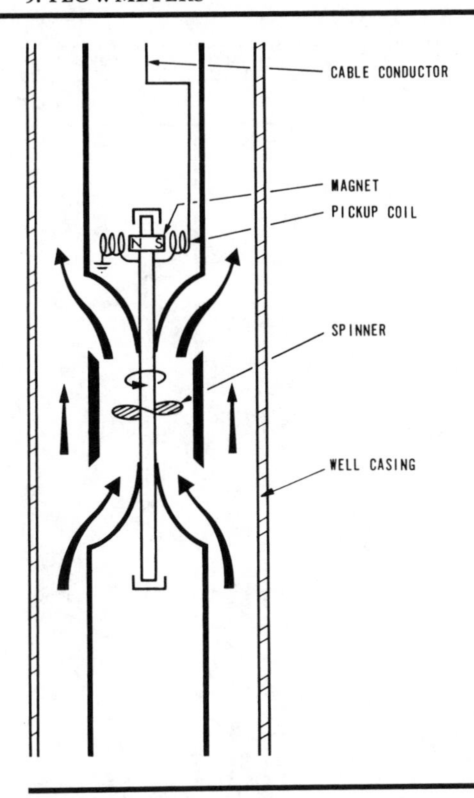

FIGURE 5.4 *Detail of Spinner Section of a Continuous Flowmeter. Courtesy Schlumberger Well Services.*

Since the continuous flowmeter log is recorded with the tool in motion, the fluid velocity active on the spinner is the algebraic sum of the cable speed *and* the fluid speed. The normal mode of operation is to move the tool against the flow. An important part of the measurement is thus the cable speed itself and this is usually recorded in Track 1 of the log. Figure 5.5 shows a typical continuous flowmeter recording.

FULL-BORE FLOWMETERS

The full-bore flowmeter is a variation of the continuous flowmeter. In the collapsed mode, the tool will fit inside the tubing. Once in the casing, the tool may be opened and spinner blades will be extended to cover a large portion of the area available for wellbore fluids. This mechanical marvel thus extends the usefulness of continuous flowmeters to cover low-flow rates and can record flows of as little as 200 B/D in 5½-in. pipe. Figure 5.6 shows the full-bore flowmeter in the closed and opened modes.

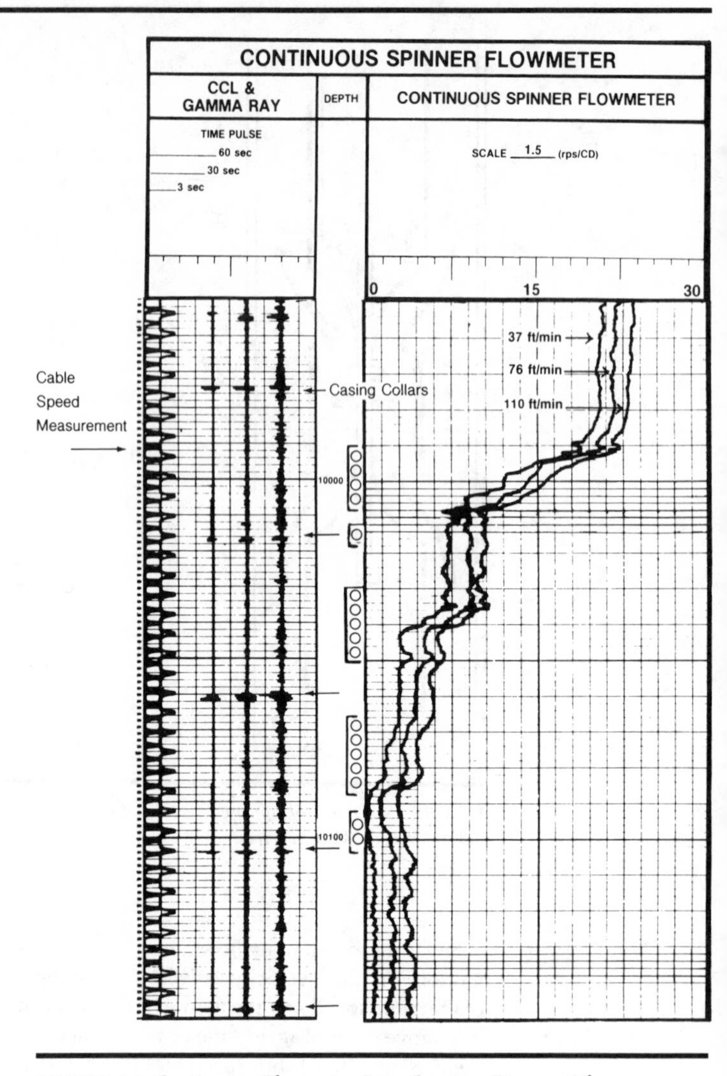

FIGURE 5.5 *Continuous Flowmeter Log. Courtesy Dresser Atlas.*

COMBINATION TOOLS

It should be noted that flowmetering tools are frequently run in combination with other sensors—temperature, fluid density, fluid pressure, etc. By a relay system, the operator may select one sensor at a time and make several different production logs on one trip in the hole. Examples of these combined tools include the PCT (production combination tool) illustrated in figure 5.7.

INTERPRETATION OF FLOWMETER SURVEYS

Interpretation of the packer flowmeter is straightforward, since flow rate is a direct function of spinner rps. Interpretation of the continu-

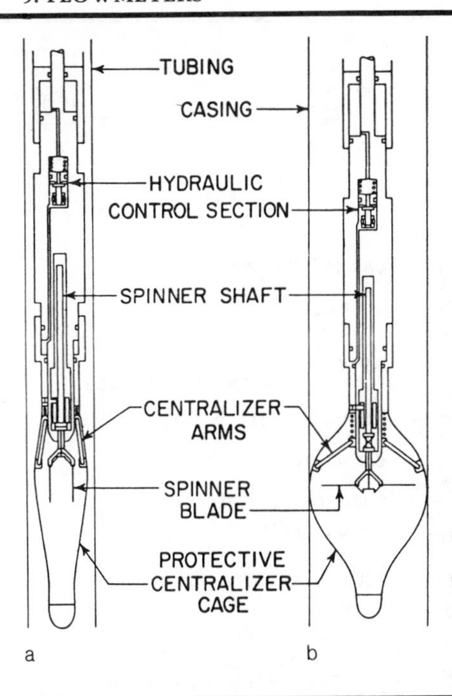

FIGURE 5.6 *Full-Bore Flowmeter.* (a) *Tool Closed,* (b) *Tool Open. Courtesy Schlumberger Well Services.*

ous flowmeters, however, is more intricate. If the spinner were frictionless, the response of the tool would be linear and depend only on the geometry of the spinner. Figure 5.8 plots rps against fluid velocity for the ideal frictionless-spinner case.

The fact that spinner mountings are not frictionless causes the real response to differ from the ideal. Additionally, the effects of the viscosity of the fluid cause further deviations from the ideal. Figure 5.9 shows the same plot of rps against fluid velocity but adds the effects of fluid viscosity. The viscous-fluid response lines are shown as being continuous; however, the friction of the spinner in its mounting results in a threshold below which the spinner will not turn. The true response therefore is as shown in figure 5.10. Note that where fluid velocities are small, regardless of direction, the spinner will not turn at all.

The generally accepted practice is to calibrate the continuous flowmeter in the well itself rather than trying to estimate the effects of viscosity and friction. The technique calls for several passes to be made in the well at different cable speeds both against and with the flow direction. A plot may then be prepared from which the analyst can deduce the fluid speed at each point in the well. Figure 5.11 illustrates the technique. The tool velocity (from the cable speed and direction recording in Track I) is plotted against rps for a single depth in the well for several passes of the tool. In this example, four passes were made going down and three coming up. The proce-

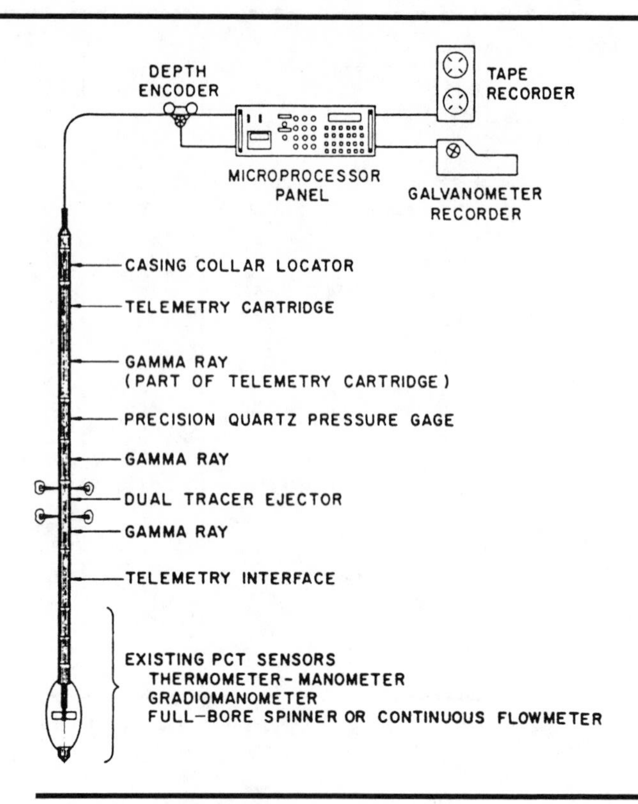

FIGURE 5.7 *Production Combination Tool. Courtesy Schlumberger Well Services.*

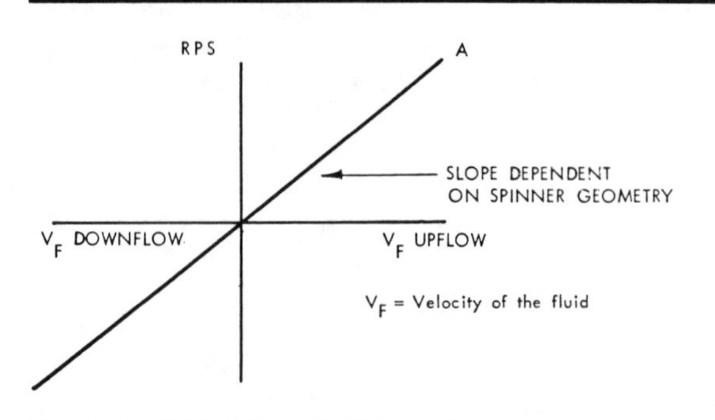

FIGURE 5.8 *Ideal Spinner Response. Courtesy Schlumberger Well Services.*

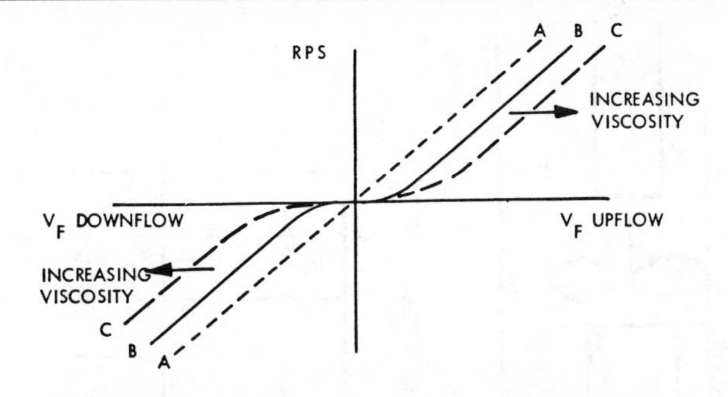

FIGURE 5.9 ***Spinner Response in Viscous Fluids. Courtesy Schlumberger Well Services.***

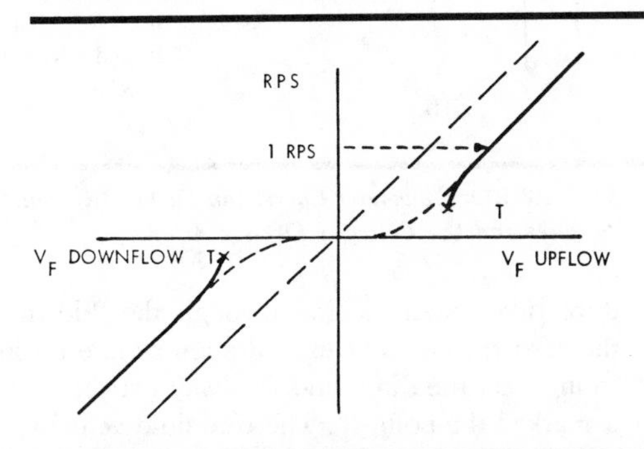

FIGURE 5.10 ***True Response of Spinner. Courtesy Schlumberger Well Services.***

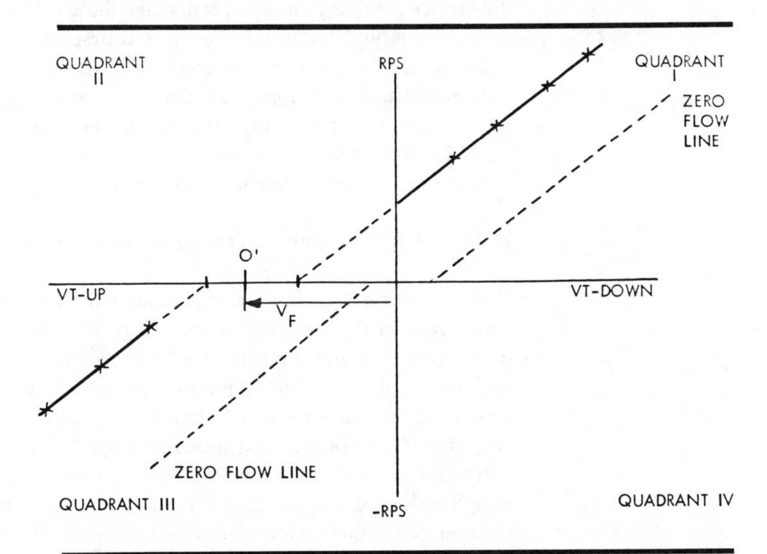

FIGURE 5.11 ***Calibration of Continuous Flowmeter. Courtesy Schlumberger Well Services.***

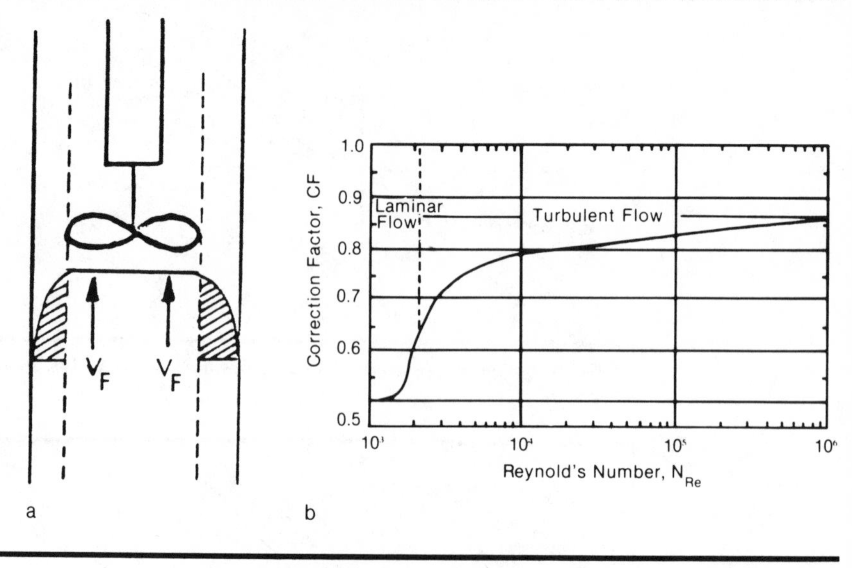

FIGURE 5.12 ***Flowmeter Correction Factor.*** **(a)** ***Courtesy Schlumberger Well Services and*** **(b)** ***Courtesy Dresser Atlas.***

dure is to extend a line through the "down" points to intersect the zero rps line *and* extend another line through the "up" points to intersect the same line. Midway between these two intersections is marked the point 0′. The true fluid velocity is then read off from zero tool velocity line to the point 0′. A graph of this sort is prepared for each point in the well for which a flow rate is required. (Normally the service company engineer will make these calibration plots. However, you should know how to do it yourself.)

Note that flow rate is still unknown. The tool has measured the fluid speed and direction. It remains to convert the speed to a volumetric flow rate by multiplying by the cross-sectional area of the pipe. Conventionally, flowmeter results are quoted as fluid speeds in feet per minute. A useful conversion is:

$$B/D = 1.4 \times \text{ft/min} \times (\text{casing ID in inches})^2.$$

Before making the conversion, remember that flowmeters are run centralized in the pipe and therefore record the speed of the fluid in the center of the pipe. Figure 5.12(a) shows the position of the tool in the flow stream together with the flow profile. The flow profile across the pipe is a function of the Reynolds number (see chap. four). Use of the fluid speed measured by a flowmeter without correction would overestimate true flow rate. The correction factor as a function of the Reynolds number is shown in figure 5.12(b). In most cases, this factor is between 0.84 and 0.85. It is applied as follows:

$$V_{\text{superficial}} = V_{\text{measured}} \times \text{correction factor}.$$

QUESTION #5.1

A flowmeter survey completed 6 passes. The cable speed, direction, and corresponding spinner rps were as follows:

Tool Speed (ft/min)		*rps*
Logging Down	10	5
	60	10
	110	15
Logging Up	−110	−5
	−160	−10
	−210	−15

a. Plot these points on linear graph paper and determine the fluid speed.
b. Use a correction factor of 0.85 and determine the flow rate in B/D for 7-in. 20-lb pipe.

Where continuous flowmeter surveys are run in bare-foot completions, a measure of the hole size is vital to the correct computation of flow rates. Such hole-size gauging can be obtained from a through-tubing caliper of the sort illustrated in figure 5.13.

RADIOACTIVE TRACERS AND THERMOMETERS

Both radioactive tracer-tool surveys and temperature logs can indicate flow rates and flow directions. However, since they each form a subject in itself they are treated separately. Chapter 6 covers the radioactive tracer logs in detail and chapter 7 the temperature logs.

MEASUREMENTS IN DEVIATED HOLES

There are grave dangers in running flowmeters in deviated holes. Controlled experiments in flow loops, where the deviated angle of the hole can be changed at will, have shown that, if the flowmeter tool lies on the low side of the pipe, it may record a downflow even in places where flow is upward. Figure 5.14 illustrates the mechanics of these erroneous measurements. The best preventive measure is to use either a packer or basket type of flowmeter.

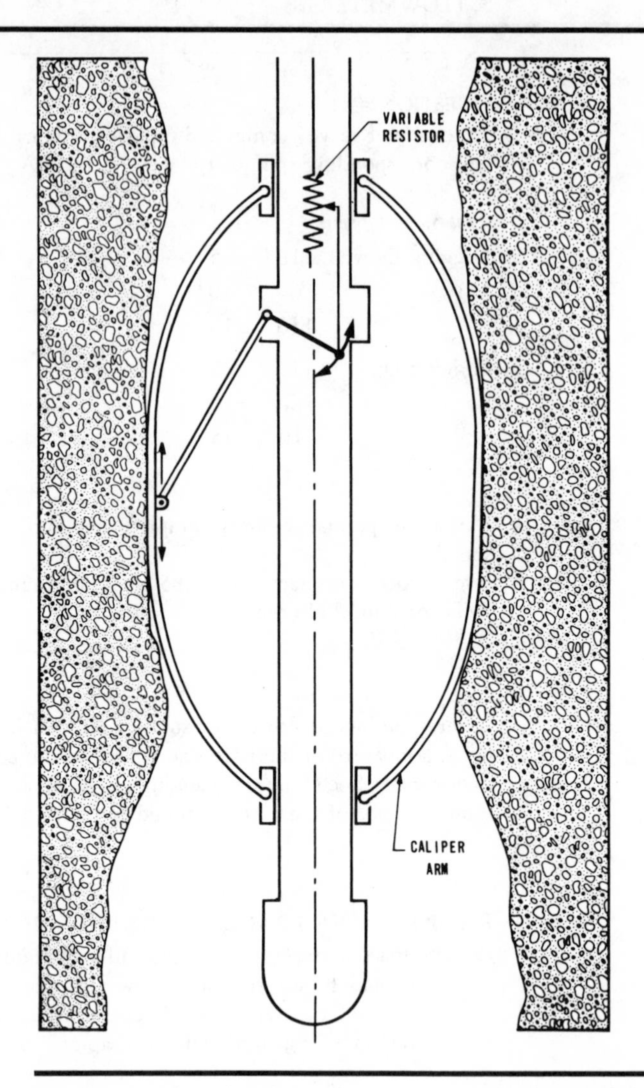

FIGURE 5.13 *Through-Tubing Caliper. Courtesy Schlumberger Well Services.*

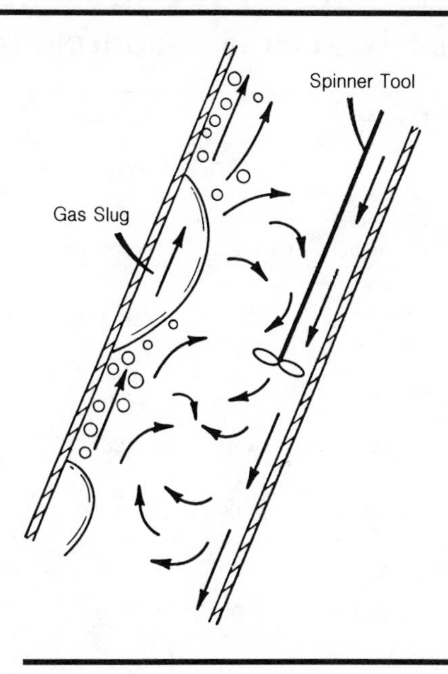

FIGURE 5.14 *Erroneous Flow Measurement in a Deviated Pipe. Reprinted by permission of the SPE-AIME from Hill and Oolman 1982, fig. 2, p. 2433. © 1981 SPE-AIME.*

APPENDIX: AVAILABLE PROFILING TOOLS

	Gearhart	Dresser	Schlumberger			Welex
	Flowmeters					
Tool	Continuous Flowmeter	Continuous Flowmeter	Continuous Flowmeter	Packer Flowmeter	Full-Bore Flowmeter	Continuous Flowmeter
Tool diameter (inches)	$1\frac{11}{16}$, $1\frac{7}{16}$	$2\frac{1}{8}$, $1\frac{11}{16}$	2, $1\frac{11}{16}$	$2\frac{1}{8}$, $1\frac{11}{16}$	Variable to casing ID	1, $\frac{11}{16}$
Maximum temperature (°F)	300	300	350	285	350	300
Maximum pressure (psi)	15,000	18,000	15,000	10,000	15,000	15,000
Minimum flow rates (BPD)	150	200	300	10	50	200
Maximum flow rates (BPD)	60,000	60,000	60,000	1,900	50,000	60,000
	Radioactive Tracer Tools					
Tool diameter (inches)		$1\frac{1}{2}$		$1\frac{11}{16}$		
Maximum temperature (°F)		350		275		
Tracer capacity (cc)				20		
Number of ejections				150–200		

BIBLIOGRAPHY

Anderson, R. A., Smolen, J. J., Laverdiere, L., Davis, J. A.: "A Production Logging Tool with Simultaneous Measurements," *J. Pet. Tech.* (Feb. 1980).

Hill, A. D., and Oolman, T.: "Production Logging Tool Behavior in Two-Phase Inclined Flow," *J. Pet. Tech.* (Oct. 1982).

Leach, B. C., Jameson, J. B., Smolen, J. J., and Nicolas, Y.: "The Full Bore Flowmeter," paper SPE 5089 presented at the SPE 49th Annual Meeting, Houston, Oct. 6–9, 1974.

Peebler, Bob: "Multipass Interpretation of the Full Bore Spinner," Schlumberger publication C-11993.

Answers to Text Question

QUESTION #5.1

a. Fluid speed = 50 ft/min
   Corrected speed = 42.5

b. Flow rate = 2471 B/D

CHAPTER SIX

RADIOACTIVE TRACER LOGS

APPLICATIONS

Radioactive tracers have three main applications:

1. determining flow rates and flow profiles
2. diagnosing completion problems
3. evaluating treatment effectiveness

The essential components of radioactive tracer logs are a radioactive material and a gamma ray detector. And the methods of using radioactive materials together with gamma ray detectors include:

Treating the well with a radioactive injectant
Using a tracer ejector tool
Monitoring the deposit of radioactive salts on the casing from water production

WELL TREATMENT

A common practice is to frac a well using specially "doped" radioactive sand. Before the frac a base gamma ray log is run. After the frac a second gamma ray log is run. Where differences in the two gamma ray logs appear, it is assumed that the formation took the frac material. Sometimes formations will be tested for injectivity before attempting a frac. In that case, after the base gamma ray log is run, injection water, containing a soluble radioactive material, will be injected. A postinject log may indicate which zones are most likely to take frac material. If a particular zone takes a disproportionate share of the injectant, there is time to place a bridge plug or perform a cement squeeze before proceeding with a frac job. Figure 6.1 illustrates a gamma ray log used to monitor well treatment.

TRACER EJECTOR TOOL

The tracer ejector tool (also known as the nuclear flolog) is illustrated in figure 6.2. A casing-collar locator helps to place the tool on depth. A reservoir of radioactive material is housed inside the tool and small quantities of it may be ejected on command by the operator at the surface. Beneath the ejector port lie one or two gamma ray detectors. There are two methods of using this tool, one known as the *velocity shot* and the other as the *timed run.*

Velocity Shot

With the velocity shot, the tool is held stationary in the wellbore and a "shot" of radioactive fluid is ejected. The two gamma ray

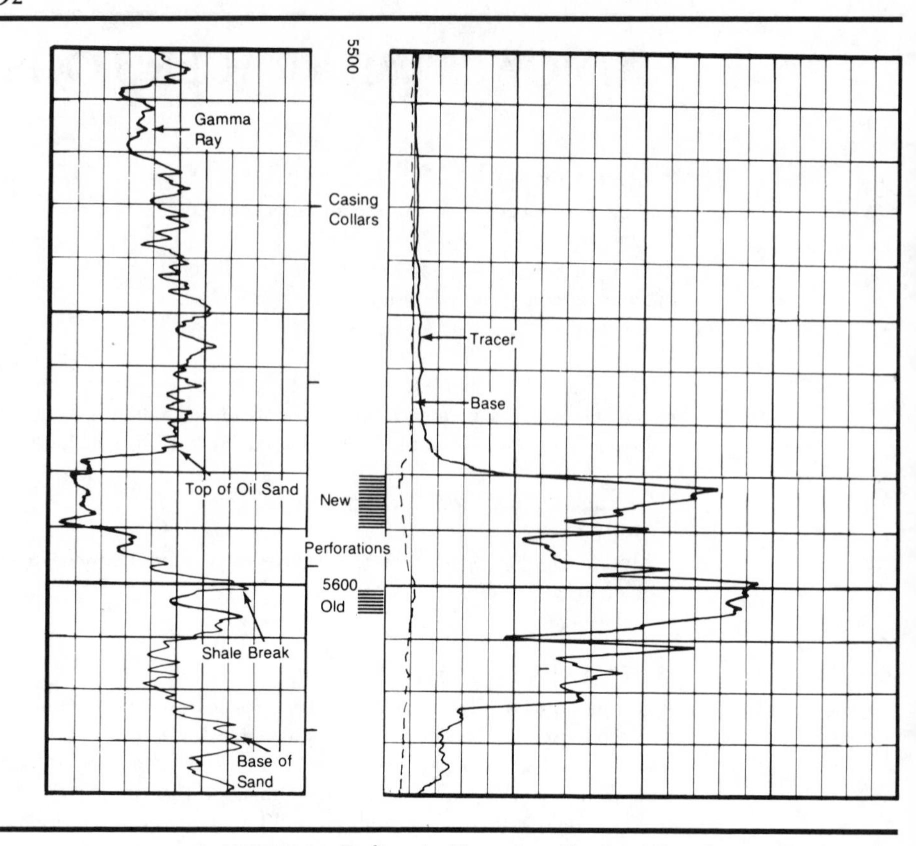

FIGURE 6.1 *Radioactive Tracer Log After Sand Frac. Courtesy Dresser Atlas.*

detectors record gamma ray intensity as a function of time. Figure 6.3 shows such a log in an injection well. Note that the passage of the radioactive slug past each detector causes a gamma ray peak. By measuring the time difference between the two peaks and knowing the spacing of the two detectors it is a simple matter to calculate the flow rate.

The factors entering into calculation of flow rate are:

Casing ID
Tool OD
Time between peaks
Distance between the detectors

It is essential to remember that the speed of the fluid in the tool/casing annulus is greater than the speed of the fluid in the undisturbed wellbore. Thus,

$$q = \frac{\text{detector spacing}}{\text{time}} \times \text{area available for flow},$$

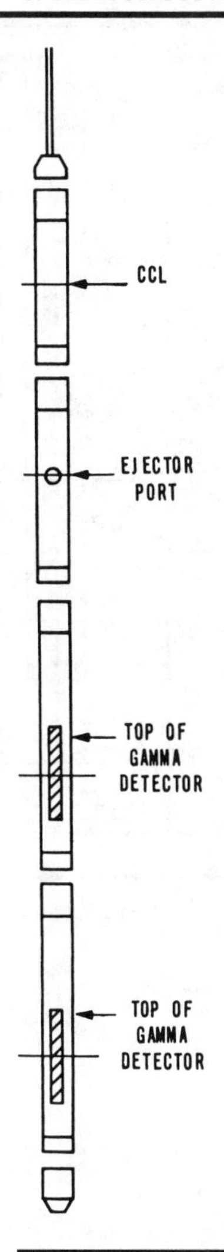

FIGURE 6.2 *The Tracer Ejector Tool. Courtesy Schlumberger Well Services.*

and

$$\text{area} = \frac{\pi}{4}[(\text{casing ID})^2 - (\text{tool OD})^2].$$

In order to express flow rates in barrels per day, some constant is required, if detector spacing, casing size, and tool OD are measured in inches and time is in seconds,

$$\text{B/D} = (\text{cubic inches/second}) \times 8.905.$$

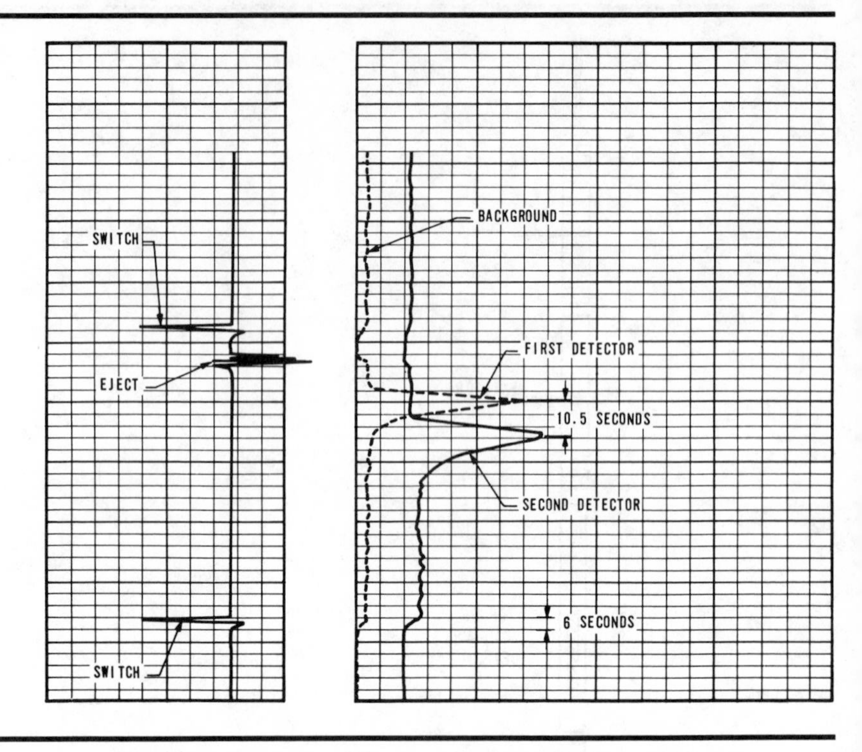

FIGURE 6.3 *Velocity Shot Log. Courtesy Schlumberger Well Services.*

QUESTION #6.1
Casing is 7-in. 26 lb.
Tool OD is $1^{11}\!/_{16}$ in.
Time between peaks = 18 s.
Detector spacing = 59 in.

Find flow rate in B/D.

Timed-Run Analysis

In the timed-run analysis, a slug of radioactive material is ejected and then a gamma ray log (vs. depth) is made repeatedly at various time intervals to trace where the radioactive slug eventually ends up. Figure 6.4 illustrates such a log.

QUESTION #6.2
Inspect the log of figure 6.4 and write down which sands actually take injection water. Explain why.

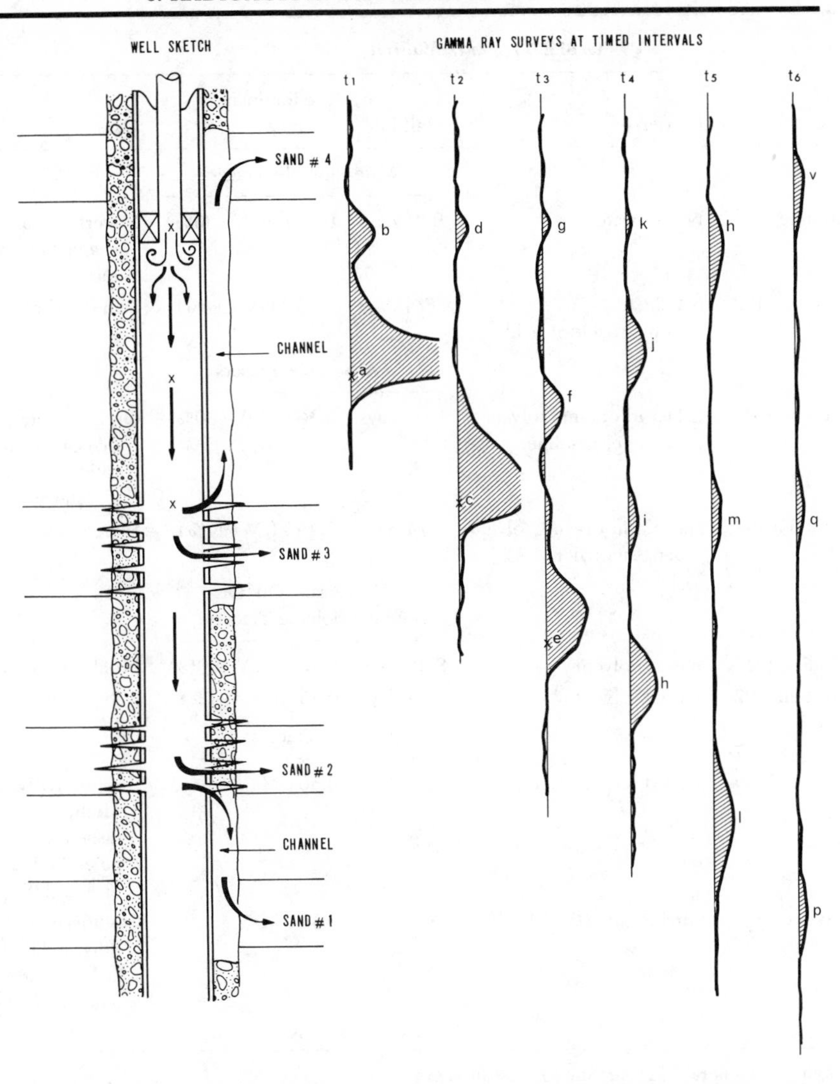

FIGURE 6.4 *Time-Run Analysis. Courtesy Schlumberger Well Services.*

CHOICE OF RADIOACTIVE TRACER MATERIALS

In general, a radioactive tracer material should be chosen that is soluble or miscible with the fluid to be profiled or traced. Since surveys may be run in oil, water, or gas flows, several different substances are required. Table 6.1 lists oil-, water-, and gas-soluble tracers. Note that the half-life is listed for each material. The half-life for most radioactive tracers is in the order of days or weeks. For safety reasons, radioactive tracers are used only in injection wells. Their use in producing wells raises problems with disposal of radioactive fluids at the surface and contamination of separator and other surface equipment.

TABLE 6.1 *Tracer Materials*

Isotope	Carrier	Half-Life	Predominant Gamma	Uses
		Water-Soluble Tracers		
Iodine-131	NaI in water	8.05 days	0.364 MeV (80%)	Waterflood profiles, channeling, cement location
Iridium-192	Na_2TrCl in hydrochloric acid	74 days	0.46 MeV (94%)	Same as above
		Oil-Soluble Tracers		
Iodine-131	C_6H_5I in organic solvent benzene, gasoline	8.05 days	0.364 MeV (80%)	Oil-injectivity profiles, oil-cement slurry location. Erratic results if water present
Iridium-192	Na_2IrCl in organic solvent benzene, xylene	74 days	0.46 MeV (94%)	Same as above
		Universal—Oil- Or Water-Soluble Tracers		
Iodine-131	Special solvent	8.05 days	0.364 MeV (80%)	Profiles in oil or water or oil–water mixtures
Iridium-192	Special solvent	74 days	0.46 MeV (94%)	
		Gas Tracers		
Iodine-131	Methyl iodide (CH_3I)	8.1 days	0.364 MeV	Gas-injectivity profiles. Methyl iodide furnished in glass ampules, boils at 72.2°C or 162.2°F
Iodine-131	Ethyl iodide (C_2H_5I)	8.1 days	0.364 MeV	Gas-injectivity profiles. Ethyl iodide furnished in glass ampules, boils at 42.5°C or 106°F

Source: General Nuclear, Inc., Houston, Texas.
Note: Aside from normal radioactivity hazard, methyl and ethyl iodides are very toxic gases with low boiling points and must be handled accordingly. Neoprene O-rings also deteriorate rapidly in the presence of these gases.

MONITORING NATURAL RADIOACTIVE DEPOSITS

In some reservoirs, the rise of the water table can be monitored very simply by running a gamma ray log at time intervals throughout the life of a well. The rising water table will make itself visible on the repeated gamma ray logs as an increase in radioactivity. It is thought that the formation water carries with it dissolved radioactive salts that deposit on the casing as precipitates, which build up over

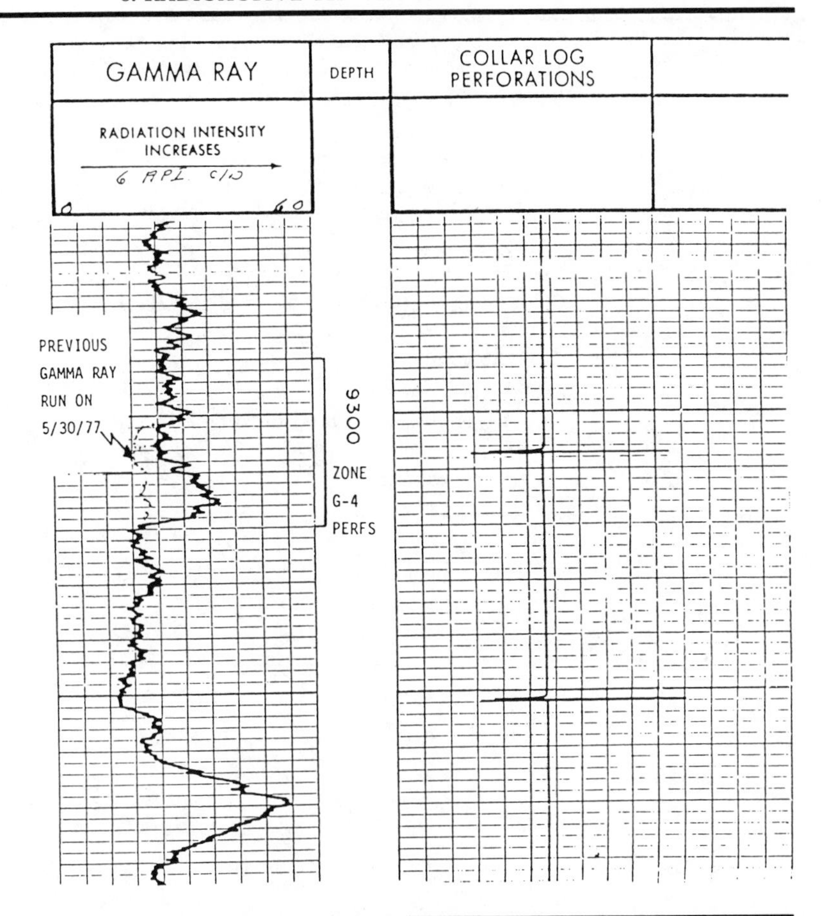

FIGURE 6.5 *Monitoring OWC Changes with a GR Log.*

a period of time. Studies suggest that a considerable volume of water production or flow past the wellbore is required before any detectable change can be seen on a GR log. Figure 6.5 shows a gamma ray log run in a well before and after production. The oil/water contact (OWC) appears to have risen from 9320 to 9302 ft in the production period.

This technique is by no means universal but should be taken advantage of when it works.

BIBLIOGRAPHY

"Interpretative Methods for Production Well Logs," second edition, Dresser Atlas (1982), section 4.

"Production Log Interpretation," Schlumberger (1973) 20–23.

Answers to Text Questions

QUESTION #6.1

Casing ID = 6.276
Tool OD = 1.6875
Area for flow = 28.7 sq in.
Speed = 59/18 = 3.28 in./s
Flow rate = 94.07 cu in./s = 837.7 B/D

QUESTION #6.2

Sands taking water:
#4 (via channel behind packer)
#3
#1 (via channel)

CHAPTER SEVEN

FLUID IDENTIFICATION

Production logging tools that can differentiate between oil, gas, and water in a producing well allow diagnosis of completion and production problems. They offer the operator assistance in planning remedial work and in monitoring reservoir performance. In particular, these tools help:

a. to pinpoint gas, oil, and water entries into, and exits from, the production string, and
b. to determine, in combination with flow measurements, how much of which fluid is produced from which zone.

TOOLS AVAILABLE

Devices used for fluid identification can be broadly categorized into two groups:

1. those that respond directly to the physical properties of wellbore fluids (such as density, dielectic constant, etc.), and
2. those that respond to the physical effects caused by the actual flow of fluids (such as temperature changes, noise, etc.).

The devices in the second group require that the well be flowing in order to function, but the devices from the first group work even if the wellbore fluids are static. Table 7.1 summarizes the fluid-typing tools available. The following devices are discussed in this chapter:

Gradiomanometer
Gamma absorption
Capacitance (dielectric)
Resonators (vibrators)
Fluid sampler
Manometer

GRADIOMANOMETER

The gradiomanometer tool (see fig. 7.1) effectively measures the pressure difference between two points by measuring the expansion or contraction of a metal bellows system filled with oil. In a vertical wellbore, the pressure difference measured may be directly converted to a pressure gradient, and hence to a fluid density in grams per cubic centimeter. In deviated wellbores, the pressure difference mea-

TABLE 7.1 *Fluid Identification Tools*

Tool Type	Measured Parameter	Service Company Name	OD (inches)	Max. Temp. (°F)	Max. Press. (psi)
Density	Pressure difference	Gradiomanometer	$1\frac{11}{16}$	350	15,000
	Gamma ray Absorption	Fluid Density Tool	$1\frac{13}{4}$, $1\frac{11}{16}$	400	15,000
	Vibration	Densimeter	—	—	—
Dielectric	Electrical capacitance	Watercutmeter Hydrolog	$1\frac{7}{16}$	300	15,000
Manometer	Pressure	Manometer	$1\frac{11}{16}$	350	15,000
HP quartz gauge	Pressure	Quartz Pressure Gauge	$1\frac{11}{16}$	300	12,000
Noise	Noise frequency spectrum	BATS	$1\frac{11}{16}$	350	15,000
		Sonan	$1\frac{25}{32}$	350	17,000
		Audio Log	$1\frac{11}{16}$	300	16,500
Thermometer	Temperature	—	—	—	—
Fluid sampler	PVT sample recovered	Borehole Fluid Sampler	$1\frac{11}{16}$	300	10,000

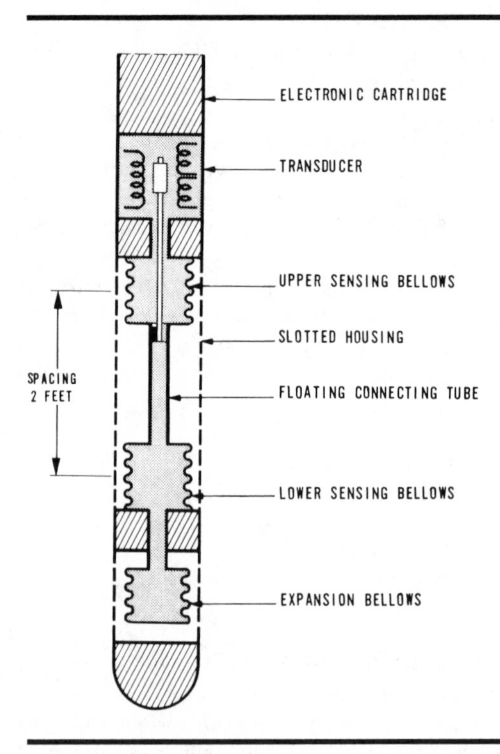

FIGURE 7.1 *Gradiomanometer Tool. Courtesy Schlumberger Well Services.*

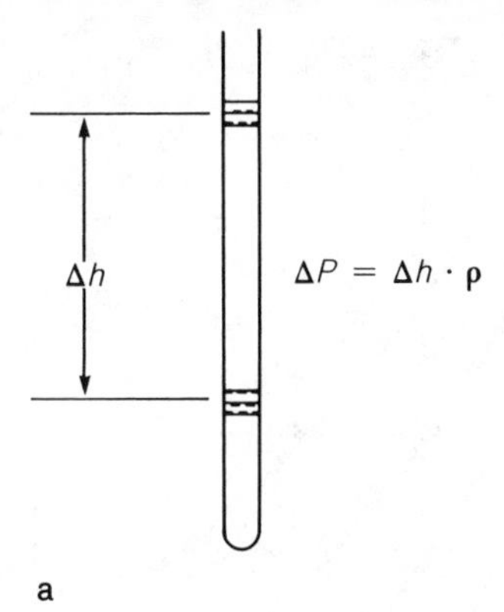

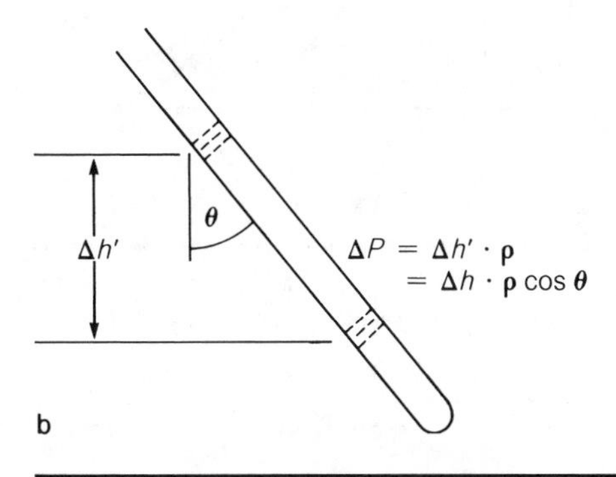

FIGURE 7.2 *Gradiomanometer Measurements in Deviated Wellbores:* (a) *Vertical Wellbore,* (b) *Deviated Wellbore.*

sured requires a correction equal to the cosine of the deviation angle in order to convert the measurement to a fluid density (see fig. 7.2). Since the measurement made is that of a pressure difference, it is assumed that the only cause of the pressure difference is the hydrostatic head of the wellbore fluid. But this assumption is not valid if fluid flow is restricted to a small cross-sectional area. In that case, the measured pressure difference will be the sum of both the hydrostatic component and a friction gradient. Figure 7.3 shows the expected friction gradient as a function of pipe ID and flow rate. In normal circumstances (5½- or 7-in. casing), the friction term is negligible.

Figure 7.4 shows a gradiomanometer recording. The tool is logged continuously up (or down) the hole and records fluid density vs. depth. The log is normally recorded with a sensitivity of 1 g/cc per log track. An amplified curve with 5 to 10 times the sensitivity is commonly recorded as well.

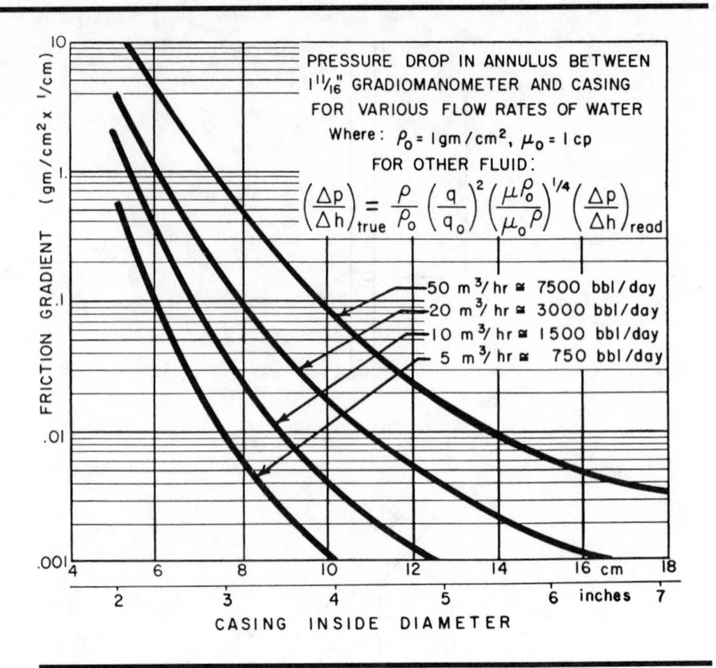

FIGURE 7.3 *Friction Gradients Caused by Fluid Flow in Small Pipe. Courtesy Schlumberger Well Services.*

QUESTION #7.1
Inspect the log shown in figure 7.4 and deduce the fluid(s) produced from each set of perforations.

If a bubble-flow regime exists in the wellbore, the interpretation of measured fluid density must take into account the holdup of the flowing mixture. In general, we may always write

$$\rho_{mix} = \rho_h y_h + \rho_l(1 - y_h),$$

where h denotes the heavier phase and l the lighter one, and y is the holdup. It follows, therefore, that if the densities of the light and heavy phases are known, the holdup may be deduced by

$$y_h = \frac{\rho_{mix} - \rho_l}{\rho_h - \rho_l}.$$

The holdup, once determined in this manner, may be used in conjunction with the holdup equation (derived in chap. 4) to deduce the flow rates of both the light and heavy phases.

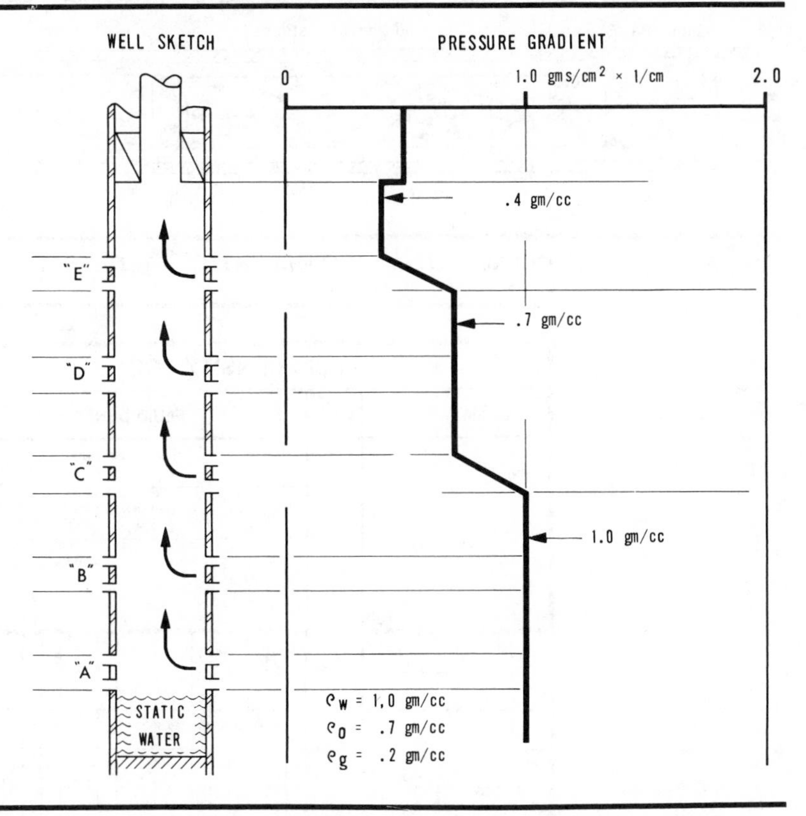

FIGURE 7.4 *Gradiomanometer Log. Courtesy Schlumberger Well Services.*

FLUID DENSITY TOOL (GAMMA RAY ABSORPTION)

The fluid density tool operates on much the same principle as the formation density tools (i.e., a source of gamma rays is positioned with respect to a detector of gamma rays so that the wellbore fluid acts as an absorber). Figure 7.5 illustrates the operating principle. A high count rate indicates a fluid of low density and a low count rate indicates a fluid of high density. Figure 7.6 shows a fluid density log.

The advantages of this type of tool over the gradiomanometer are that (a) its measurement is not affected by the wellbore deviation angle, or by friction effects. However, since it is a tool relying on a statistical radioactive decay, the readings are subject to statistical variations. It should also be noted that the measured quantity is the average density of the flowing mixture and thus subject to the same holdup effects as the gradiomanometer.

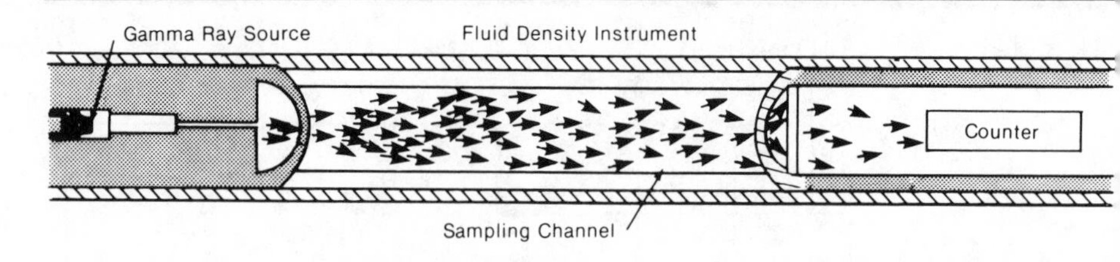

FIGURE 7.5 *The Fluid Density Tool. Courtesy Dresser Atlas.*

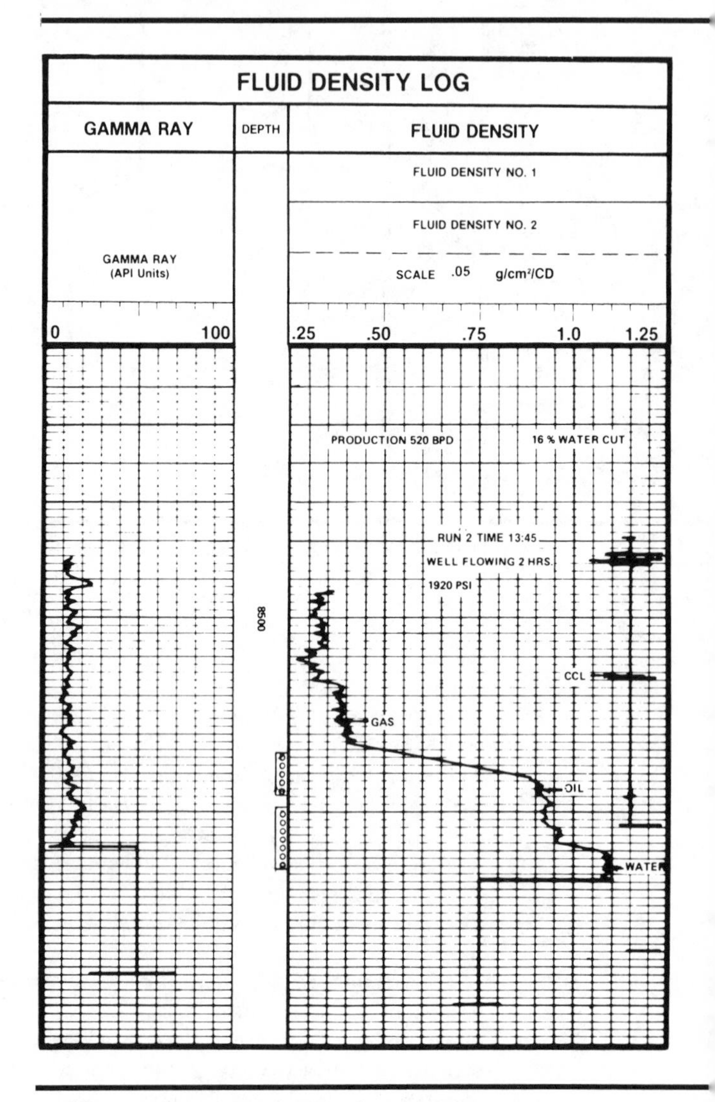

FIGURE 7.6 *A Fluid Density Log. Courtesy Dresser Atlas.*

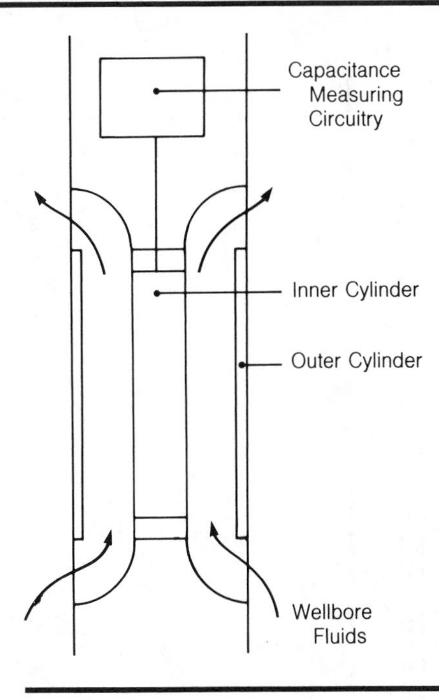

FIGURE 7.7 *Capacitance Watercut Meter.*

CAPACITANCE (DIELECTRIC) TOOLS

Another group of tools widely used to distinguish water from hydrocarbons depend for their operation on the difference between the dielectric constant of water (~80) and that of oil or gas (~6). A simple way to find the dielectric constant of a fluid is to use the fluid itself as the dielectric between the plates of a capacitor. The capacitance of the resulting capacitor may be found by classical methods such as including it in an RC network and finding the resonant frequency.

A conventional design is shown in figure 7.7. Two cylindrical metal tubes are arranged so that wellbore fluids flow through the annular space between them. The raw readings of such a device are in terms of a frequency. Each tool will have a calibration graph to convert a measured frequency to a *watercut value.* These tools behave well provided that the continuous phase is oil. In practice, the measurement may become unreliable if the watercut in the flowing mixture exceeds 30%.

RESONATOR (VIBRATOR)

Tools that measure fluid density by means of mechanical vibration have been in use for some time. They are designed so that the flow stream is directed to pass through a hollow metal cylinder con-

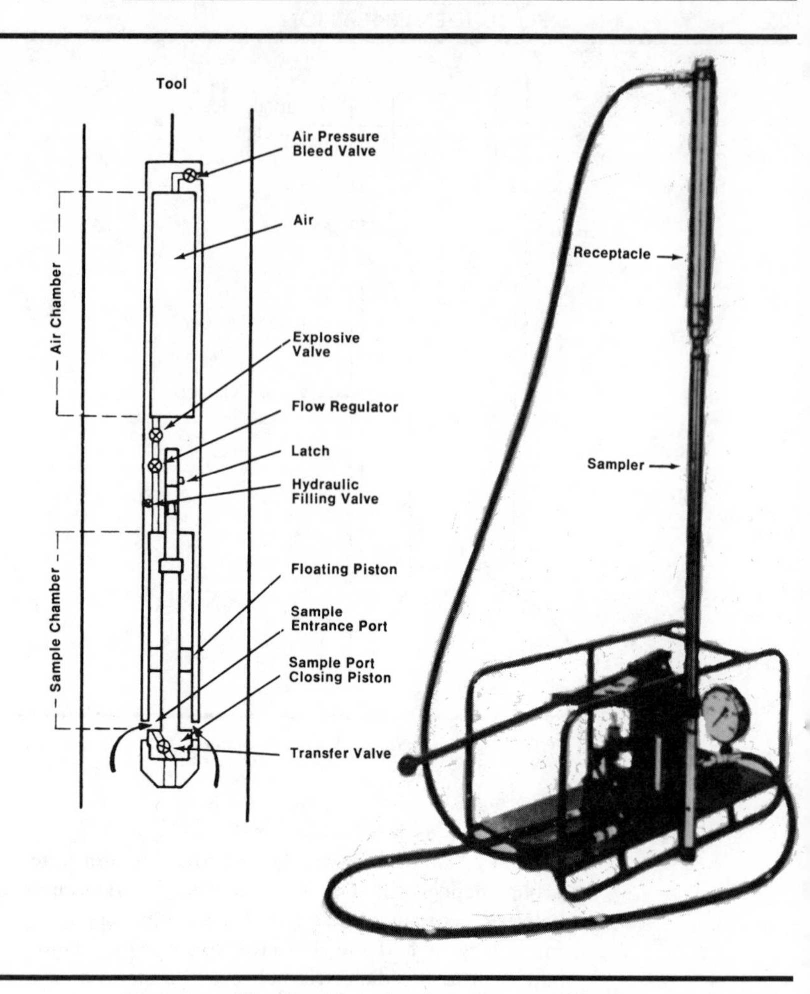

FIGURE 7.8 *The Fluid Sampler Tool. Courtesy Schlumberger Well Services.*

taining radial blades in line with the fluid stream. The cylinder is set into circular vibration by an electromagnetic driving system and oscillates at a natural frequency that depends on the density of the fluids passing through the cylinder. The raw reading of the tool is a frequency. Each tool has a calibration chart to convert from frequency to a *density index.* The main problem with tools of this sort is their tendency to plug up with sand or other grains carried in the flow stream.

FLUID SAMPLER

The fluid sampler is analogous to a formation tester, the difference being that the latter takes samples of the formation fluids but the

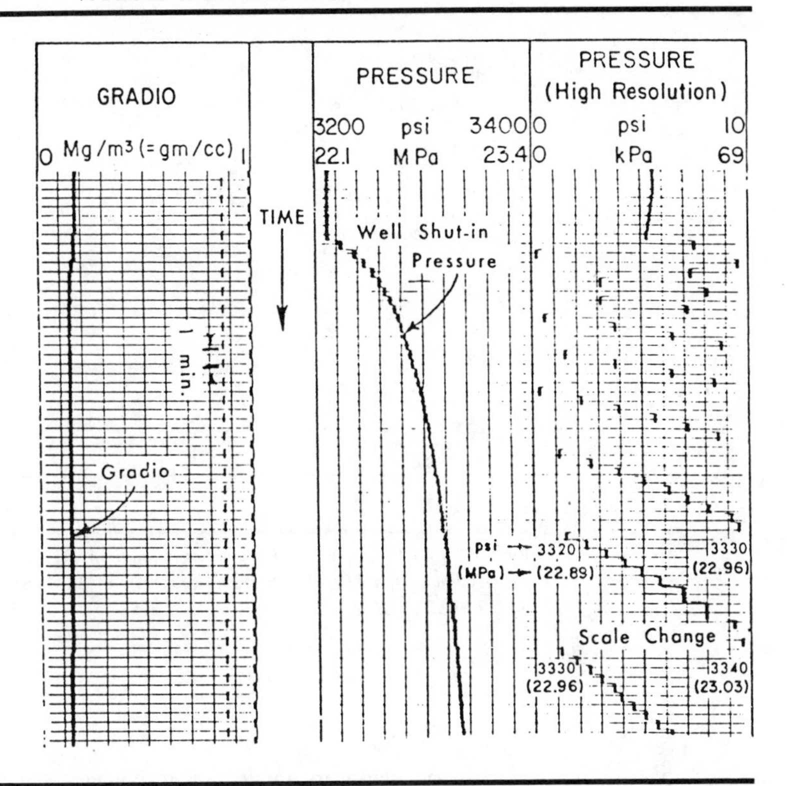

FIGURE 7.9 *Pressure Buildup Recording with a Manometer. Reprinted by permission of the SPE-AIME from Anderson et al. 1980, fig. 5A, p. 195. © 1980 SPE-AIME.*

former takes samples of the wellbore fluids. Figure 7.8 shows the tool. It is run into the well to the required depth and then actuated by the operator on the surface. A valve opens and a sample chamber fills with the fluids in the wellbore. The valve is then closed and the sample is recovered and analyzed at the surface. For this purpose, a miniseparator is used to gauge the relative amounts of oil, water, and gas recovered.

MANOMETER

The manometer is a pressure gauge. It may be used in two ways, either to provide a continuous recording of pressure against depth or to provide a recording of pressure versus time at some fixed point in the well. When used in the continuous logging mode, the manometer will show a pressure gradient reflecting the density of the fluid in the well. Changes in this gradient indicate density changes in the well fluids and thus serve to pick up fluid entries. When

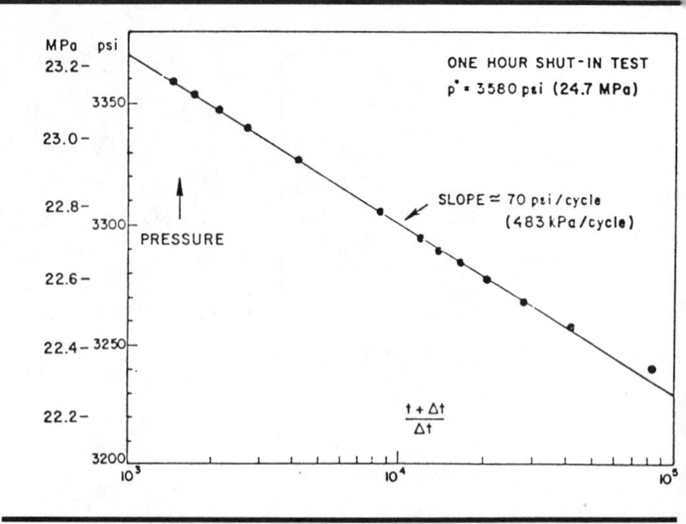

FIGURE 7.10 *"Horner" Plot from Log Shown in Figure 7.9. Reprinted by permission of the SPE-AIME from Anderson et al. 1980, fig. 5B, p. 195. © 1980 SPE-AIME.*

used in the stationary mode, the manometer can be used in the familiar mode of a pressure bomb. Either the well can be shut in for a pressure buildup test, or the well can be flowed on different chokes for a step-rate test. Figure 7.9 shows a pressure buildup test recorded with a manometer, and figure 7.10 a pressure buildup plot used to interpret the data. Two types of gauges are available, a conventional one with about 3% accuracy and a quartz-crystal gauge (HP) with an accuracy of 0.025%.

OTHER MEASUREMENTS

The use of temperature logs to detect which type of fluid is flowing is covered in chapter 8. The use of sibilation or noise logs is covered in chapter 9. These logs are mentioned here to assure that they are remembered in the correct perspective as fluid-identifying devices.

Many service companies offer tools that combine numerous sensors in one tool string. This simplifies the practical problem of obtaining logs when working against wellhead pressure with a riser and pressure-control equipment, since all the required measurements can be made on one trip in the hole without time delays caused by frequent tool changes. A schematic of such a combination tool is given in figure 7.11 and an example log is shown in figure 7.12.

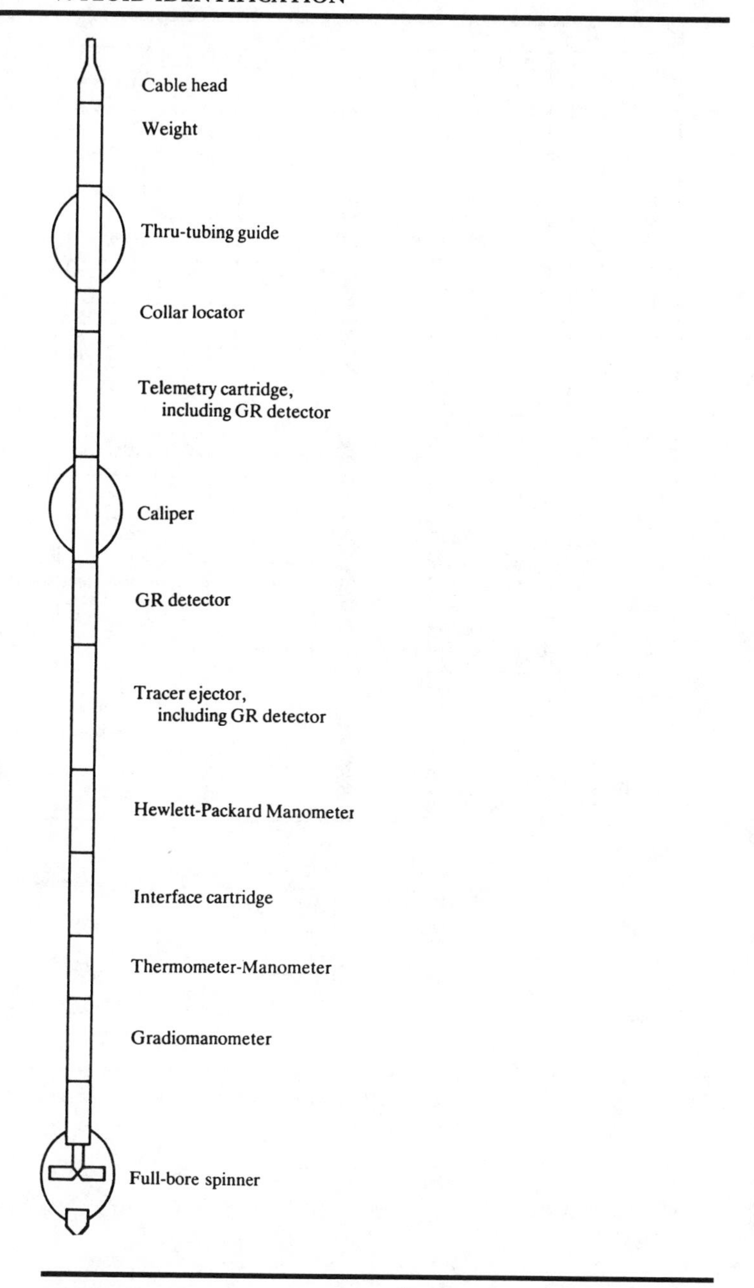

FIGURE 7.11 *The Production Logging Tool (PLT). Courtesy Schlumberger Well Services.*

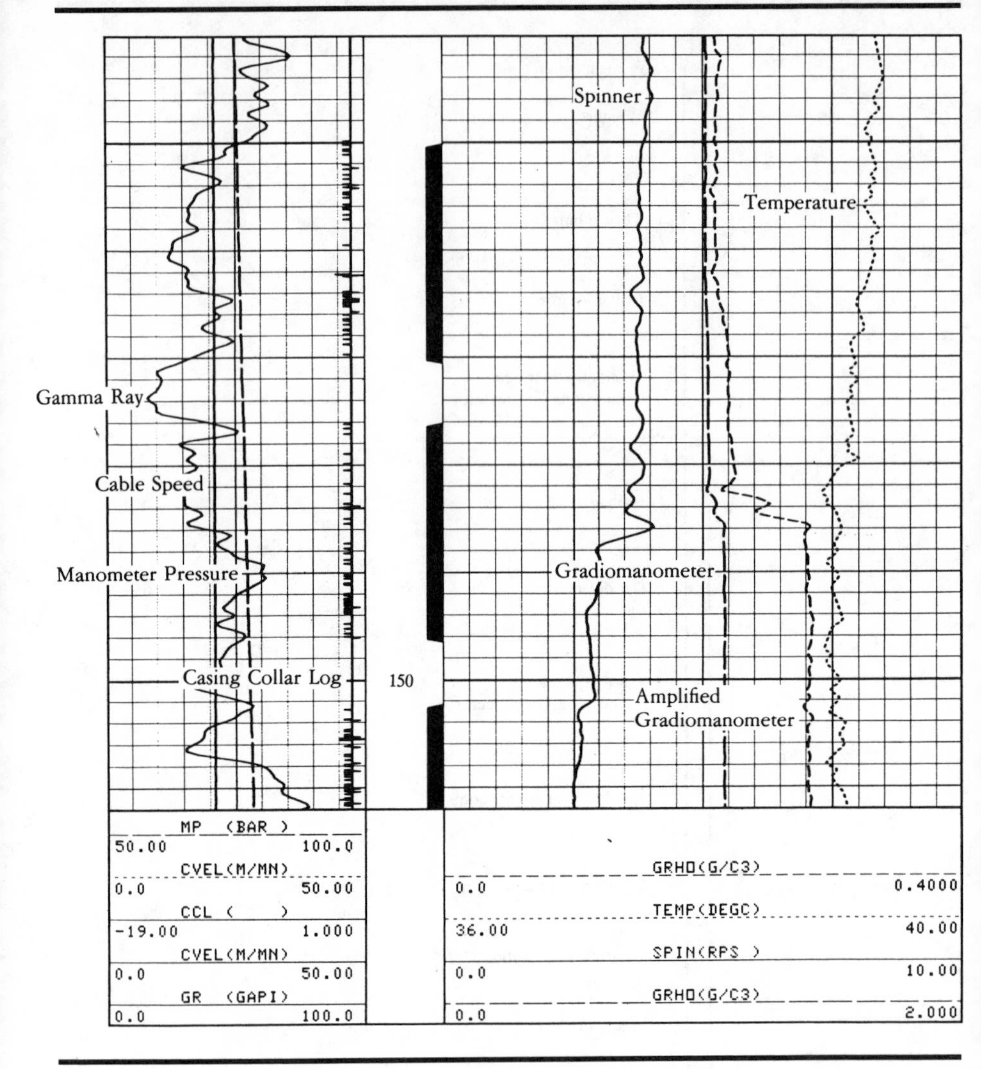

FIGURE 7.12 *Production Logs Run with Combination Tool. Courtesy Schlumberger Well Services.*

BIBLIOGRAPHY

Anderson, R. A., Smolen, J. J., Laverdiere, L., and Davis, J. A.: "A Production-Logging Tool with Simultaneous Measurements," paper SPE 7447 presented at the SPE 53rd Annual Technical Conference and Exhibition, Houston, Oct. 1–4, 1978.; *J. Pet. Tech.* (Feb. 1980).

Answers to Text Question

QUESTION #7.1

Intervals A and B: water column, either static or moving

C: hydrocarbon entry, possibly with some water

D: no entry or mixed entry

E: gas or gas and liquid entry

CHAPTER EIGHT

TEMPERATURE LOGGING

Among the reasons for using temperature logs are:

1. to find the cement top after a recent cementing operation
2. to find a lost circulation zone in a currently drilling well
3. to find fluid entry and exit points in production and injection wells

Temperature sensors are temperature sensitive resistors. In order to react quickly to temperature changes, temperature logging tools are designed with the element exposed directly to the wellbore fluids. For this reason, they are delicate instruments easily damaged by physical abuse or by the junk normally found at the bottom of a well.

Though sensitive enough to detect temperature changes of 0.5°F, these devices are not normally very accurate in absolute readings; and the error in any temperature recorded is probably in the ±5°F range. Figure 8.1 illustrates a temperature logging tool and a typical temperature log.

FUNDAMENTALS

Undisturbed formation temperature increases predictably with depth. This increase in temperature with depth is known as the *geothermal gradient* (G) and is usually in the range of 1 to 2°F per 100 ft. Figure 8.2 is a useful guide to geothermal gradients—the temperature at any depth may be extrapolated using the relationship

$$T_{form} = T_{surf} + \text{depth} \cdot G.$$

Obviously, the actual temperature on the surface fluctuates seasonally so the value used for the *surface temperature* is actually the mean annual surface temperature and will be in range of 60 to 70°F. Seasonal near-surface temperature changes do not penetrate very deeply into the ground and for well logging purposes can be ignored. One exception is the permafrost zone that can exist near polar regions.

Estimation of formation temperature from openhole logs can be made provided it is borne in mind that at the time a logging run is made the wellbore is cooler than the surrounding formation, due to mud circulation. If several logging runs are made in the same hole, undisturbed formation temperature can be estimated from a plot of temperature against time. The method of Dowle and Cobb

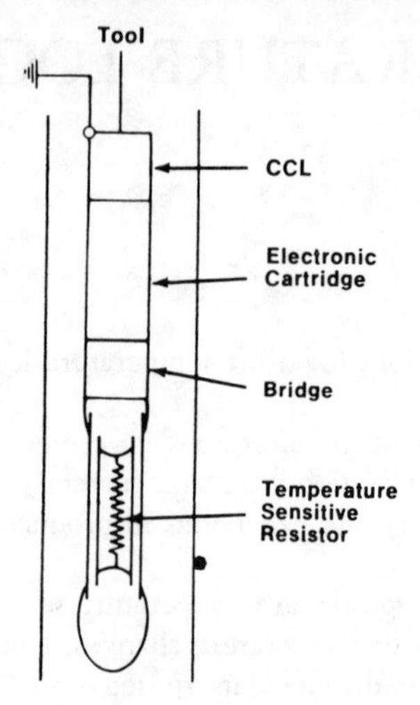

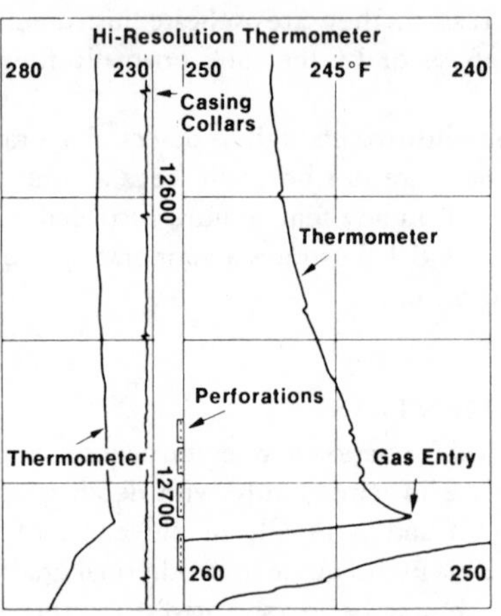

FIGURE 8.1 *High-Resolution Thermometer and a Typical Temperature Log. Courtesy Schlumberger Well Services.*

is recommended. If t_k is the circulation time and Δt is the time since circulation stopped, then a plot of the observed temperature at time Δt against $(t_k + \Delta t)/\Delta t$ on a log scale should give a straight line with an intercept at $(t_k + \Delta t)/\Delta t = 1$ equal to the undisturbed formation temperature, T_i. Figure 8.3 illustrates the method.

QUESTION #8.1
Depth: 16,200 ft.
Drilling stopped: 00:30 hr.
Circulation stopped: 4:00 hr.
Circulation time: 3.5 hr.

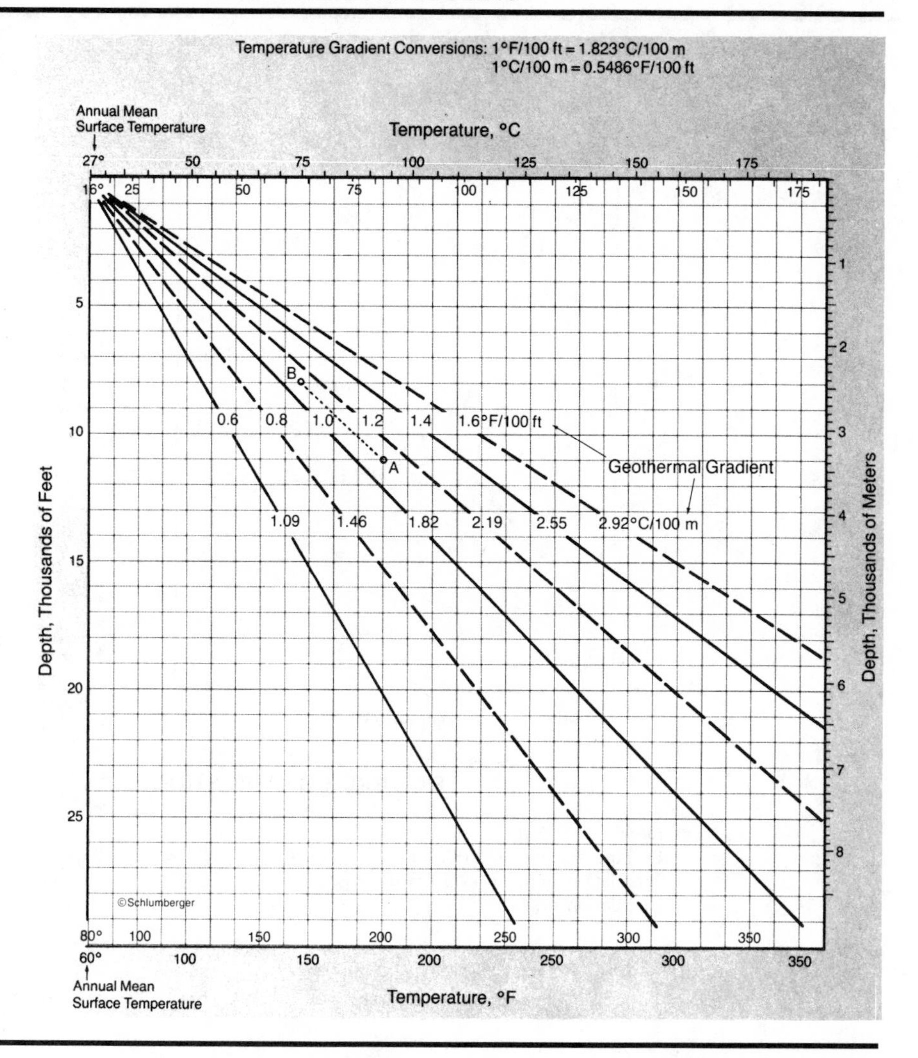

FIGURE 8.2 *Geothermal Gradients. Courtesy Schlumberger Well Services.*

Three log runs were made, the corresponding times and temperatures were:

Tool	*Time Off Bottom*	*Time since Circulation Stopped*	*Temp. (°F)*
Induction	12:15	8:15	241
Density	15:00	11:00	257
Dipmeter	17:30	13:30	262

Plot temperature vs. $(t_k + \Delta t)/\Delta t$ and deduce static formation temperature.

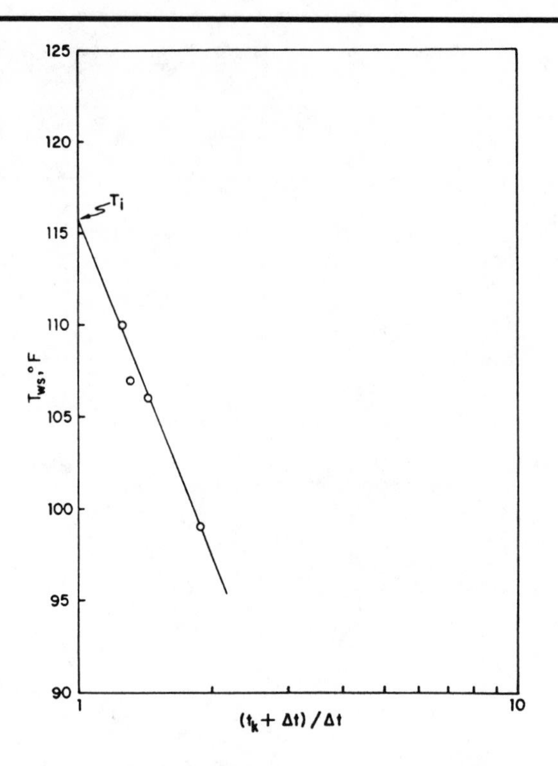

BOTTOMHOLE TEMPERATURE EXTRAPOLATION

Depth—7,646 ft
Drilling stopped—22:00/2nd
Circulation stopped—2:30/3rd
Circulation time—4½ hr

Tool	Thermometer Depth (ft)	Time Off Bottom	Time Since Circulation Stopped (hr)	Temperature (°F)
Sonic	7,608	07:36/3rd	5:06	99
DIL	7,608	12:48/3rd	10:18	106
FDC	7,620	14:29/3rd	14:29	107
SNP	7,620	20:37/3rd	18:07	110

FIGURE 8.3 *Determination of Static Formation Temperature. Reprinted by permission of the SPE-AIME from Dowdle and Cobb 1975, fig. 4, p. 1329.*

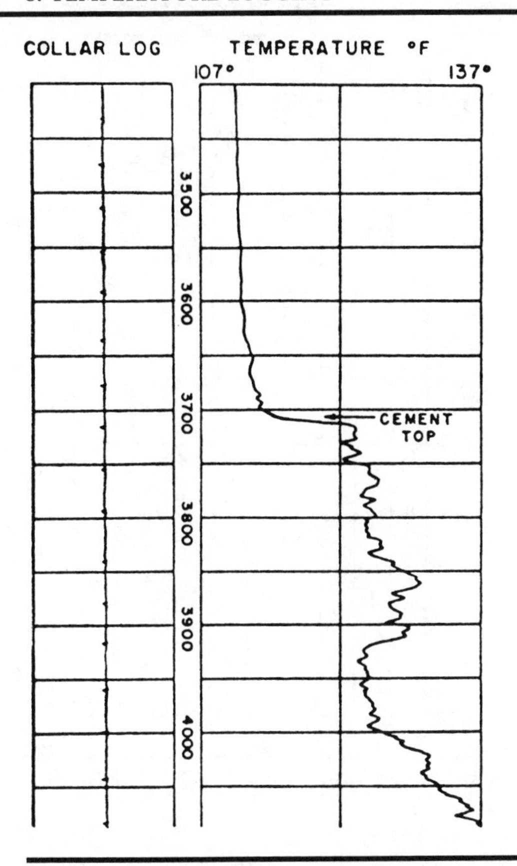

FIGURE 8.4 *Cement-Top Pick from Temperature Log.*

CEMENT-TOP EVALUATION

Although no longer widely used for finding cement tops in recently cemented wells, the temperature log may be used for this purpose—its advantage being that it is cheap and demands less rig time. Its disadvantage is that it gives no indication of the cement quality or the ability of the cement job to make a hydraulic seal. The principle involved is the exothermic chemical reaction that takes place while cement is curing. The heat given off raises the temperature in and around the borehole at those places where cement is placed. Thus, a marked drop in temperature may be expected at the cement top. Figure 8.4 shows a temperature log run in a recently cemented well and the corresponding pick for the cement top.

LOST-CIRCULATION ZONES

In the event that circulation is lost in a currently drilling well, a temperature log can be a useful indicator of the thief zone in question. Just below the point of lost circulation, the mud in the hole is likely

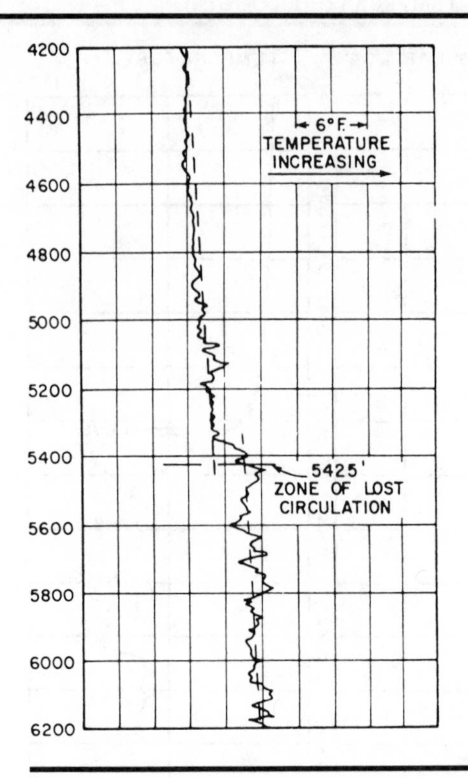

FIGURE 8.5 *Lost-Circulation Zone Detection with a Temperature Log.*

to have been stagnant for some time and therefore to have assumed a higher temperature than the mud in the column, which is still free to circulate. Thus, a temperature discontinuity will exist across the lost-circulation zone. Figure 8.5 illustrates the effect.

TEMPERATURE PROFILES IN PRODUCTION AND INJECTION WELLS

General

If left undisturbed, the temperature in a wellbore will assume the ambient temperature of the surrounding formations; and a log of temperature against depth will indicate the geothermal gradient. However, if the well is flowing, either due to production or injection of fluids, then the observed temperature profile will depart from the geothermal gradient. This surprisingly simple rule should always be borne in mind when analyzing temperature logs. Look for departures from the geothermal gradient as the prime indicator of fluid movement. Once a particular flow regime has reached thermal equilibrium, the difference between the observed temperature in the

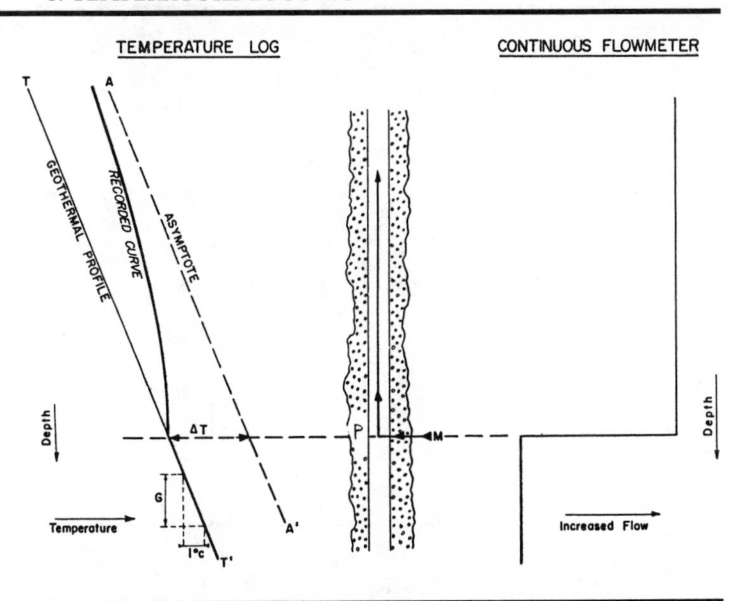

FIGURE 8.6 *Liquid-Production Temperature Profile. Courtesy Schlumberger Well Services.*

borehole and the geothermal gradient is related to the mass flow by the equation

$$\Delta T = bM/G,$$

where:

ΔT = the temperature difference,
b = a constant that depends on the physical characteristics of the fluid produced and on the thermal conductivity of the formation,
M = the mass flow rate, and
G = the geothermal gradient.

Thus, other things being equal, the ΔT is proportional to the weight of fluid produced or injected per unit time.

Liquid Production

Figure 8.6 shows a temperature profile for a single-point entry of liquid production. Things to note include:

a. Below the production point, the temperature profile follows the geothermal profile.
b. At the production point, the temperature profile is vertical.
c. Above the production point, the temperature profile asymptotically approaches a new gradient offset from the geothermal by an amount ΔT.

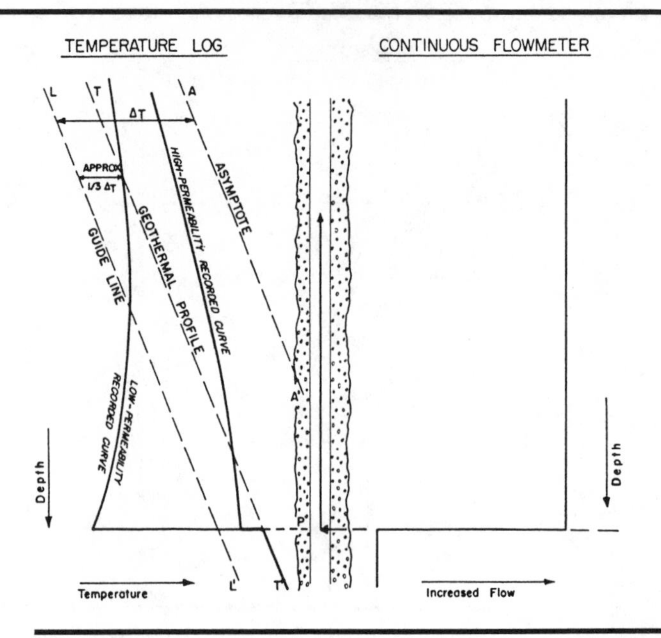

FIGURE 8.7 *Gas-Production Temperature Profiles. Courtesy Schlumberger Well Services.*

Gas Production

Figure 8.7 shows temperature profiles for a single-point entry of gas. Things to note include:

a. Below the production point, the temperature profile follows the geothermal profile.
b. At the production point, the temperature profile is *horizontal* and shows a marked cooling effect due to gas expansion from reservoir pressure to well flowing pressure,
c. Above the production point, the temperature rises, crosses the geothermal, and approaches an asymptote offset ΔT above the geothermal.

Two traces are shown on figure 8.7—one for a high-permeability formation and one for a low-permeability formation. Note that the initial cooling effect at the point of production is less for the high-permeability formation than for the low.

Water Injection

Figure 8.8 illustrates the temperature profiles to be expected in a water-injection well. Things to note are:

a. Depending on the temperature of the injected water relative to the undisturbed formation temperature, the temperature profile above the injection point may show an increase or a decrease with depth.

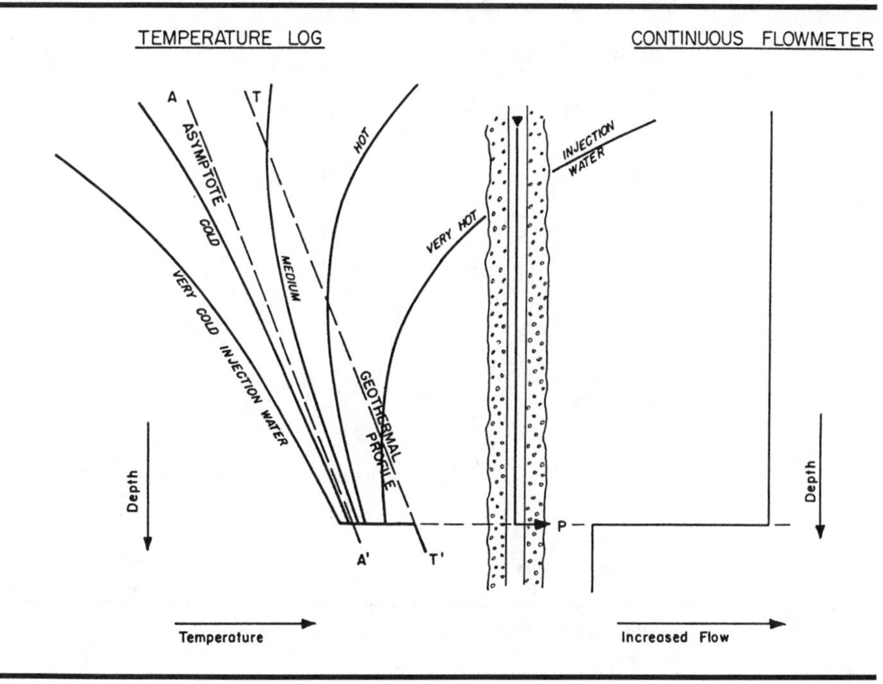

FIGURE 8.8 *Water-Injection Temperature Profiles. Courtesy Schlumberger Well Services.*

b. At the injection point, the temperature profile is horizontal.
c. Below the injection point, the temperature profile returns to geothermal.

Gas Injection

The temperature profiles for a gas-injection well (fig. 8.9) are entirely similar to those for a water-injection well, and the same observations apply.

Further examples of liquid and gas production and injection profiles—together with the effects of casing leaks, casing-formation annulus flow, etc.—are given in a paper entitled "Temperature Logs in Production and Injection Wells," by A. Poupon and J. Loeb. The reader is encouraged to read this paper in its entirety.

LOGGING TECHNIQUES

Several special logging techniques can improve interpretation of temperature profiles. These include:

1. shut-in temperature surveys
2. differential temperature logs
3. radial differential temperature logs

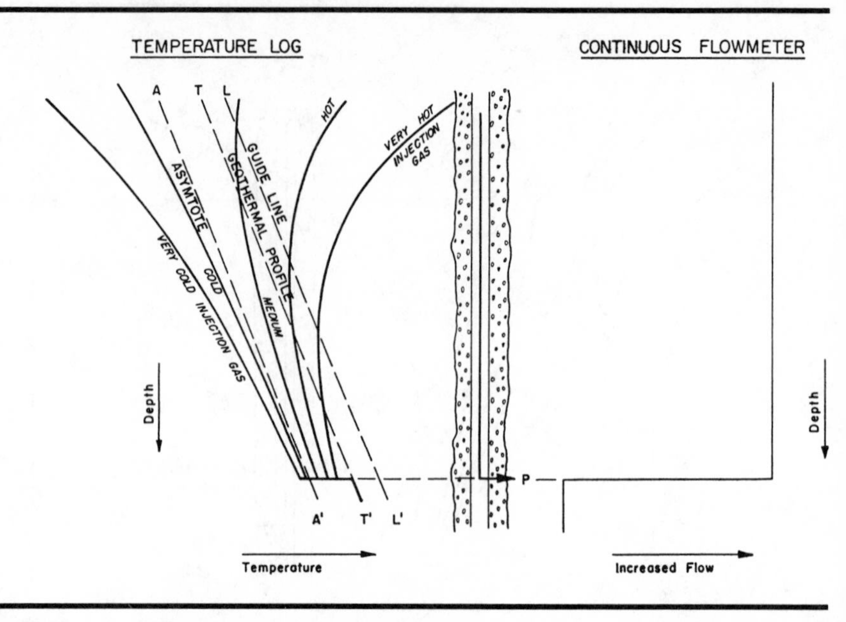

FIGURE 8.9 *Gas-Injection Temperature Profiles. Courtesy Schlumberger Well Services.*

Shut-in Temperature Surveys

If more than one zone is taking water in an injection well, it is sometimes difficult to judge from the flowing temperature profile which zone is taking what percentage of the injected total. Two techniques are offered here. The first simply relates the ΔT value to the volumetric flow. Figure 8.10 illustrates the concept. In the figure, T_G is the undisturbed formation temperature and T_h is the temperature observed in the borehole.

A second useful technique is to stop injection altogether and repeat the temperature profile several times at various time intervals, such as at 3, 6, 12, and 24 hours. Or, depending on the local conditions, at 12, 24, and 48 hours. Zones that were taking relatively cool injection water will remain cooler than surrounding formations for a relatively long time and will be visible on the repeat profiles. Figure 8.11 shows such an example.

Differential-Temperature Surveys

A differential-temperature log is a recording made, versus depth, of the difference in temperature between two points in the well. Effectively, the trace is a differentiation of the temperature curve itself. In practice this is accomplished either by having matched sensors on the logging probe some short distance apart or by memorizing the temperature taken at one point and comparing it to the temperature taken at some other point, such as 1 ft deeper, or shallower, depending on the direction of logging. Differentiation can be on a

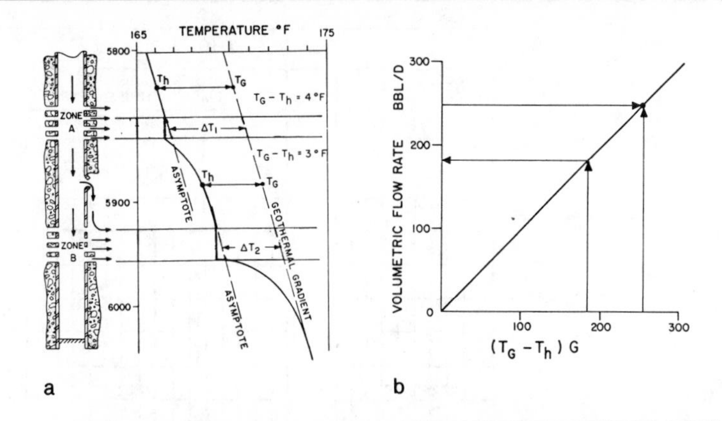

FIGURE 8.10 *Water Injection: (*a*) Temperature Profile, (*b*) Flow Rate vs.* Δ T. *Reprinted by permission of the SPE-AIME from Witterholt and Tixier 1972. © 1972 SPE-AIME.*

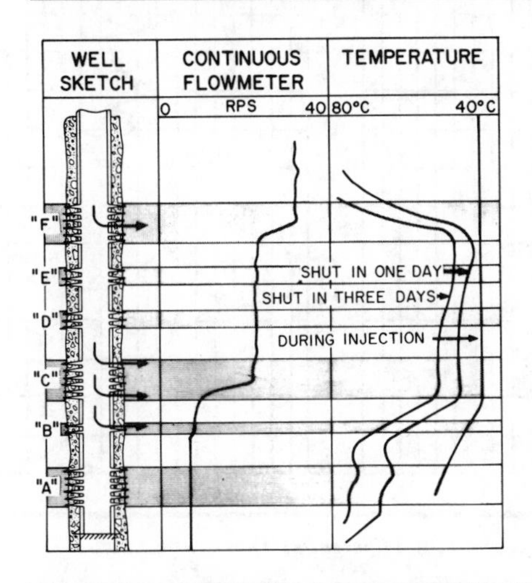

FIGURE 8.11 *Shut-In Temperature Survey. Reprinted by permission of the SPE-AIME from Witterholt and Tixier 1972. © 1972 SPE-AIME.*

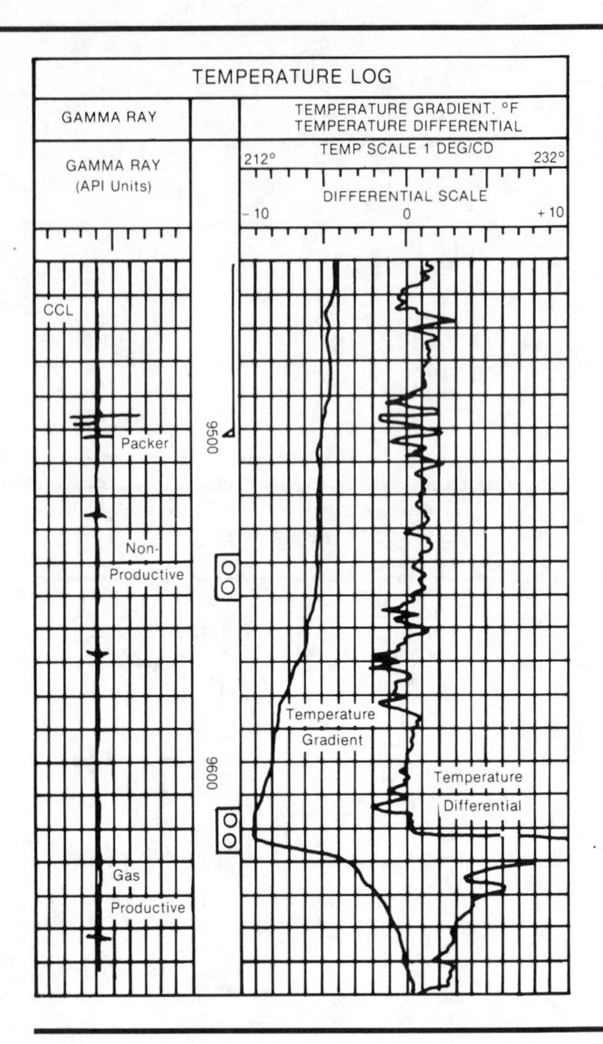

FIGURE 8.12 *Differential-Temperature Log. Courtesy Dresser Atlas.*

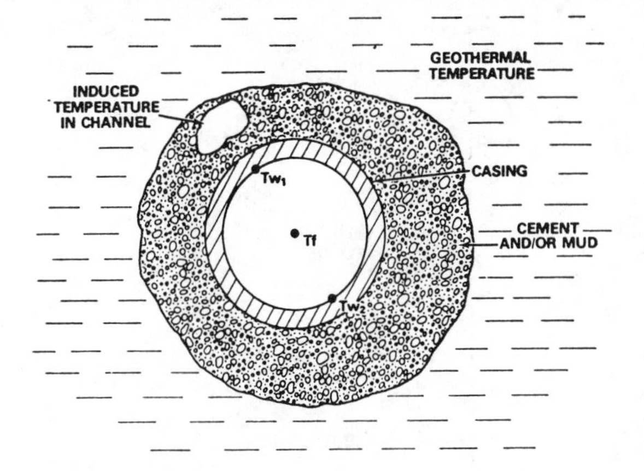

FIGURE 8.13 *Temperatures In and Around a Cased Well. Reprinted by permission of the SPE-AIME from Cooke 1979, fig. 1, p. 677.* © *1979 SPE-AIME.*

depth or a time basis. Provided that the logging speed remains constant, both methods produce the same answer. Figure 8.12 shows a conventional temperature survey together with a differential-temperature curve. Note that where even slight changes in temperature occur the differential curve accentuates the occurrence. Where the temperature does not change with depth, the differential curve is likewise unchanged.

Radial Differential-Temperature Tool

The radial differential-temperature tool (RDT) was designed to detect channels behind pipe. The operating principle relies on the probability that the temperature in the channel is different from the temperature in the surrounding formation. If fluid is channeling from above or below it is probable that such a temperature difference will be present. Figure 8.13 shows a plan view of a channel. The temperature on the side of the casing near the channel, T_{w1}, is likely to be different from the temperature of the casing opposite the channel, T_{w2}. Thus, a temperature sensor held stationary at a given depth, but free to rotate through 360°, should observe a temperature fluctuation if a channel with fluid flowing in it is present. The method used to make the horizontal scan is shown in figure 8.14. Anchor springs hold the tool in the casing and a rotation motor is actuated to cause the RDT sensor to scan round the casing. The resulting log is shown in figure 8.15. Note that the log is a record of temperature versus degrees of rotation round the casing. This particular survey with measurements made at 6400, 6440, 6500,

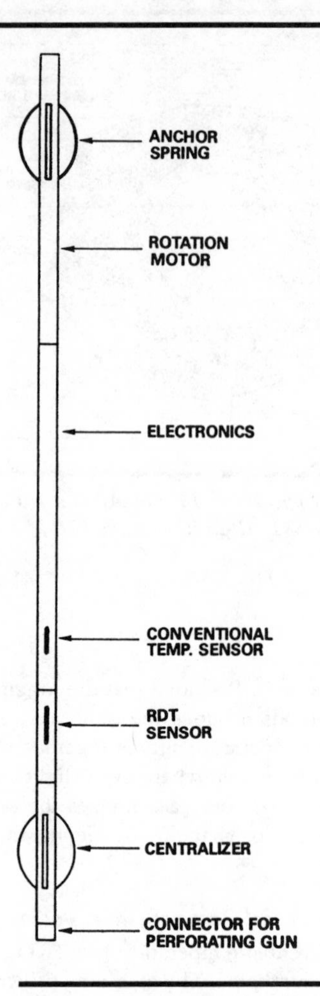

FIGURE 8.14 *RDT Tool. Reprinted by permission of the SPE-AIME from Cooke 1979, fig. 2, p. 677. © 1979 SPE-AIME.*

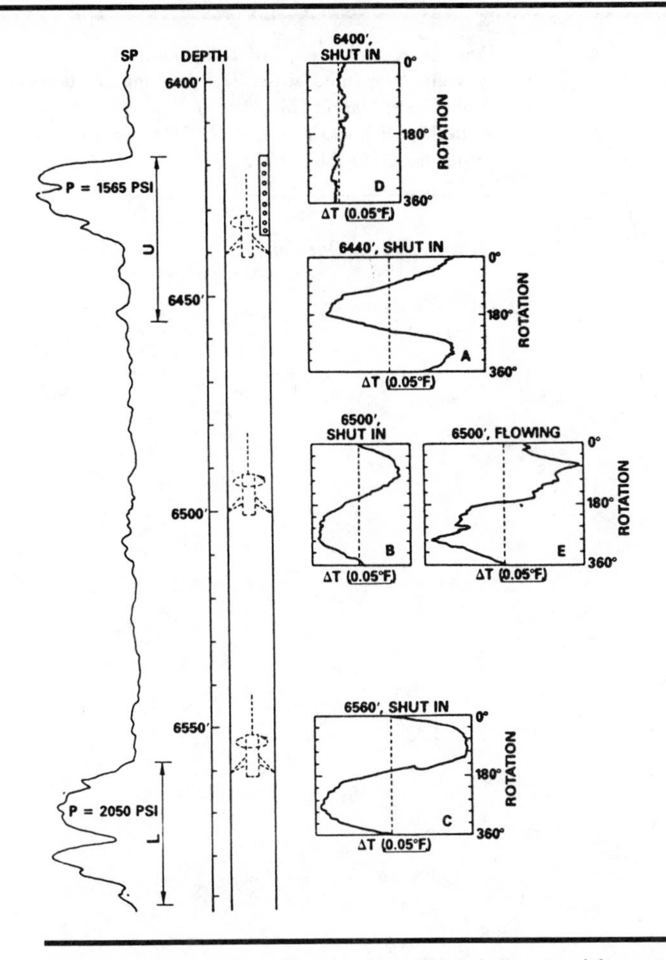

FIGURE 8.15 *RDT Scan Showing Gas Channel. Reprinted by permission of the SPE-AIME from Cooke 1979, fig. 3, p. 678. © 1979 SPE-AIME.*

and 6560 ft shows that gas is channeling from the lower sand (marked "L" on the figure) to the upper sand (marked "U"). For remedial action, the tool carries a perforating gun that can shoot squeeze perforations directly into the channel once it is detected.

BIBLIOGRAPHY

Cooke, Claude E.: "Radial Differential Temperature (RDT) Logging—A New Tool for Detecting and Treating Flow Behind Casing," paper SPE 7558 presented at the SPE-AIME 53rd Annual Technical Conference and Exhibition, Houston, Oct. 1–3, 1978; *J. Pet. Tech.* (June 1979).

Dowdle, W. L. and Cobb, W. M.: "Static Formation Temperature from Well Logs—An Empirical Method," *J. Pet. Tech.* (November 1975).

Poupon, A. and Loeb, J.: "Temperature Logs in Production and Injection Wells," presented at the 27th Meeting of European Assn. of Expl. Geophysicists, Madrid (May 1965).

Witterholt, E. J. and Tixier, M. P.: "Temperature Logging in Injection Wells," SPE 4022 (October 1972).

Answer to Text Question

QUESTION #8.1

297°F

CHAPTER NINE

NOISE LOGGING

Noise logging is applicable whenever fluid flow, either in the borehole or in the casing formation annulus, produces a detectable noise. Detection is by means of a microphone suspended in the well. Experience teaches that analysis of noise is a refined technique for deducing the source of the noise. Student life in the low-rent district is enough to turn even the untrained ear into a veritable Sherlock Holmes, distinguishing the characteristic sounds of flushing toilets from draining bath tubs. The human ear and brain perform this function well by making an amplitude-frequency analysis of the total audible spectrum. Tools for well-flow analysis have to perform a similar function in order to earn their keep. Only by this kind of frequency analysis can the hiss of gas be distinguished from the gurgle of liquids.

TOOLS AVAILABLE

Many service companies offer noise logging service, under a number of trade names such as:

Sonan Log	(Dresser)
Audio Log	(Schlumberger)
Borehole Audio Tracer Survey (BATS)	(Gearhart)
Noise Log	(McCullough)
Borehole Sound Survey	(Dia-Log)

In general, measurements are made at preselected stations in the well. At each station, the amplitude of the noise in a number of frequency bands is determined and plotted on the log. Subsequently, these individual station readings may be joined together by straight lines to give the appearance of a continuous log. Figure 9.1 illustrates such a log. The total noise amplitude is generated by flow from formation A via the casing/formation annulus to formation C. Note the increase in noise at restriction B.

OPERATING PRINCIPLE

Through controlled experiments it is possible to derive noise amplitude-frequency spectra that are characteristic of fluid flow regimes. For example, figure 9.2 shows the spectrum for 70 BWPD expanding across a 90 psi differential into a channel behind the pipe. Note that the majority of the noise energy is concentrated in the frequency

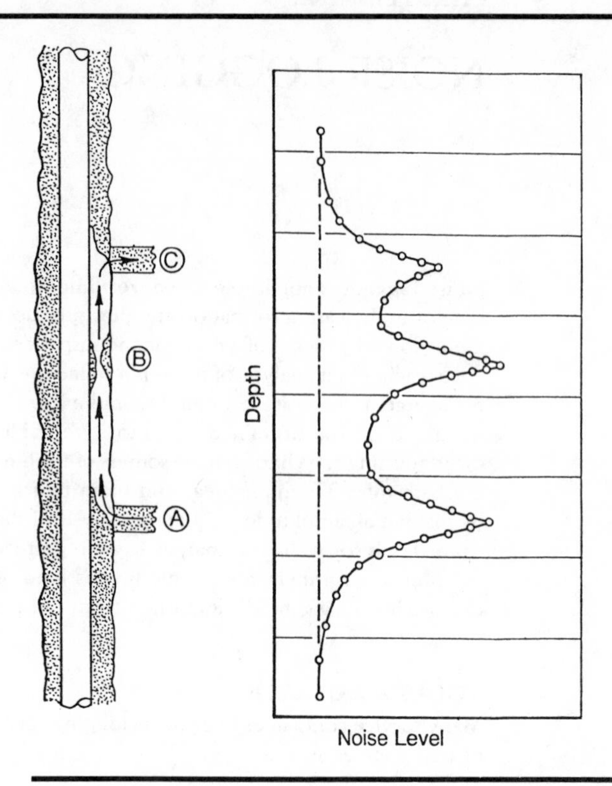

FIGURE 9.1 *Noise Log.*

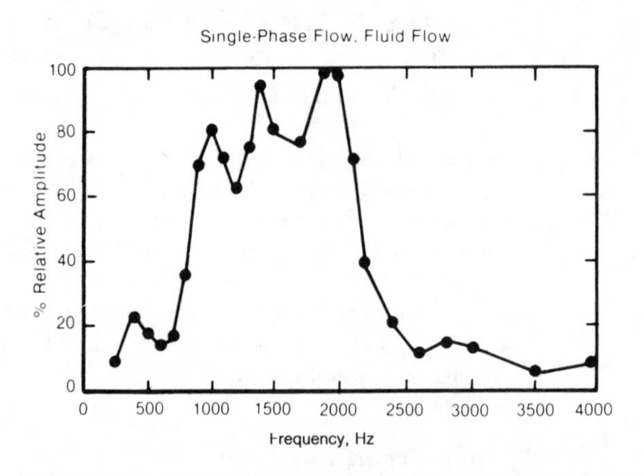

FIGURE 9.2 *Noise Spectrum—Single-Phase Water Flow. Courtesy Dresser Atlas.*

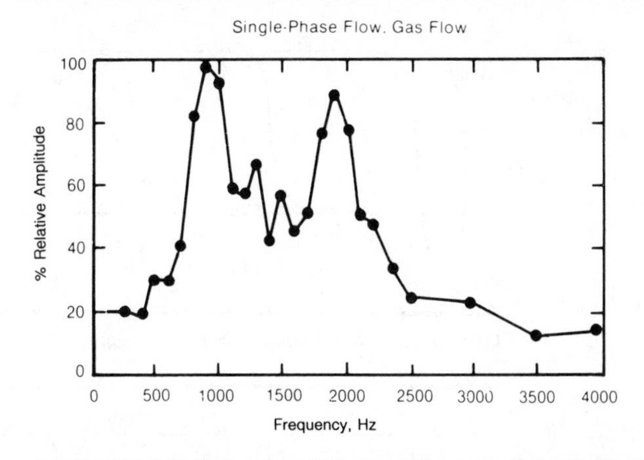

FIGURE 9.3 *Noise Spectrum—Single-Phase Gas Flow. Courtesy Dresser Atlas.*

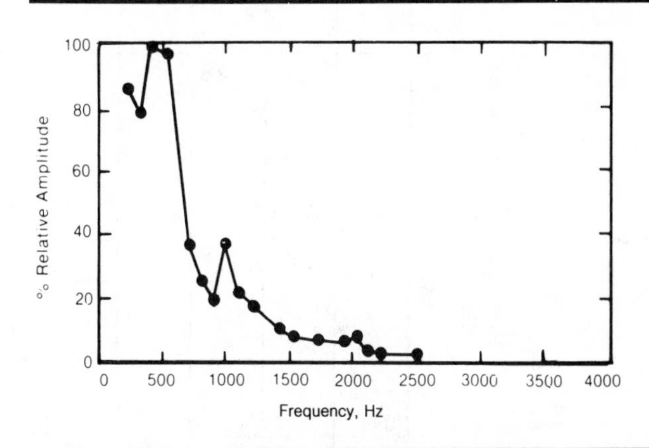

FIGURE 9.4 *Noise Spectrum—Two-Phase Flow (Gas/Water). Courtesy Dresser Atlas.*

range from 800 to 2000 Hz. By contrast, figure 9.3 shows the spectrum for sound emanating from 3.8 Mcf/D of gas expanding across a pressure differential of 10 psi into a gas-filled channel behind the pipe. Note the two peaks at 800 and 1800 Hz. Figure 9.4 shows a very different spectrum, obtained when 0.3 Mcf/D of gas expands into a water-filled channel. Here the peak noise is at less than 500 Hz.

In order to distinguish these spectra from one another, the total signal is filtered through bandpass filters that allow the frequencies ≥200+ Hz, ≥600+ Hz, ≥1000+ Hz, and ≥2000+ Hz to pass. The choice of limits on these bands varies somewhat between different service companies. Since the 200-Hz filter allows all frequencies

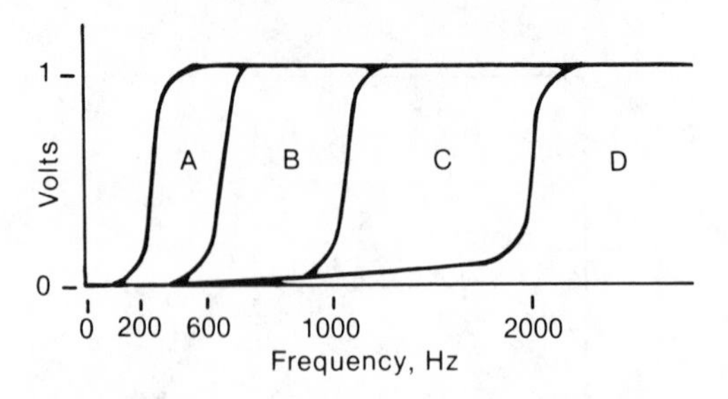

FIGURE 9.5 *Filter Response in Noise Logging. Courtesy Dresser Atlas.*

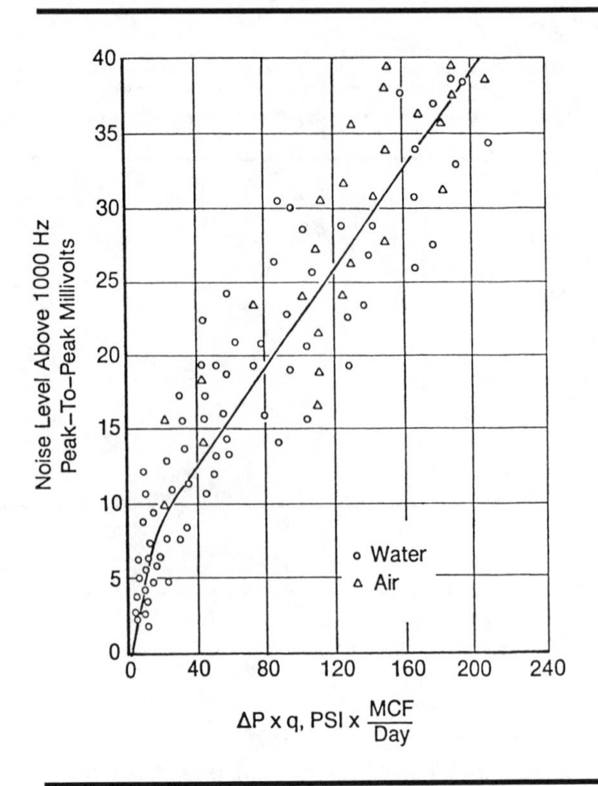

FIGURE 9.6 *Noise vs. Normalized Flow Rate (Single-Phase Gas Flow).*

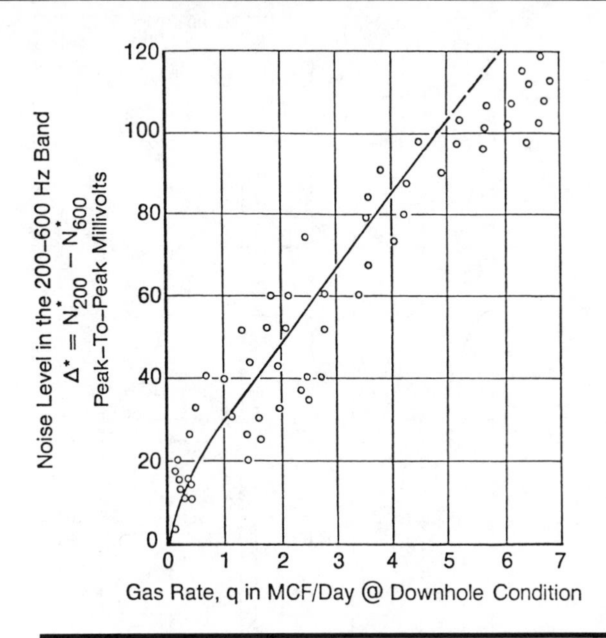

FIGURE 9.7 *Noise vs. Gas Flow Rate (Two-Phase Flow).*

above 200 Hz to pass, this filter normally gives the highest amplitude. The 2000-Hz filter, which only allows frequencies above 2000 Hz to pass, normally gives the lowest amplitude. Figure 9.5 illustrates the bandpass filter response.

INTERPRETATION

The interpretation of noise logs is an empirical art governed by common sense and a body of experimental data. In general, the noise level in the low-frequency bands correlates fairly well with a normalized gas flow rate, the normalization factor depending on the pressure drop. Figure 9.6 shows such a correlation. In two-phase flow, a slightly different correlation can be made (see figure 9.7).

Bear in mind that sounds carry great distances in cased wells. Any surface noise should be eliminated before attempting a noise survey. Pump jacks, motors, etc., can generate noise unrelated to fluid flow in the well and confuse interpretation of noise logs. Two general rules apply: (1) Changes in noise level indicate a change in volumetric flow rate. (2) Changes in the relative noise levels in different frequency bands indicate changes in the phase make-up of the fluid mixture.

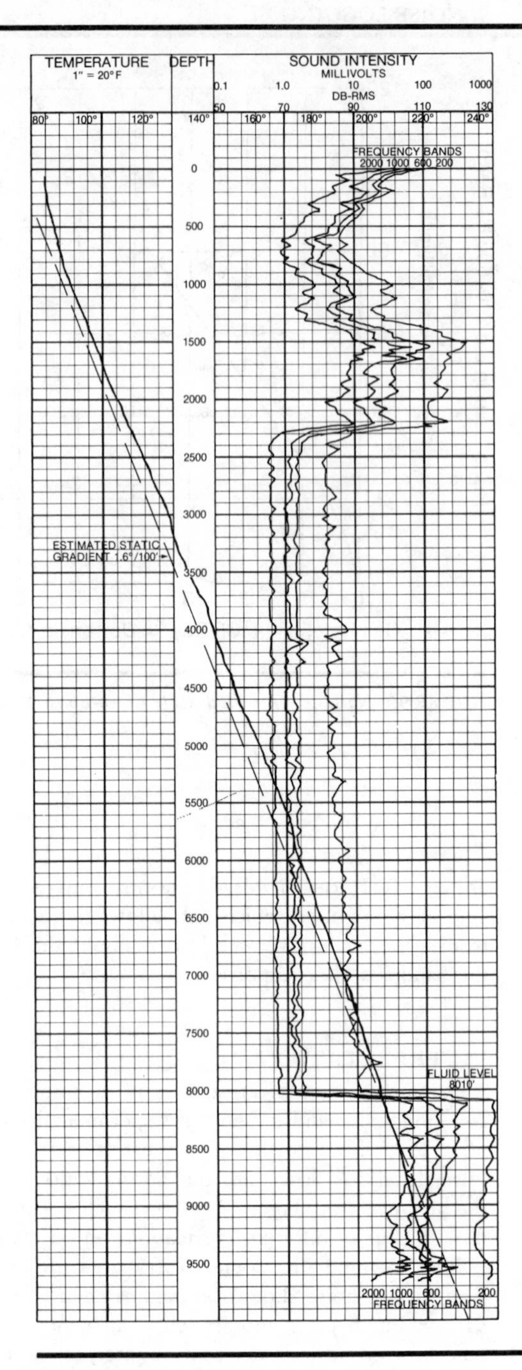

FIGURE 9.8 *Combination Noise and Temperature Log. Courtesy the DIA-LOG Company.*

QUESTION #9.1

Figure 9.8 shows a combination temperature and noise log. Inspect the log. Deduce:

a. the type of flow from 8000 to 9500 ft, and
b. the type of flow from 2250 to 8000 ft.
c. What happens at 2250 ft?

BIBLIOGRAPHY

Britt, E. L.: "Theory and Application of the Borehole Audio Tracer Survey," SPWLA *Trans.,* 17th Annual Logging Symposium, Denver, Jun. 9–12, 1976.

"General Catalog," the DIA-LOG Company.

"An Introduction to Sonan Logging," Technical Memorandum 9385, Dresser Atlas.

Robinson, W. S.: "Field Results from the Noise-Logging Technique," *J. Pet. Tech.* (Nov. 1976); presented at the SPE 49th Annual Meeting held in Houston, Oct. 6–7, 1974.

"Noise Logging Service," brochure no. 75M/5-76/C & A, N. L. McCullough, Houston.

Answers to Text Question

QUESTION #9.1

a. From 8000 to 9500 ft, probably gas percolating up through a static oil or water column.
b. From 2250 to 8000 ft, probably gas flowing in casing above the static fluid column top at 8000 ft.
c. At 2250 ft, gas enters tubing.

CHAPTER TEN

INTERPRETATION

HOLDUP EQUATIONS

The basics of biphasic flow have been covered in chapter 4. In this chapter, the practical applications of the holdup equation will be covered. The purpose is to equip the analyst with a practical tool with which to analyze a combination of a flowmeter log and a holdup log (whether this be from a gradiomanometer or other tool). The methods described will allow the analyst to calculate the flow rate for each phase of a biphasic mixture at each point in the well and hence to deduce the production of each phase from each perforated interval. This is an essential step in quantitative analysis without which remedial action cannot be planned properly.

Figure 10.1 depicts the simultaneous flow of two phases in a vertical pipe. One phase is referred to as the *heavy* phase and the other as the *light* phase. Depending on the mixture these may be:

Mixture	*Light Phase*	*Heavy Phase*
Oil and water	Oil	Water
Oil and gas	Gas	Oil
Gas and water	Gas	Water

The holdup equation states that the density of a flowing biphasic mixture is given by

$$\rho_m = y_h \times \rho_h + (1 - y_h)\rho_l,$$

where:

ρ_m is the density of the mixture,
ρ_h is the density of the heavy phase, and
ρ_l is the density of the light phase.

The holdup of the heavy phase is therefore given by

$$y_h = \frac{\rho_m - \rho_l}{\rho_h - \rho_l}.$$

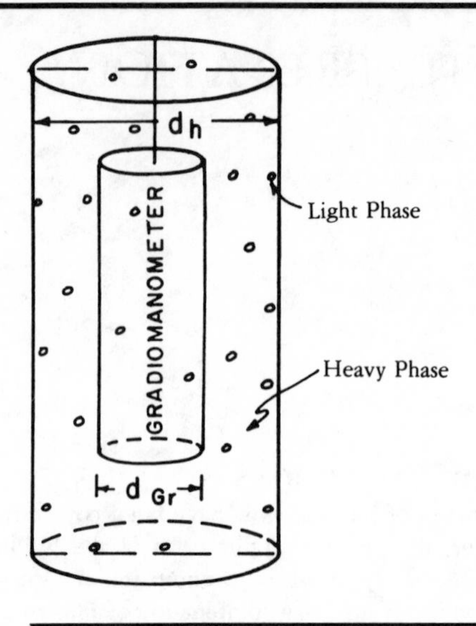

FIGURE 10.1 *Biphasic Flow. Courtesy Schlumberger Well Services.*

QUESTION #10.1
Oil and water flow together and have a density of 0.8 g/cc.
If the oil has a density of 0.6 g/cc and the water a density of 1.0 g/cc, what is the water holdup in this flow mixture?

The holdup only tells us what fraction of the cross-sectional area available for fluid flow is occupied by a particular phase. In order to tell the actual flow rates of each phase we need additional data, namely, total flow rate, the pipe cross-sectional area, the tool diameter, and the interphase slip velocity. In general, the equations will be

$$Q_t = Q_h + Q_l,$$
$$Q_h = y_h A V_h,$$
$$Q_l = y_l A V_l,$$
$$V_l = V_h + V_s, \text{ and}$$
$$1 = y_h + y_l,$$

where:

Q_t = total flow rate,
Q_h = heavy-phase flow rate,
Q_l = light-phase flow rate,
A = cross-sectional area,
V_h = heavy-phase velocity,

V_l = light-phase velocity, and
V_s = slip velocity.

The known quantities are normally Q_t, A, V_s, and y_h. The unknown quantities are Q_h and Q_l. With some minor algebraic manipulation an expression can be derived for Q_h or Q_l as follows:

$$Q_h = Q_t - Q_l = Q_t - (1 - y_h) \cdot A \cdot (V_h + V_s).$$

But

$$V_h = Q_h / y_h A,$$
$$\therefore Q_h = Q_t - (1 - y_h) \cdot A \cdot (Q_h / y_h A + V_s),$$

whence

$$Q_h = y_h [Q_t - (1 - y_h) \cdot A \cdot V_s].$$

Once Q_h is determined, Q_l can be found from

$$Q_l = Q_t - Q_h.$$

PRACTICAL APPLICATIONS

In order to apply this equation to find the volumetric flow rates of the two phases, consistency of units of measurement must be maintained. If flow rates are to be expressed in barrels per day then the product $A \cdot V_s$ must also be in barrels per day. However A, the cross-sectional area of the pipe available for flow, will normally be quoted either in square inches or square centimeters, and V_s, the slip velocity, will normally be quoted in feet per minute or centimeters per second. Therefore, a conversion constant is required to convert the area × speed product into a flow rate in B/D. Thus, in practical terms, the flow-rate equation reduces to

$$Q_h = y_h [Q_t - (1 - y_h) \cdot A \cdot V_s \cdot 1.781],$$

where A is in sq in. and V_s is in ft/min, or to

$$Q_h = y_h [Q_t - (1 - y_h) \cdot A \cdot V_s \cdot 0.5433],$$

where A is in cm² and V_s is in cm/s.

Note that the cross-sectional area referred to is the net area available for flow in the casing/tool annulus. Thus, both the ID of the casing and the OD of the tool must be known in order to solve the flow equation accurately. This net area may be calculated from

$$A = \pi \left[\left(\frac{\text{casing ID}}{2} \right)^2 - \left(\frac{\text{tool OD}}{2} \right)^2 \right].$$

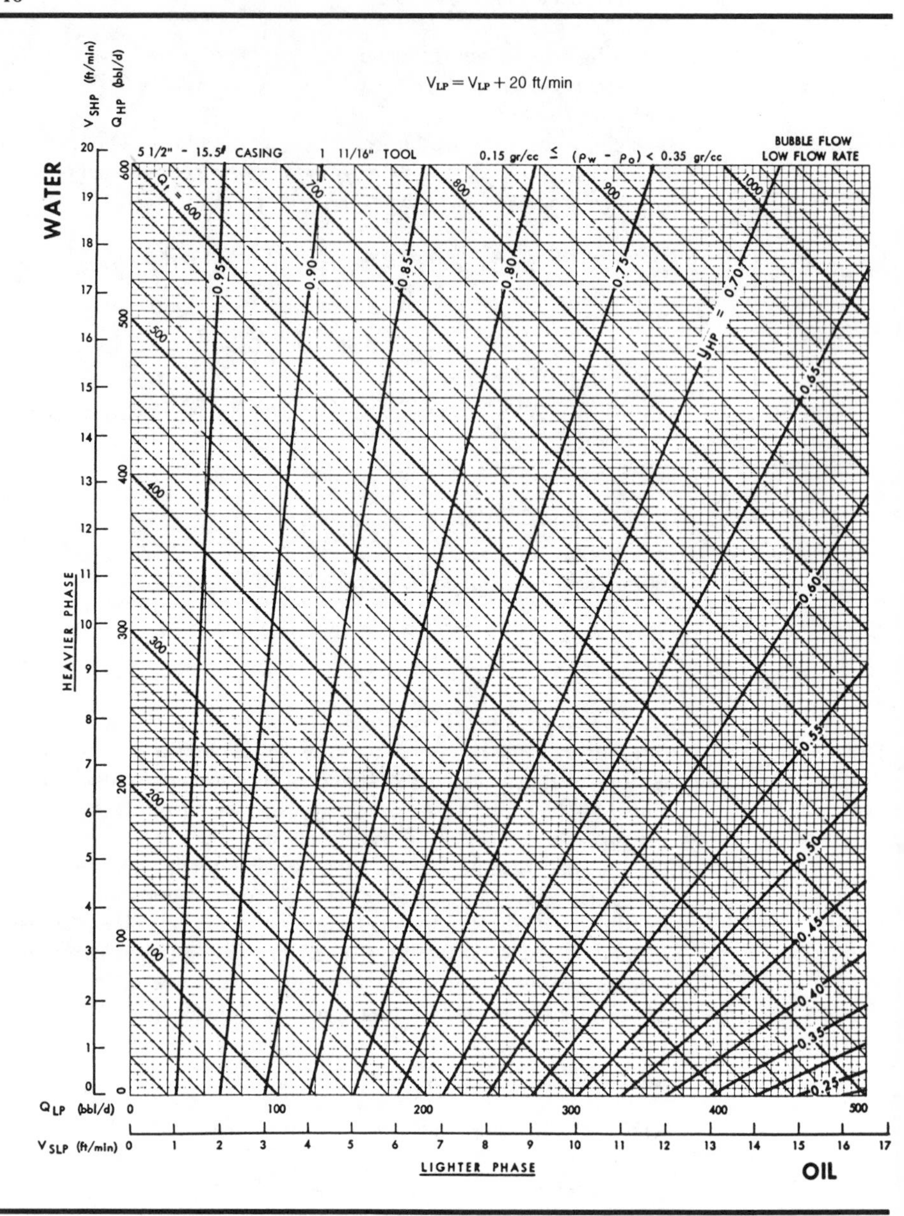

FIGURE 10.2 *Flow Rate and Holdup Chart. Courtesy Schlumberger Well Services.*

This type of calculation may be performed either using a calculator or, graphically, using a chart of the type shown in figure 10.2. On the graph, the *Y*-axis plots the heavy-phase flow rate and the *X*-axis the light-phase flow rate. In figure 10.2, there are two series of lines. One set (running diagonally from NW to SE) corresponds to constant total flow rate. Another set (running diagonally from NE to SW) corresponds to constant values of y_h.

QUESTION #10.2

Oil and water flow in 5-in. 15.5-lb casing which has an ID of 4.95 in. The slip velocity, V_s, is 20 ft/min. The water holdup has been found from a gradiomanometer survey to be 0.7. Assume that a $1^{11}\!/_{16}$-in. OD tool was used to make the measurements, and that the total flow rate is 1000 B/D.

a. What is the water flow rate?
b. What is the oil flow rate?

Note also that the chart shown in figure 10.2 is specifically built for a given set of values for casing size and weight (and therefore casing ID), tool OD, and slip velocity, V_s. If the analyst chooses to use the graphical method, an appropriate chart should be obtained from the relevant service-company publication to suit the particular circumstances.

FLOWMETER AND GRADIOMANOMETER COMBINATIONS

As an example of the practical method to be employed, the log shown in figure 10.3 will be considered. This well produces oil and water from two perforated intervals. Total flow rate is 850 B/D (at downhole conditions) with 485 BOPD and 365 BWPD. Downhole fluid densities are 1.05 g/cc for the water and 0.8 g/cc for the oil. The objective of the analysis is to deduce the oil and water flow rates at each station where a total flow rate has been determined and thereby to determine the oil and water production rates from each perforated interval. Notice that the gradiomanometer log (fig. 10.3) has both a direct recording (the solid trace) and an amplified trace (dotted). It is customary to use the amplified trace since it allows better resolution. However, since use of the amplified trace requires a rescaling in terms of heavy-phase holdup, the amplified gradiomanometer trace is usually read in terms of chart divisions.

In order to rescale the gradiomanometer, two points on the log need to be considered. At Station #4, an assumption is made that the reading of 17.0 divisions corresponds to 100% water, for which, obviously, $y_h = 1.0$. At Station #1, the reading of 5.5 divisions corresponds to some value of y_h that can be back-calculated from

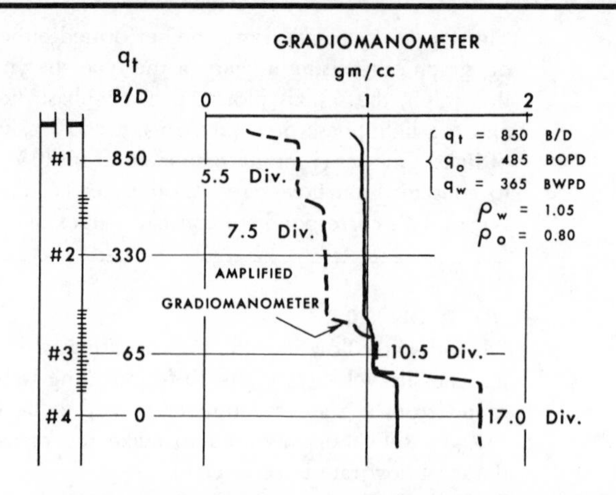

FIGURE 10.3 *Gradiomanometer Log in a Well Producing Oil and Water. Reprinted by permission of the SPE-AIME from Curtis 1967. © 1967 SPE-AIME.*

the known flow rates, casing and tool sizes, and slip velocity. Mathematically,

$$y_h = \frac{-(Q_t - AV_s) \pm \sqrt{(Q_t - AV_s)^2 + 4Q_h AV_s}}{2AV_s}.$$

In this case, the product AV_s in B/D is computed as

$$AV_s = \frac{\pi(4.95^2 - 1.6875^2)}{4} \cdot 20 \cdot 1.781$$

$$= 605.81 \text{ B/D}.$$

The heavy-phase holdup at Station #1 is then calculated as

$$y_h = \frac{-244.19 \pm 971.65}{2 \cdot 605.81} = 0.6.$$

An alternative method of finding y_h above all the perforated intervals is to use the holdup and flow rate chart, as shown in figure 10.4. By plotting the point corresponding to 485 BOPD and 365 BWPD the holdup of 0.6 is directly determined.

The gradiomanometer may now be recalibrated in terms of y_h using the two calibration points determined, that is:

y_h = 1.0 when gradio reads 17.0 divisions, and
y_h = 1.6 when gradio reads 5.5 divisions.

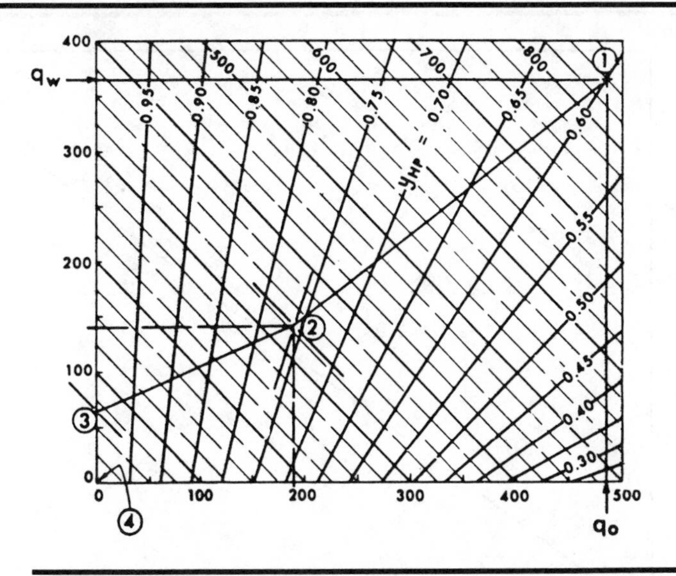

FIGURE 10.4 *Use of Holdup and Flow Rate Chart. Reprinted by permission of the SPE-AIME from Curtis 1967. © 1967 SPE-AIME.*

Thus, for any gradio reading, y_h may be calculated for this particular log by

$$y_h = 0.41 + \frac{\text{gradio divisions}}{28.75}.$$

A more straightforward method is to simply plot a recalibration graph on a sheet of linear graph paper as shown in figure 10.5.

Everything is now in place to proceed with station-by-station analysis. The procedure will be as follows for each station:

1. Read gradiomanometer log.
2. Convert reading to y_h (holdup value).
3. Plot total flow rate and y_h on the holdup–flowrate graph.
4. Read the water and oil flow rates on the X and Y axes.
5. Repeat process for next station.

In order to keep track of the log readings and the calculated answers it is useful to work with a table of the sort shown in table 10.1. Note that the production rate from any set of perforations can be deduced from the oil and flow rate found above and below those perforations. A useful way of presenting the results is shown in figure 10.6. Flow rate is plotted against depth for both oil and water and one phase or the other is shaded or color coded. Coded

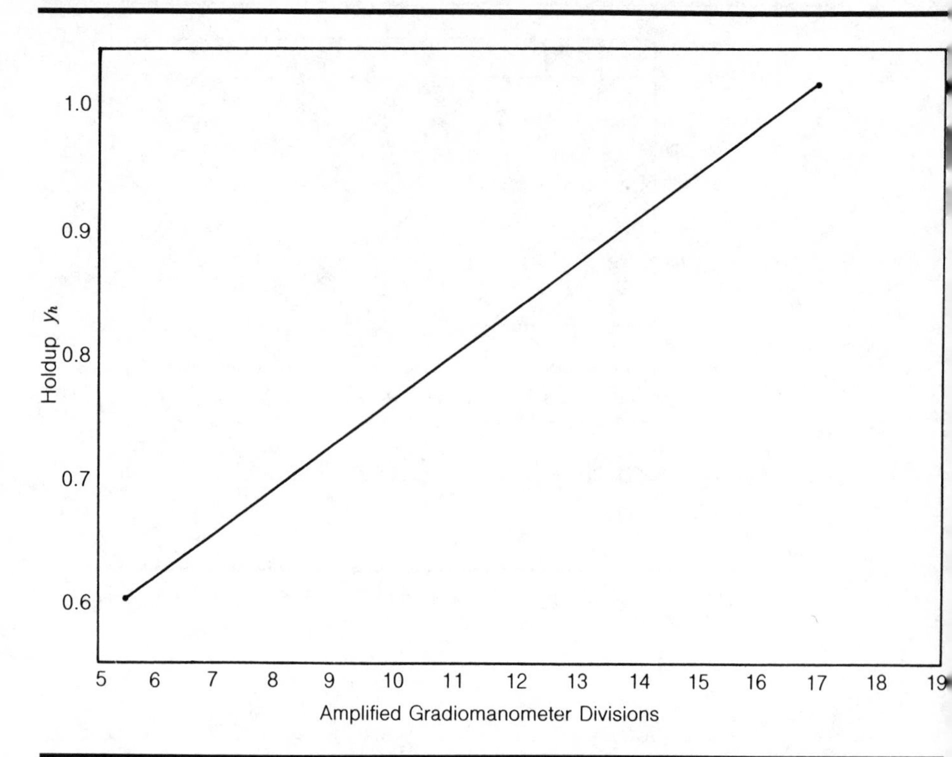

FIGURE 10.5 *Recalibration of Gradiomanometer Log in Terms of* y_h.

TABLE 10.1 *Tabulation of Log Data and Calculations*

Station #	Total Flow Rate B/D	Amplified Gradio Divisions	y_h	Q_w	Q_o	Production BWPD	Production BOPD
1	850	5.5	0.6	365	485		
						278	242
2	330	7.5	0.67	87	243		
						87	243
4	0	17.0	1.0	0	0		

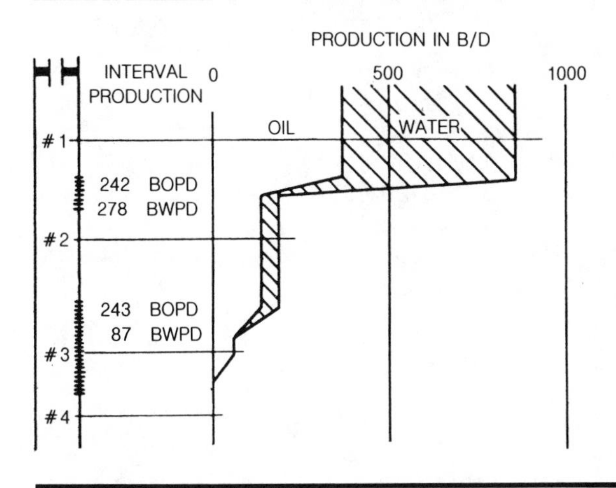

FIGURE 10.6 ***Presentation of Results of Gradiomanometer Analysis. Reprinted by permission of the SPE-AIME, adapted from Curtis 1967. © 1967 SPE-AIME.***

bargraphs may also be used to show the production from each set of perforations.

BIBLIOGRAPHY

Curtis, M. R.: "Flow Analysis in Producing Wells," paper SPE 1908 presented at the SPE 42nd Annual Meeting, Houston, October 1967.

Nicolas, Y. and Witterholt, E. J.: "Measurements of Multiphase Fluid Flow," paper SPE 4023 presented at the SPE 47th Annual Meeting, San Antonio, Tex., Oct. 8–11, 1972.

Answers to Text Questions

QUESTION #10.1

$y_h = 0.5$

QUESTION #10.2

a. 572.8 BWPD
b. 427.2 BOPD

CHAPTER ELEVEN

THE GAMMA RAY LOG

Gamma ray logs are used for three main purposes:

1. correlation
2. evaluation of the shale content of a formation
3. mineral analysis

The gamma ray log measures the natural gamma ray emissions from subsurface formations. Since gamma rays can pass through steel casing, measurements can be made in both open and cased holes. In other applications, induced gamma rays are measured (e.g., in pulsed neutron logging), but that will not be discussed in this chapter.

Figure 11.1 shows a typical gamma ray log. It is normally presented in Track 1 on a linear grid and is scaled in API units, which will be defined later. On this grid, gamma ray activity increases from left to right. Modern gamma ray tools are in the form of double-ended subs that can be sandwiched into practically any logging tool string; thus, the gamma ray can be run with practically any tool available.

Gamma ray tools consist of a gamma ray detector and the associated electronics for passing the gamma ray count rate to the surface. Table 11.1 lists some of the common tool sizes, ratings, and applications.

ORIGIN OF NATURAL GAMMA RAYS

Gamma rays originate from three sources in nature: the radioactive elements in the uranium group, the thorium group, and potassium. Uranium 235, uranium 238, and thorium 232 all decay, via a long chain of daughter products, to stable lead isotopes as illustrated in figure 11.2. An isotope of potassium, K40, decays to argon, giving off a gamma ray as shown in figure 11.3. Note that each type of decay is characterized by a gamma ray of a specific energy (wave length) and that the frequency of occurrence for each specific energy is different. Figure 11.4 shows this relationship between gamma ray energy and frequency of occurrence. This is an important concept, since it is used as the basis for measurement in the natural gamma spectroscopy tools.

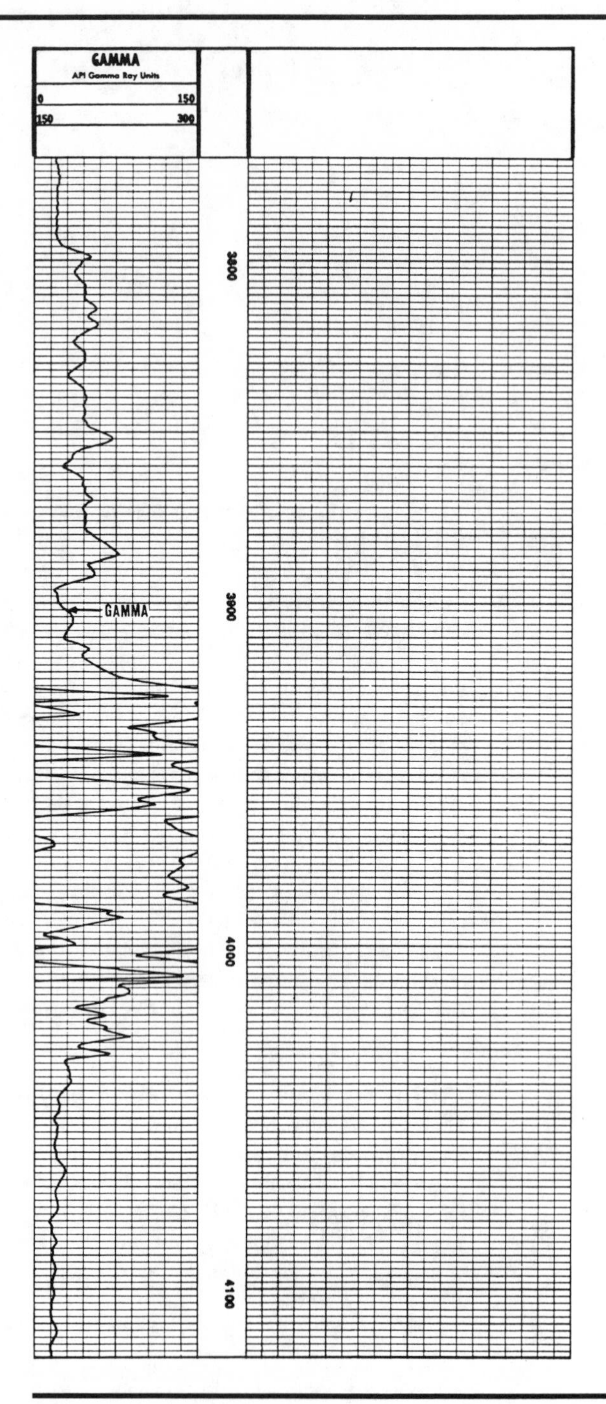

FIGURE 11.1 *Typical Gamma Ray Log. Courtesy WELEX, a Halliburton Company.*

TABLE 11.1 *Scintillation Gamma Tool (SGT) and Natural Gamma Tool (NGT)*

OD (inches)	Temperature (°F)	Pressure (psi)	Application
3⅝	350	20,000	Openhole (SGT)
2¾	500	25,000	HEL* and through drillpipe
1 11/16	350	20,000	Through tubing
3⅝	350	20,000	Lithology (NGT)
	275	10,000	Lithology + P_e (CSNG-Z)†
	400	20,000	Lithology (CSNG)

* HEL = Hostile-environment logging.
† Mark of Welex, a Halliburton Company.

ABUNDANCE OF NATURALLY OCCURRING RADIOACTIVE MINERALS

An average shale contains 6 ppm uranium, 12 ppm thorium, and 2% potassium. Since the various gamma ray sources are not all equally effective, it is more informative to consider this mix of radioactive materials on a common basis, for example, by reference to potassium equivalents (i.e., the amount of potassium that would produce the same number of gamma rays per unit of time). Reduced to a common denominator, the average shale contains uranium equivalent to 4.3% potassium, thorium equivalent to 3.5% potassium, and 2% potassium. An *average* shale is hard to find. Shale is a mixture of clay minerals, sand, silts, and other extraneous materials; thus, there can be no "standard" gamma ray activity for shale. Indeed, the main clay minerals vary enormously in their natural radioactivity. Kaolinite has no potassium, whereas illite contains between 4 and 8% potassium. Montmorillonite contains less than 1% potassium. Occasionally, natural radioactivity may be due to the presence of dissolved potassium or other salts in the water contained in the pores of the shale.

OPERATING PRINCIPLE OF GAMMA RAY TOOLS

Traditionally, two types of gamma ray detectors have been used in the logging industry: Geiger-Mueller and scintillation detectors. Today, practically all gamma ray tools use scintillation detectors containing a sodium iodide crystal (fig. 11.5). When a gamma ray strikes the crystal, a single photon of light is emitted. This tiny flash of light then strikes a photo cathode made from cesium antimony or silver magnesium. Each photon, when hitting the photo cathode, releases a bundle of electrons. These in turn are accelerated in an electric field to strike another electrode producing an even bigger bundle (a shower) of electrons. This process is repeated through a number of stages until a final electrode conducts a small current

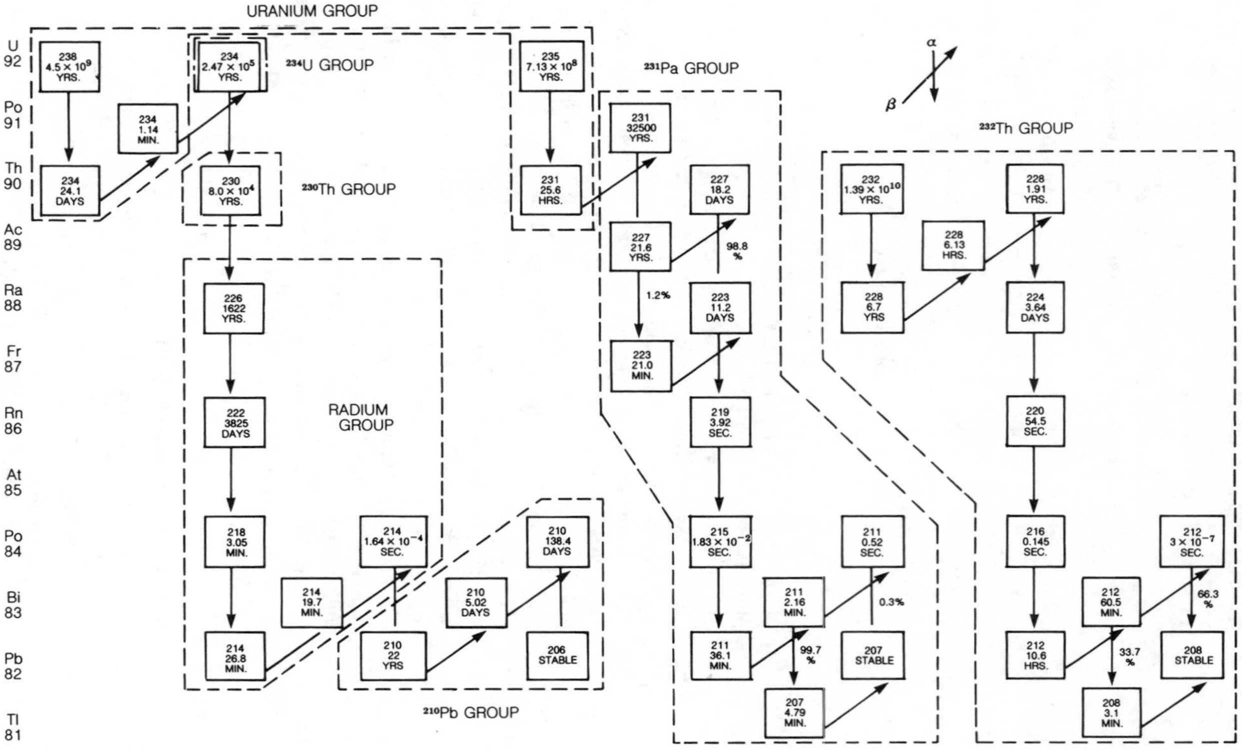

FIGURE 11.2 *Classification of the Radioactive Disintegration Series.*

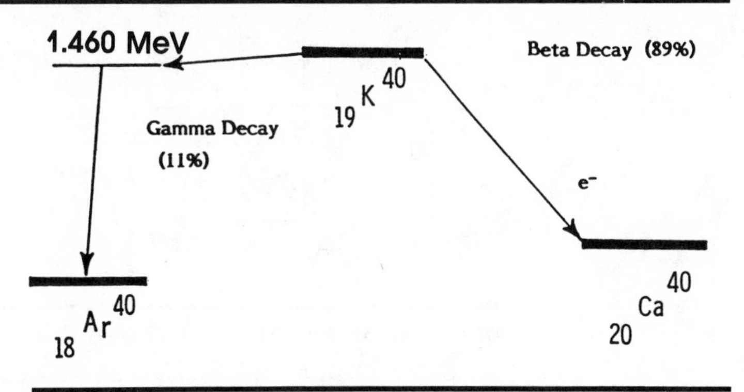

FIGURE 11.3 *Decay Scheme of $_{19}K40$.*

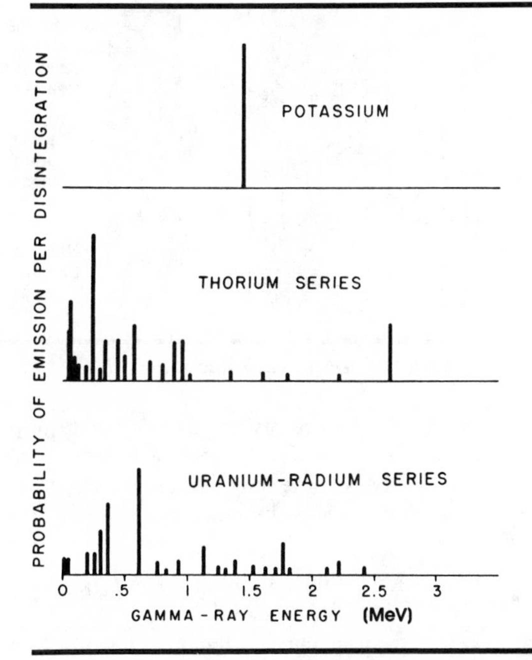

FIGURE 11.4 *Gamma Ray Emission Spectra of Radioactive Minerals. From Tittman 1956.*

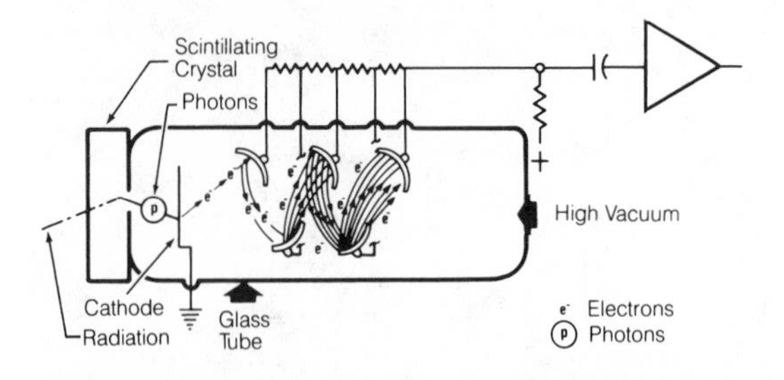

FIGURE 11.5 *Scintillation Counter. Courtesy Gearhart Industries, Inc.*

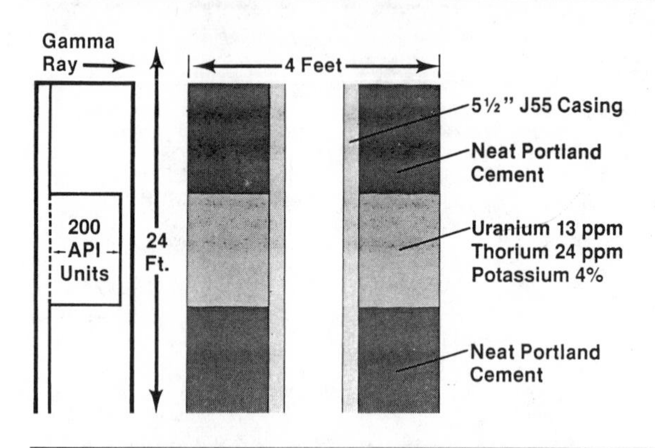

FIGURE 11.6 *API Gamma Ray Standard.*

through a measure resistor to give a voltage pulse that signals that a gamma ray struck the sodium iodide crystal. The system has a very short dead time and can register many counts per second without becoming swamped by numerous signals.

CALIBRATION OF GAMMA RAY DETECTORS AND LOGS

One of the problems of gamma ray logging is the choice of a standard calibration system, since all logging companies use counters of different sizes and shapes encased in steel housings with varying characteristics. On very old logs, the scale might be quoted in micrograms of radium per ton of formation. For many reasons this was found to be an unsatisfactory method of calibration for gamma ray logs, so an API standard was devised. A test pit (installed at the University of Houston) contains an "artificial shale," as illustrated in figure 11.6. A cylinder, 4 ft in diameter and 24 ft long contains a central 8-ft section consisting of cement mixed with 13 ppm uranium, 24 ppm thorium, and 4% potassium. On either side of this central section are 8-ft sections of neat portland cement. This sandwich is cased with 5½-in. J55 casing. The API standard defines the difference in gamma ray count rate between the neat cement and the radioactively doped cement as 200 API units. Any logging service company may place its gamma ray tool in this pit to make a calibration.

Field calibration is performed using a portable jig that contains a radioactive *pill.* The pill is a 0.1 millicurie source of radium 226. When placed 53 in. from the center of the gamma ray detector, it produces a known increase over the background count rate. This increase is equivalent to a known number of API units, depending on the tool type and size and the counter it encloses.

TIME CONSTANTS

All radioactive processes are subject to statistical variations. For example, if a source of gamma rays emits an average of 100 gamma rays each second over a period of hours, the source will emit 360,000 gamma rays per hour (100/s × 60 s × 60 min). However, if the count is measured for 1 second, the actual count might be less than 100 or more than 100. Thus, a choice must be made. Gamma rays can be counted for a very short interval of time, resulting in a poor estimate of the real count rate; or the gamma rays can be counted for a long time, resulting in a more accurate estimate of the count rate at the expense of an inordinately long time period. In order to average out the statistical variations, various *time constants* may be selected according to the radioactivity level measured. The lower the count rate, the longer the time constant required for adequate averaging of the variations.

In the logging environment, gamma rays can be counted for a short period of time (e.g., 1 second) with the recognition that during that time period, the detector will have moved past the formation whose activity is being measured. Thus, the logging speed and the time interval used to average count rates are interrelated. The following rules of thumb are generally recognized:

Logging Speed	*Time Constant*
3600 ft/hr	1 second
1800 ft/hr	2 second
1200 ft/hr	3 second
900 ft/hr	4 second

At very slow logging speeds (900 ft/hr = 1.5 ft/s) and long time constants, a more accurate measurement of absolute activity is obtained at the expense of good bed resolution. At high logging speeds and short time constants, somewhat better bed resolution is obtained at the expense of absolute accuracy. At some future time, when the efficiency of gamma ray detectors and their associated electronics improve by one or two orders of magnitude, the use of a time constant will be obsolete except in the cases of very, very inactive formations with intrinsically low gamma ray count rates.

To illustrate this interdependence of logging speed and time constant, figure 11.7 shows the same formation logged at two different speeds. On the first run, the logging speed was 80 ft/min and the

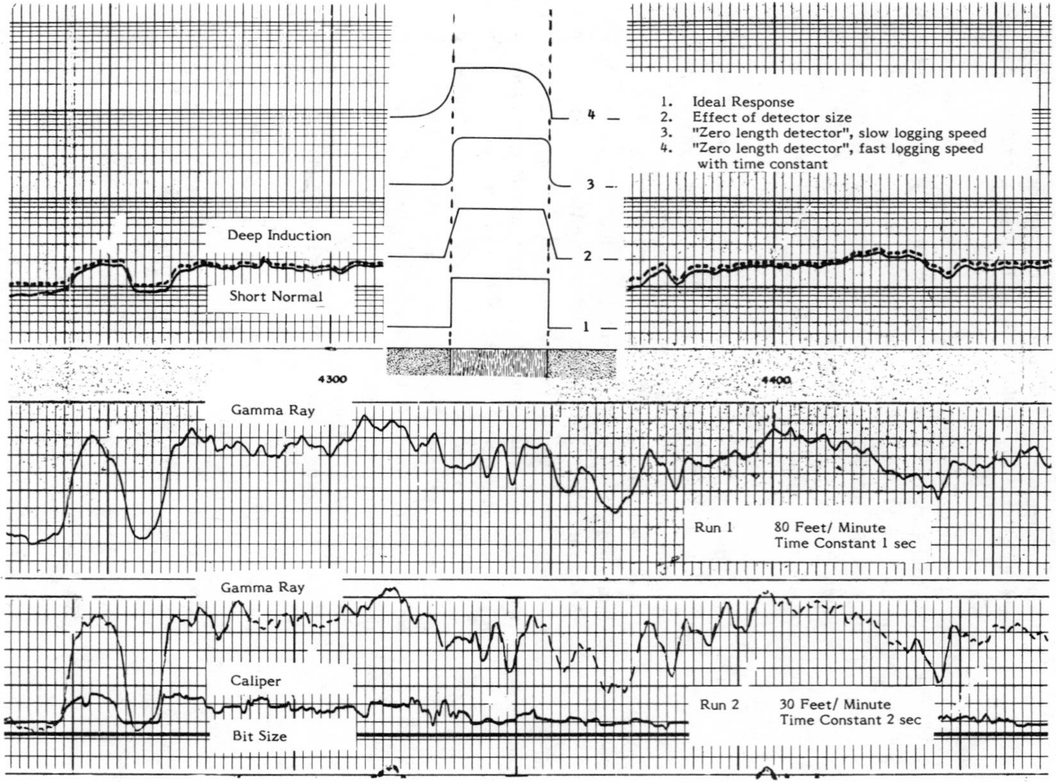

FIGURE 11.7 *Effects of Logging Speed and Time Constant on Gamma Ray Log.*

time constant 1 second. On the second run, the speed was 30 ft/min and the time constant was 2 seconds. Note the differences in both statistics and bed resolution between the two runs.

PERTURBING EFFECTS ON GAMMA RAY LOGS

Gamma ray logs are subject to a number of perturbing effects including:

Sonde position in the hole (centering/eccentering)
Hole size
Mud weight
Casing size and weight
Cement thickness

Since there are innumerable combinations of hole size, mud weights, and tool positions, an arbitrary standard set of conditions is defined as a 3⅝-in. OD tool eccentered in an 8-in. hole filled with 10-lb mud. A series of charts exist for making the appropriate corrections. Figure 11.8 applies to logs run in openhole and corrects for hole size and mud weight. Figure 11.9 applies to logs run in cased hole and corrects for casing and cement as well. Note that if a gamma ray log is run in combination with a CNL-FDC tool, it is run eccentered. If it is run with a laterolog or an induction log, it will be centered, most of the time.

QUESTION #11.1

Use figure 11.8 to estimate GR_{cor} under the following conditions:

GR_{Log} reads 67 API units.
Hole size = 8 in.
Mud weight = 16 lb/gal.
Tool is centered.

Find GR_{cor} = ______.

ESTIMATING SHALE CONTENT FROM GAMMA RAY LOGS

Since it is common to find radioactive materials associated with the clay minerals that constitute shales, it is a commonly accepted practice to use the relative gamma ray deflection as a shale-volume indicator. The simplest procedure is to rescale the gamma ray log between its minimum and maximum values from 0 to 100% shale. A number of studies have shown that this is not necessarily the best method, and alternative relationships have been proposed. To further explain these methods, the *gamma ray index* is defined as a linear rescaling of the GR log between GR_{min} and GR_{max} such that

$$\text{gamma ray index} = \frac{GR - GR_{min}}{GR_{max} - GR_{min}}$$

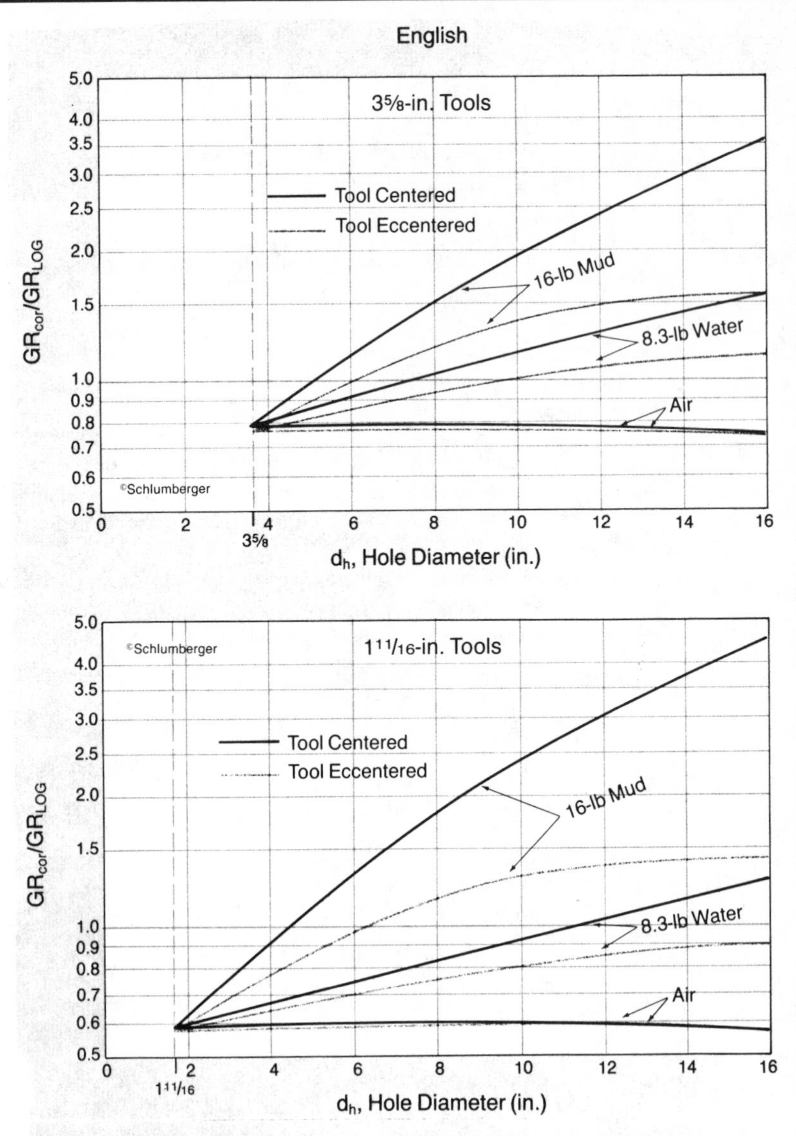

FIGURE 11.8 ***Gamma Ray Corrections for Hole Size and Mud Weight. Courtesy Schlumberger Well Services.***

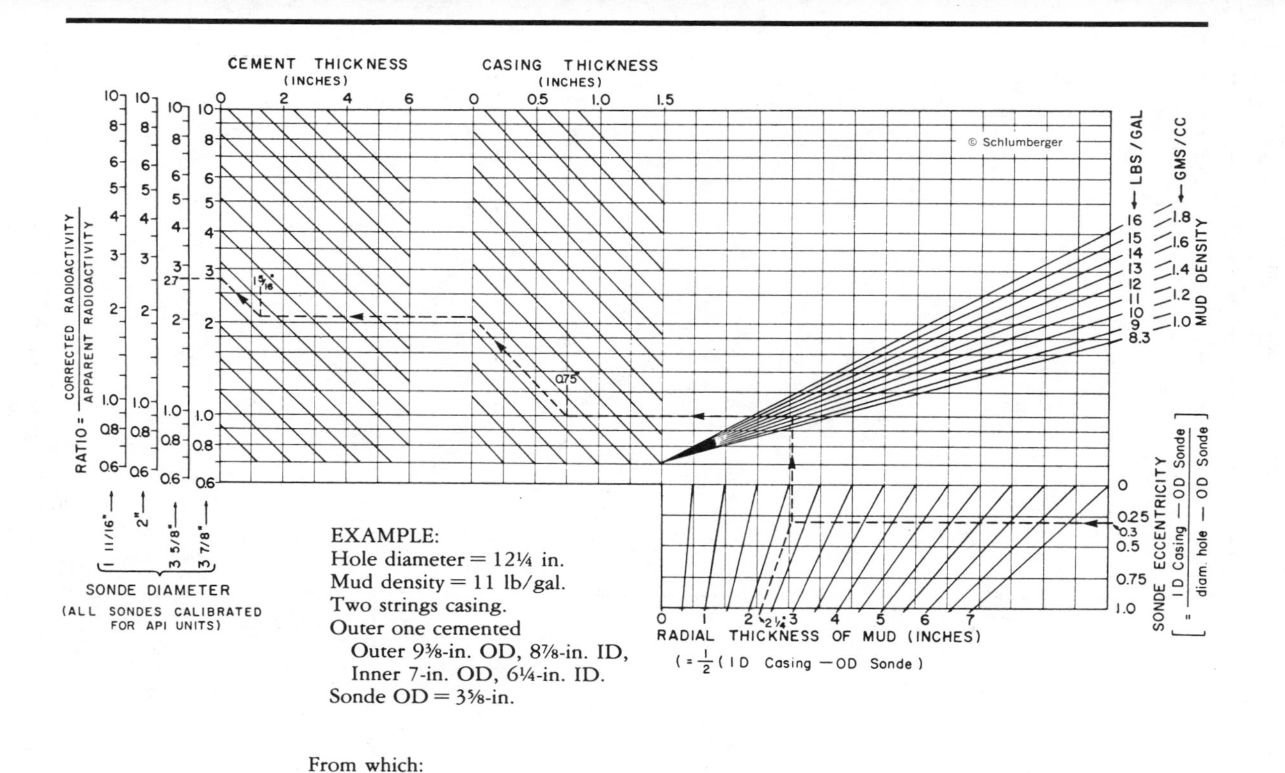

EXAMPLE:
Hole diameter = 12¼ in.
Mud density = 11 lb/gal.
Two strings casing.
Outer one cemented
Outer 9⅜-in. OD, 8⅞-in. ID,
Inner 7-in. OD, 6¼-in. ID.
Sonde OD = 3⅝-in.

From which:
Cement thickness = ½ (12¼ − 8⅝) = 1 5/16 in.
Casing thickness = ½ (9⅜ − 8⅞ + 7 − 6¼) = ⅝ in.
Mud thickness = ½ (8⅞ − 7 + 6¼ − 3⅝) = 2¼ in.
Sonde eccentricity = (6¼ − 3⅝)/(12¼ − 3⅝) = 0.30.

FIGURE 11.9 *Gamma Ray Corrections for Cased Holes. Courtesy Schlumberger Well Services.*

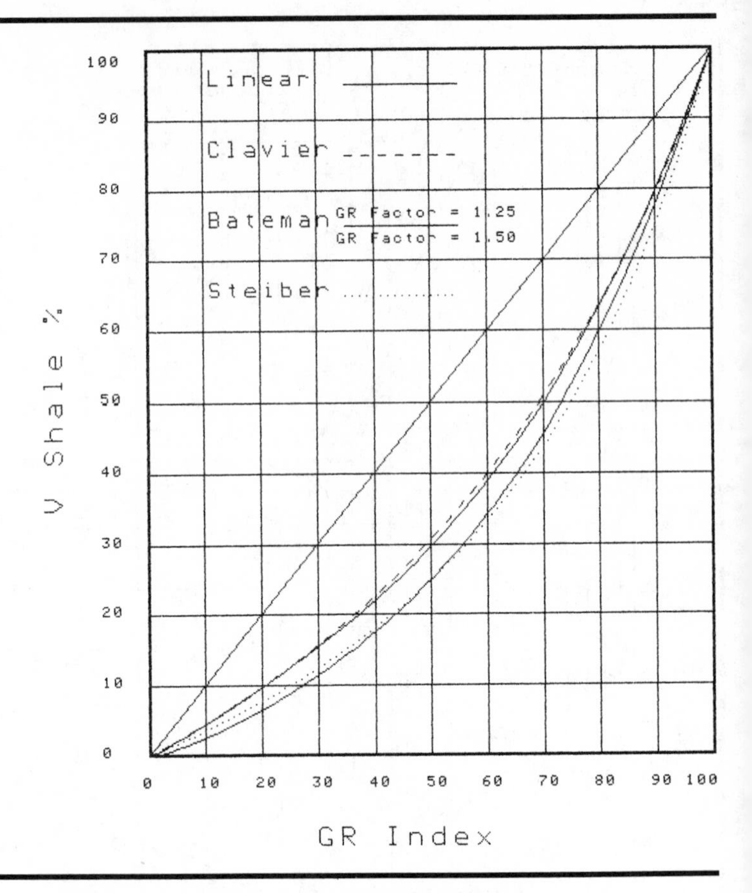

FIGURE 11.10 V_{sh} *as a Function of Gamma Ray Index.*

If this index is called *X,* then the alternative relationships can be stated in terms of *X* as follows:

Relationship	*Equation*
Linear	$V_{sh} = X$
Clavier	$V_{sh} = 1.7 - \sqrt{[3.38 - (X + 0.7)^2]}$
Steiber	$V_{sh} = \dfrac{0.5\,X}{(1.5 - X)}$
Bateman	$V_{sh} = X^{(X + GR\ \mathrm{factor})}$

where the *GR* factor is a number chosen to force the result to imitate the behavior of either the Clavier or the Steiber relationship. Figure 11.10 illustrates comparatively the difference between these alternative relationships.

QUESTION #11.2

On the gamma ray log shown in figure 11.11, choose a value for GR_{min}, GR_{max} and then compute V_{sh} in Sand C using the linear, Clavier, and Steiber methods.

GAMMA RAY SPECTROSCOPY

Each radioactive decay produces a gamma ray that is unique. These various gamma rays have characteristic energy levels and occur in characteristic abundance, as expressed in counts per time period. The simple method of just counting how many gamma rays a formation produces can be carried a step further to count how many gamma rays from each energy group it produces. The spectrum produced when the number of occurrences is plotted against the energy group is characteristic of the formation logged.

Figure 11.12 shows such a spectrum, where energies from 0 to approximately 3 MeV have been split into 256 specific energy bins. The number of gamma rays in each bin is plotted on the *Y*-axis. This spectrum can be thought of as a mixture of the three individual spectra belonging to uranium, thorium, and potassium. Some unique mixture of these three radioactive families would have the same spectrum as the observed one. The trick is to find a quick and easy method of discovering that unique mixture. Fortunately, on-board computers in logging trucks are capable of quickly finding a "best fit" and producing continuous curves showing the concentration of U, Th, and K.

Figure 11.13 illustrates a gamma ray spectral log. Both total gamma ray activity (SGR) and a uranium-free version of the total activity are displayed in Track 1. Units are API. In Tracks 2 and 3, the concentration of U, Th, and K are displayed. Depending on the logging service company, the units may be in counts/s, ppm, or percent.

QUESTION #11.3

In the example shown in figure 11.13, determine which element is responsible for the high activity seen on the total gamma ray intensity curve at the point marked A.

INTERPRETATION OF NATURAL GAMMA RAY SPECTRA LOGS

The interpretation of natural gamma ray spectra logs is a new art and, as such, is still developing. Two general techniques are in use. One is the use of the uranium curve as an indicator of fractures. This is more fully described by Fertl et al. Another technique is to

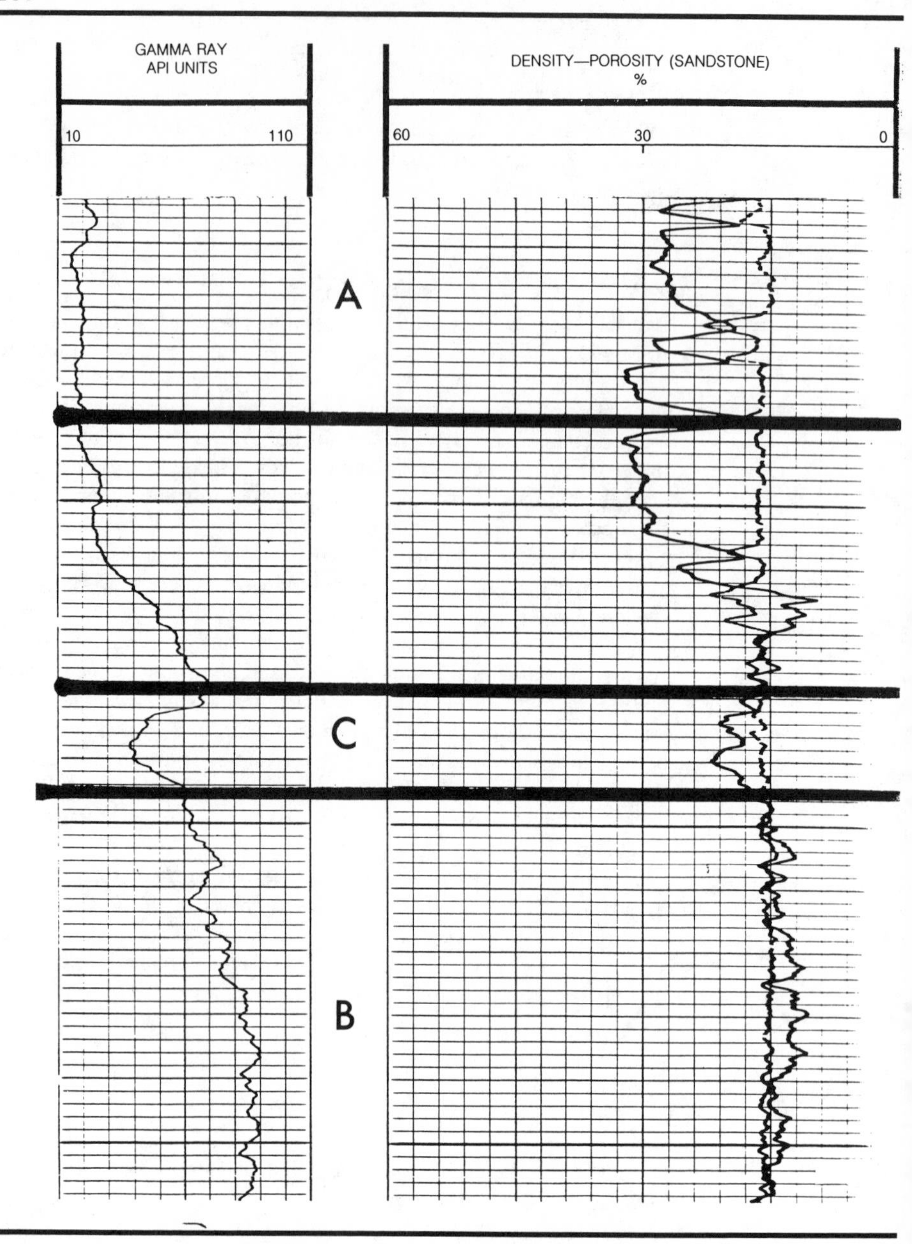

FIGURE 11.11 *Estimation of Shale Content from Gamma Ray Log.*

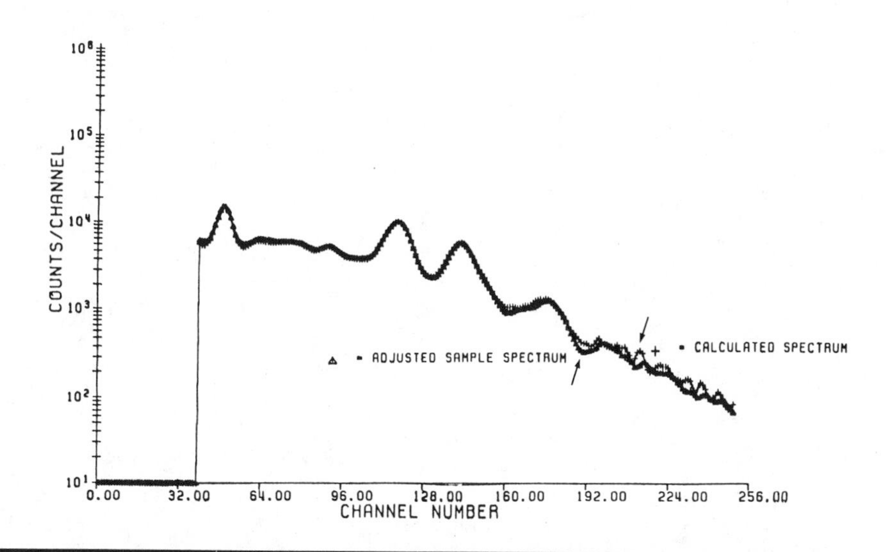

FIGURE 11.12 *Gamma Ray Spectrometry.*

apply the U, Th, and K concentrations in combination with other log data to determine lithology and clay type, as described by Marett et al. Still another approach could be called the geochemical method as described by Hassan et al. (1976). Figure 11.14 illustrates the variation of the thorium/potassium ratio in a number of minerals ranging from potassium-feldspar to bauxite. Figure 11.15 maps a number of radioactive minerals as a function of their thorium and potassium content. Other elemental ratios are also useful indicators. For example, a low U/Th ratio indicates reduced black shales. Uranium by itself may indicate a high organic carbon content, which in turn may indicate the presence of gas. If additional data are available, for example, the photoelectric absorption coefficient (Pe) obtained from the Litho-Density tool, plots of the sort shown in figure 11.16 can be made to assist in mineral identification.

Field presentations of gamma ray spectra can assist the analyst in the task of mineral identification by offering curve plots with ratios of the three components (U, Th, and K) already computed. Figure 11.17 gives an example of one such presentation. Track 1 shows total gamma ray together with a uranium-free curve. Track 2 gives three ratios, uranium/potassium, thorium/uranium and thorium/potassium. Track 3 gives a coded display on which the coded area represents the formations with both the highest potassium and the highest thorium content.

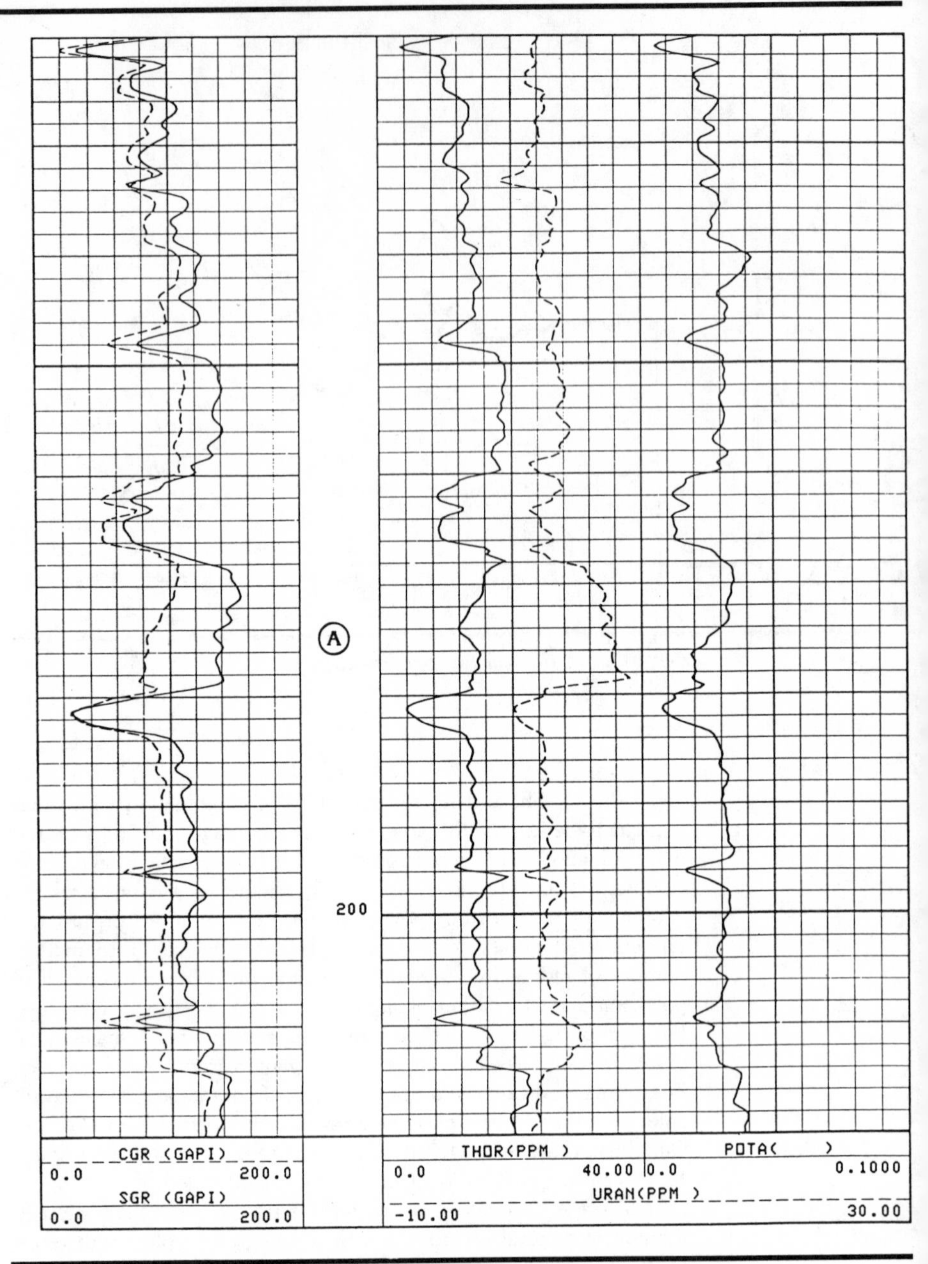

FIGURE 11.13 *Gamma Ray Spectral Log. Courtesy Schlumberger Well Services.*

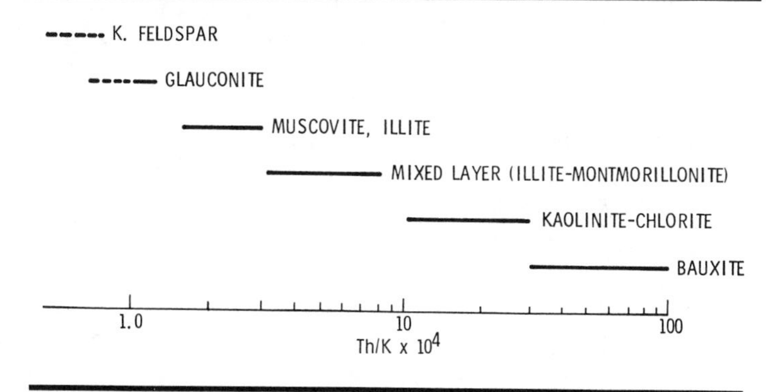

FIGURE 11.14 *Thorium/Potassium Ratios for Various Minerals. Reprinted by permission of the SPWLA from Hassan et al. 1976.*

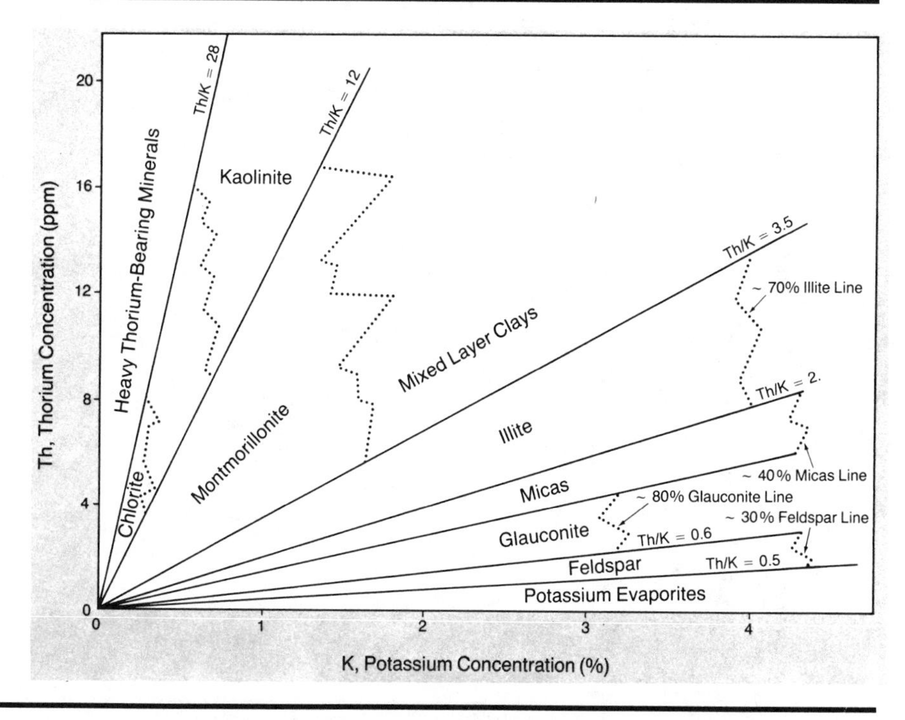

FIGURE 11.15 *Thorium/Potassium Crossplot for Mineral Identification. Courtesy Schlumberger Well Services.*

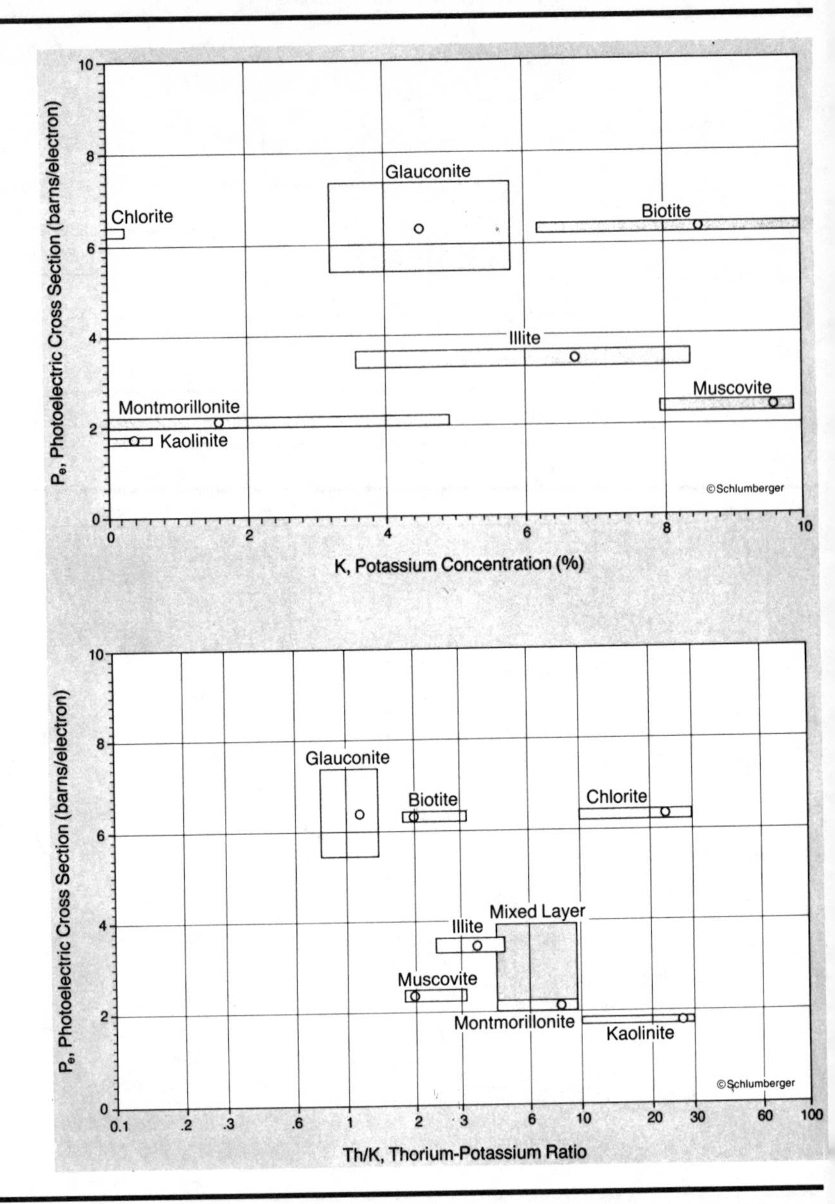

FIGURE 11.16 *Photoelectric Absorption Coefficient (Pe) and Natural Gamma Ray Spectra Crossplots. Courtesy Schlumberger Well Services.*

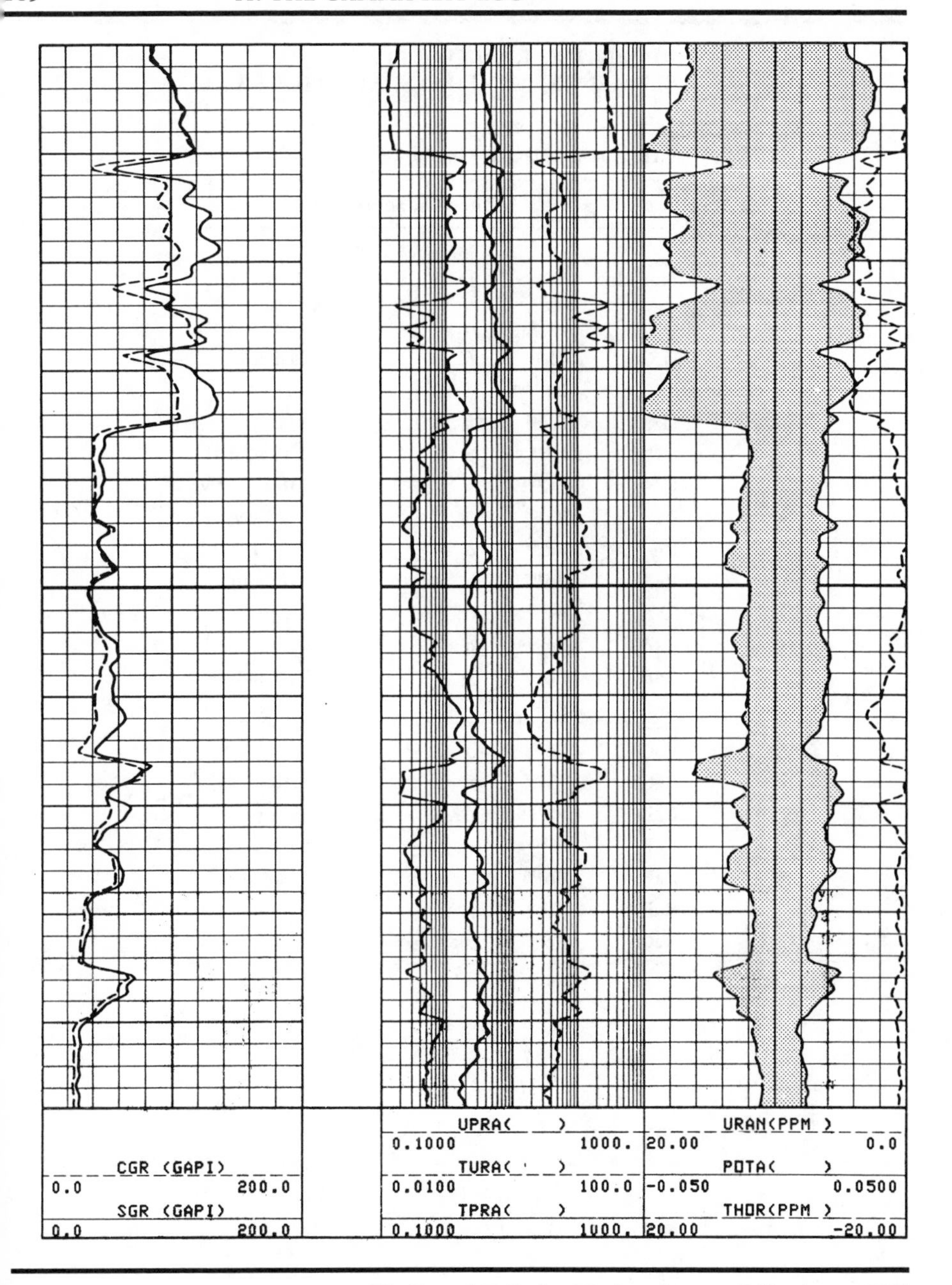

FIGURE 11.17 *Th, K, and U Ratios Display. Courtesy Schlumberger Well Services.*

APPENDIX: RADIOACTIVE ELEMENTS, MINERALS, AND ROCKS

TABLE 11A.1 *Natural Gamma Ray Emitters*

Uranium Series			
Nuclide		Mode of Disintegration	Half-Life
UI	$_{92}U^{238}$	α	4.51×10^{4} yr
UX_1	$_{90}Th^{234}$	β	24.1 d
UX_2	$_{91}Pa^{234m}$	β, IT	1.18 min
UZ	$_{91}Pa^{234}$	β	6.66 hr
UII	$_{92}U^{234}$	α	2.48×10^{4} yr
Io	$_{90}Th^{230}$	α	8.0×10^{4} yr
Ra	$_{88}Ra^{226}$	α	1620 yr
Rn	$_{86}Em^{222}$	α	3.82 d
RaA	$_{84}Po^{218}$	α, β	3.05 min
RaA′	$_{85}At^{218}$	α, β	2 s
RaA″	$_{86}Em^{218}$	α	1.3 s
RaB	$_{82}Pb^{214}$	β	26.8 min
RaC	$_{83}Bi^{214}$	α, β	19.7 min
RaC′	$_{84}Po^{214}$	α	1.6×10^{-4} sec
RaC″	$_{81}TI^{210}$	β	1.32 min
RaD	$_{82}Pb^{210}$	β	19.4 yr
RaE	$_{83}Bi^{210}$	α, β	5.01 d
RaF	$_{84}Po^{210}$	α	138.4 d
RaE′	$_{81}TI^{206}$	β	4.2 min
RaG	$_{82}Pb^{206}$	Stable	

TABLE 11A.1 (*Continued*)

Thorium Series			
Nuclide		Mode of Disintegration	Half-Life
Th	$_{90}Th^{232}$	α	1.42×10^{10} yr
$MsTh_1$	$_{88}Ra^{228}$	β	6.7 yr
$MsTh_2$	$_{89}Ac^{228}$	β	6.13 hr
RdTh	$_{90}Th^{228}$	α	1.91 yr
ThX	$_{88}Ra^{224}$	α	3.64 d
Tn	$_{86}Em^{220}$	α	51.5 s
ThA	$_{84}Po^{216}$	α	0.16 s
ThB	$_{82}Pb^{212}$	β	10.6 hr
ThC	$_{83}Bi^{212}$	α, β	60.5 min
ThC′	$_{84}Po^{212}$	α	0.30 μs
ThC′	$_{81}T^{208}$	β	3.10 min
ThD	$_{82}Pb^{208}$	Stable	

TABLE 11A.2 *Gamma Ray Lines* in the Spectra of the Important Naturally Occurring Radionuclides*

Nuclide	Gamma Ray Energy, MeV	Number of Photons per Disintegration in Equilibrium Mixture
Bi^{214}(Rac)	0.609	0.47
	0.769	0.05
	1.120	0.17
	1.238	0.06
	1.379	0.05
	1.764	0.16
	2.204	0.05
$T4^{208}$(ThC′)	0.511	0.11
	0.533	0.28
	2.614	0.35
K^{40}	1.46	0.11

* With intensities greater than 0.05 photons per disintegration and energies greater than 100 kev.

TABLE 11A.3 *Thorium-Bearing Minerals*

Name	Composition	ThO_2 Content (%)
Thorium minerals		
Cheralite	$(Th,Ca,Ce)(PO_4SiO_4)$	30, variable
Huttonite	$ThSiO_4$	81.5 (ideal)
Pilbarite	$ThO_2 \cdot UO_3 \cdot PbO \cdot 2SiO_2 \cdot 4H_2O$	31, variable
Thorianite	ThO_2	Isomorphous series to UO_2
Thorite[a]	$ThSiO_4$	25 to 63–81.5 (ideal)
Thorogummite[a]	$Th(SiO_4)_{1-x}(OH)_{4-x}$; $x < 0.25$	24 to 58 or more
Thorium-bearing minerals		
Allanite	$(Ca,Ce,Th)_2(Al,Fe,Mg)_3Si_3O_{12}(OH)$	0 to about 3
Bastnaesite	$(Ce,La)Co_3F$	Less than 1
Betafite	About $(U,Ca)(Nb,Ta,Ti)_30_9 \cdot nH_20$	0 to about 1
Brannerite	About $(U,Ca,Fe,Th,Y)_3Ti_5O_{16}$	0 to 12
Euxenite	$(Y,Ca,Ce,U,Th)(Nb,Ta,Ti)_2O_5$	0 to about 5
Eschynite	$(Ce,Ca,Fe,Th)(Ti,Nb)_2O_6$	0 to 17
Fergusonite	$(Y,Er,Ce,U,Th)(Nb,Ta,Ti)O_4$	0 to about 5
Monazite[b]	$(Ce,Y,La,Th)PO_4$	0 to about 30; usually 4 to 12
Samarskite	$(Y,Er,Ce,U,Fe,Th)(Nb,Ta)_2O_6$	0 to about 4
Thucholite	Hydrocarbon mixture containing U, Th, rare earth elements	
Uraninite	UO_2 (ideally) with Ce, Y, Pb, Th, etc.	0 to 14
Yttrocrasite	About $(Y,Th,U,Ca)_2(Ti,Fe,W)_4O_{11}$	7 to 9
Zircon	$ZrSiO_4$	Usually less than 1

Source: After Frondel, C., 1956, in Page, L. R., Stocking, H. E., and Smith, H. D., Jr., U.S. Geol. Survey Prof. Papers no. 300.
[a] Potential thorium ore minerals.
[b] Most important commercial ore of thorium. Deposits are found in Brazil, India, USSR, Scandinavia, South Africa, and U.S.A.

TABLE 11A.4 *Uranium Minerals*

Name	Composition
Autunite	$Ca(UO_2)_2(PO_4)_2$ 10–12H_2O
Tyuyamunite	$Ca(UO_2)_2(VO_4)_2$ 5–8H_2O
Carnotite	$K_2(UO_2)_2(UO_4)_2$ 1–3H_2O
Baltwoodite	U-silicate high in K
Weeksite	U-silicate high in Ca

TABLE 11A.5 *Potassium, Uranium, and Thorium Distribution in Rocks and Minerals*

	K (%)	U (ppm)	Th (ppm)
Accessory minerals			
Allanite		30–700	500–5000
Apatite		5–150	20–150
Epidote		20–50	50–500
Monazite		500–3000	2.5×10^4–20×10^4
Sphene		100–700	100–600
Xenotime		500–3, 4×10^4	Low
Zircon		300–3000	100–2500
Andesite (av.)	1.7	0.8	1.9
A., Oregon	2.9	2.0	2.0
Basalt			
Alkali basalt	0.61	0.99	4.6
Plateau basalt	0.61	0.53	1.96
Alkali olivine basalt	<1.4	<1.4	3.9
Tholeiites (orogene)	<0.6	<0.25	<0.05
(non orogene)	<1.3	<0.50	<2.0
Basalt in Oregon	1.7	1.7	6.8
Carbonates			
Range (average)	0.0–2.0(0.3)	0.1–9.0(2.2)	0.1–7.0(1.7)
Calcite, chalk, Limestone, dolomite (all pure)	<0.1	<1.0	<0.5
Dolomite, West Texas (clean)	0.1–0.3	1.5–10	<2.0
Limestone (clean)			
Florida	<0.4	2.0	1.5
Cretaceous trend, Texas	<0.3	1.5–15	<2.0
Hunton lime, Okla.	<0.2	<1.0	<1.5
West Texas	<0.3	<1.5	<1.5
Clay minerals			
Bauxite		3–30	10–130
Glauconite	5.08–5.30		
Bentonite	<0.5	1–20	6–50
Montmorillonite	0.16	2–5	14–24
Kaolinite	0.42	1.5–3	6–19
Illite	4.5	1.5	
Mica			
Biotite	6.7–8.3		<0.01
Muscovite	7.9–9.8		<0.01
Diabase, Va.	<1.0	<1.0	2.4
Diorite, quartzodiorite	1.1	2.0	8.5
Dunite, Wa.	<0.02	<0.01	<0.01
Feldspars			
Plagioclase	0.54		<0.01
Orthoclase	11.8–14.0		<0.01
Microcline	10.9		<0.01
Gabbro (mafic igneous)	0.46–0.58	0.84–0.9	2.7–3.85

TABLE 11A.5 (*Continued*)

	K (%)	U (ppm)	Th (ppm)
Granite (silicic igneous)	2.75–4.26	3.6–4.7	19–20
Rhode Island	4.5–5	4.2	25–52
New Hampshire	3.5–5	12–16	50–62
Precambrian (Okla.) Minnesota, Col. Tex.)	2–6	3.2–4.6	14–27
Granodiorite	2–2.5	2.6	9.3–11
Colorado, Idaho	5.5	2.–2.5	11.0–12.1
Oil shales, Colorado	<4.0	up to 500	1–30
Periodite	0.2	0.01	0.05
Phosphates		100–350	1–5
Rhyolite	4.2	5	
Sandstones, range (av.)	0.7–3.8(1.1)	0.2–0.6(0.5)	0.7–2.0(1.7)
Silica, quartz, quartzite, (pure)	<0.15	<0.4	<0.2
Beach Sands, Gulf Coast	<1.2	0.84	2.8
Atlantic Coast (Fla., N.C.)	0.37	3.97	11.27
Atlantic Coast (N.J., Mass.)	0.3	0.8	2.07
Shales			
"Common" shales [range (av.)]	1.6–4.2(2.7)	1.5–5.5(3.7)	8–18(12.0)
Shales (200 samples)	2.0	6.0	12.0
Schist (biotite)		2.4–4.7	13–25
Syenite	2.7	2500	1300
Tuff (feldspatic)	2.04	5.96	1.56

TABLE 11A.6 *Geological Significance of Natural Gamma Ratios*

Ratios	Remarks
Thorium/uranium (Th/U)	In *sedimentary* rocks, Th/U varies with depositional environment Th/U > 7: continental, oxidizing environment, weathered soils, etc. < 7: marine deposits, gray and green shales, graywackes < 2: marine black shales, phosphates. In *igneous* rocks, high Th/U indicative of oxidizing conditions by magma before crystallization and/or extensive leaching during postcrystallization history Source rock potential estimates of argillaceous sediments (shales) Major geologic unconformities Distance to ancient shore lines or location of rapid uplift during time of deposition Stratigraphic correlations, transgression vs. regression, oxidation vs. reduction regimes, etc.
Uranium/potassium (U/K)	Source rock potential of argillaceous sediments Stratigraphic correlations Unconformities, diagenetic changes in argillaceous sediments, carbonates, etc. Frequent correlation with vugs and natural fracture systems in subsurface formations, including localized correlation with hydrocarbon shows on drilling mud logs and core samples both in clastic and carbonate reservoirs
Thorium/potassium (Th/K)	Recognition of rock types of different facies Paleographic and paleoclimatic interpretation of facies characteristics Depositional environments, distance from ancient shore lines, etc. Diagenetic changes of argillaceous sediments Clay typing: Th/K increases from glauconite → muscovite → illite → mixed-layer clays → kaolinite → chlorite → bauxite Correlation with crystallinity of illite, average reflectance power, paramagnetic electronic resonance

BIBLIOGRAPHY

Fertl, W. H., and Frost, E., Jr.: "Experiences with Natural Gamma Ray Spectral Logging in North America," paper SPE 11145 presented at the SPE 57rd Annual Technical Conference and Exhibition, New Orleans, Sept. 25–29, 1982.

Fertl, W. H., Stapp, W. L., Vaello, D. B., and Vercellino, W. C.: "Spectral Gamma Ray Logging in the Texas Austin Chalk Trend," *J. Pet. Tech.* (March 1980); presented at the SPE 53rd Annual Technical Conference and Exhibition, Houston, Oct. 1–4, 1978.

Frondel, C., 1956, in L. R. Page, H. E. Stocking, and H. B. Smith, U.S. Geol. Survey Prof. Papers No. 300.

Gadeken, L. L., Arnold, D. M., and Smith, H. D., Jr.: "Applications of the Compensated Spectral Natural Gamma Tool," paper presented at the 25th Annual SPWLA Symposium in New Orleans, June 1984.

"Gamma Ray Spectral Data Assists in Complex Formation Evaluation," Dresser Atlas, publication REP 06/80 5M 3335, Houston (February 1979).

Hassan, M., Hossin, A., and Combaz, A.: "Fundamentals of the Differential Gamma Ray Log–Interpretation Technique," paper presented at the SPWLA 17th Annual Logging Symposium, Denver, June 9–12, 1976.

Kokesh, F. P.: "Gamma Ray Logging," *Oil and Gas J.* (July 1951).

Marett, G., Chevalier, P., Souhaite, P., Suau, J.: "Shaly Sand Evaluation Using Gamma Ray Spectrometry Applied to the North Sea Jurassic," SPWLA 17th Annual Symposium, Denver, June 1976.

Quirein, John A., Gardner, John S., and Watson, John T.: "Combined Natural Gamma Ray Spectral-Litho-Density Measurements Applied to Complex Lithologies," paper SPE 11143 presented at the SPE 57th Annual Technical Conference and Exhibition, New Orleans, Sept. 25–29, 1982.

Smith, H. D., Jr., Robbins, C. A., Arnold, D. M., and Deaton, J. G.: "A Multi-Function Compensated Spectral Natural Gamma Ray Logging System," paper SPE 12050 presented at the SPE 58th Annual Technical Conference and Exhibition, San Francisco, Oct. 5–8, 1983.

"Spectralog," Dresser Atlas, publication 3334, Houston (1981).

Tittman, J.: "Radiation Logging Lecture I: Physical Principle" and "Lecture II: Applications," Petroleum Engineering Conference on the Fundamental Theory and Quantitative Analysis of Electric and Radioactivity Logs, the University of Kansas (1956).

Answers to Text Questions

QUESTION #11.1

$GR_{cor} = 100$

QUESTION #11.2

$GR_{min} = 20$

$Gr_{max} = 85$

In sand C $GR = 40$

GR index $= 0.3077$

V_{sh} linear $= 30.77\%$

V_{sh} Clavier $= 16.23\%$

V_{sh} Steiber $= 12.90\%$

QUESTION #11.3

Uranium

CHAPTER TWELVE

PULSED NEUTRON LOGGING

Pulsed neutron logs provide a means of evaluating a formation after the well has been cased. It is of particular value for:

1. evaluating old wells, where the original openhole logs are inadequate or nonexistent
2. monitoring reservoir performance over an extended period of time
3. monitoring the progress of secondary and tertiary recovery projects
4. formation evaluation, as a last resort should the drillpipe become stuck

It is the most widely used logging method in cased holes at the present time. Other nuclear measurements are currently being developed, which may give superior results. These include carbon/oxygen logs and activation logs. Presently, however, conventional pulsed neutron logging provides the most direct method of measuring formation properties in cased holes.

TOOLS AVAILABLE

Four pulsed neutron tools are currently available:

1. Dual Detector Neutron Lifetime Log (DNLL): available in 3⅝- or 1¹¹⁄₁₆-in. OD versions.
2. Dual Spacing Thermal Decay Time Log (TDT-K): available in 1¹¹⁄₁₆-in. OD
3. Dual Spacing Thermal Decay Time Log (TDT-M): a new version of the TDT-K
4. Thermal Multigate Decay Log (TMD)

All tools attempt to measure the same formation parameters, although their operating systems are all slightly different.

PRINCIPLE OF MEASUREMENT

Regardless of the tool used, the principle of measurement remains the same. A neutron generator is turned on for a very short period of time. As a result, a burst of neutrons leaves the tool; and, since neutrons can easily pass through both the steel housing of the tool and the tubing/casing, a cloud of neutrons is formed in the formation. Fast neutrons soon became "thermalized" by collisions with atoms in the formation. The most effective thermalizing agent is the hydro-

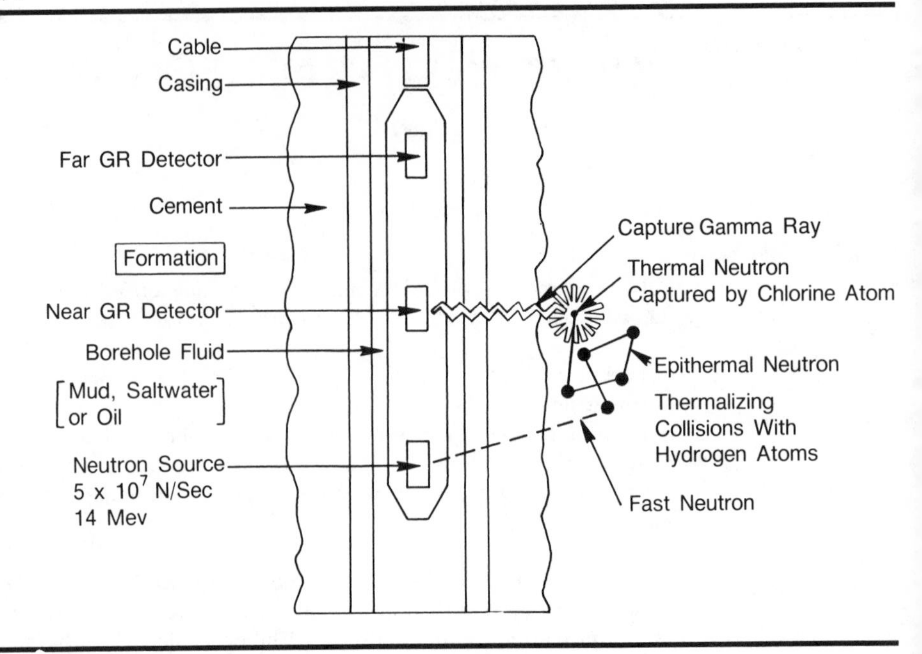

FIGURE 12.1 *Generalized Pulsed Neutron Tool*

gen present in the pore space in the form of water or hydrocarbon. Once in the thermal state, a neutron is liable to be captured. The capture process depends on the capture cross section of the formation. In general, chlorine dominates the capture process. Since chlorine is present in formation water in the form of salt (NaCl), the ability of the formation to capture thermal neutrons reflects the salt content and hence the water saturation. The capturing of a thermal neutron by a chlorine atom gives rise to a capture gamma ray. Pulsed neutron tools therefore monitor these capture gamma rays. Thus, the common elements of all commercial pulsed neutron tools are (1) a pulsed neutron generator, and (2) two gamma ray detectors at different distances from the neutron generator. A generalized neutron tool is shown in figure 12.1.

The cloud of neutrons produced by the initial neutron burst from the generator results in a cloud of thermal neutrons in the vicinity of the tool. This cloud dies away due to capture by chlorine atoms or other neutron absorbers in the formation. If there is plenty of chlorine present (i.e., high water saturation), then the cloud of thermal neutrons disappears quite quickly. If, however, hydrocarbons are present (i.e., low water saturation), then the cloud of thermal neutrons decays much more slowly. The rate of decay is measured by monitoring how many capture gamma rays enter the gamma ray counter(s) as a function of time. Figure 12.2 plots the relative counting rate on the Y-axis and the time (microseconds) since the initial burst of fast neutrons on the X-axis. Note that after a few hundred

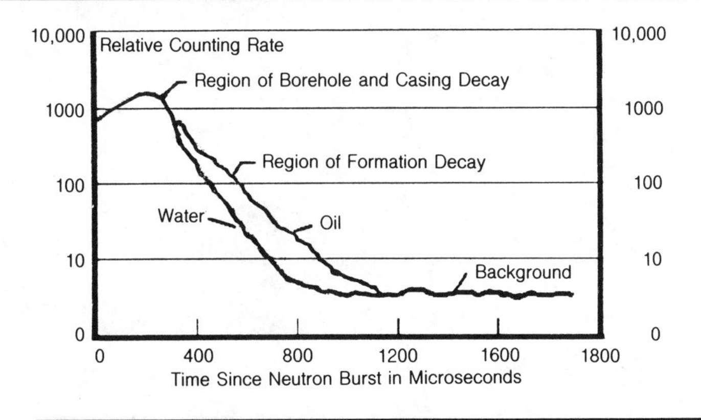

FIGURE 12.2 ***Thermal-Neutron Decay Curves for Oil- and Water-Bearing Formations.***

microseconds a straight-line portion of the decay curve develops. Note also how the water line has a steeper slope than the oil line. Note, too, that at later times the background gamma ray count rate remains substantially constant.

The Y-axis on figure 12.2 is logarithmic, but the time scale (X-axis) is linear. Thus, the straight-line portions of the curve represent exponential decay. If N is the number of gamma rays observed at time t and N_0 is the number observed at $t = 0$, then

$$N = N_0 e^{-t/\tau},$$

where τ (greek letter tau) is the time constant of the decay process. Tau is measured in units of time. It is convenient to quote values of tau in microseconds ($1\ \mu s = 10^{-6}$ s). The capture cross section of the formation, the property of interest, is directly related to tau by the equation

$$\Sigma = 4550/\tau,$$

where Σ (greek letter sigma) is the capture cross section measured in capture units (cu). Thus, the essence of measuring Σ is to first find the straight-line portion of the capture gamma ray decay, and then to measure its slope. This is accomplished in different ways by the various commercially available tools.

TDT-K

The TDT-K tool uses a gating system to find and measure the slope of the straight-line portion of the decay curve. The timing of the gates is related, by a feedback mechanism, to the actual value of tau. That is, the gates are not fixed in time but "slide" back and forth depending on the capture cross section of the formation as

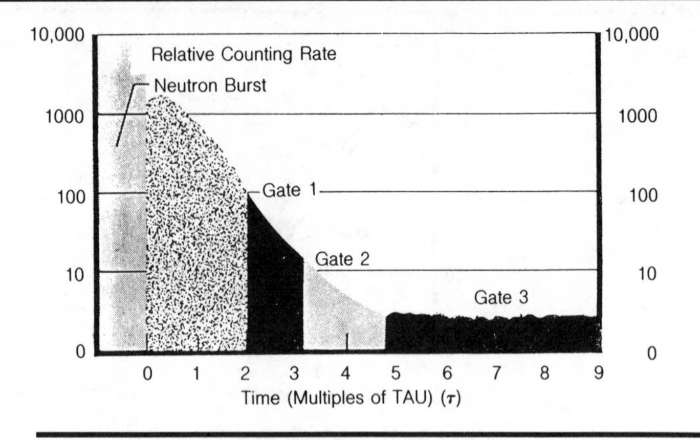

FIGURE 12.3 *TDT-K Gating System.*

measured during the preceding duty cycle. Figure 12.3 illustrates the TDT-K gate system:

Gate 1 is open between 2τ and 3τ.
Gate 2 is open between 3τ and 5τ.
Gate 3 is open between 5τ and 9τ.

Gate 3 measures the background gamma ray count rate. This constant is then subtracted from the rates found when Gates 1 and 2 are open, and the corrected counts are used to find τ and hence Σ. The feedback loop that attempts to maintain the slope-measuring process on the straight-line portion of the curve is controlled so that

$$\frac{N_1 - \frac{1}{2}N_3}{N_2 - \frac{2}{3}N_3} = 2,$$

where N_1, N_2, and N_3 are the count rates from Gates 1, 2, and 3, respectively. The ratio curve is derived from the count rates during appropriate gates for the near- and far-spacing counts.

DNLL

By contrast, the neutron lifetime log uses fixed gates and operates on a duty cycle of 1000 μs. Table 12.1 details the timing and measurements for the various gates and detectors.

1. During Sequence No. 1, pulses are counted by a count-rate meter in the logging panel at the surface and presented on the log format as a monitor curve that is an indication of source output.

2. During Sequence No. 3, and No. 4, pulses are counted by a count-rate meter in the logging panel at the surface presented on the log as the long- and short-space curves and are then fed into

TABLE 12.1 *DNLL Gating System*

Sequence No.	Time (seconds)	Detector	Operation
1	0–20	Monitor	Source on output 14 MeV neutrons
2	20–100	Short-space Long-space	Quiescent period
3	100–300	Short-space	Short-space measurement
4	100–1000	Long-space	Long-space measurement
5	300–400	Short-space	Quiescent period
6	400–600	Short-space	Gate #1 measurement
7	600–700	Short-space	Quiescent period
8	700–900	Short-space	Gate #2 measurement
9	900–1000	Short-space	Quiescent period
10	1000	Short-space Long-space	Sequence repeats

the digital ratio computer card. The ratio card takes the ratio of the short-space and long-space counts and computes the (SS/LS) ratio curve, which is presented on the log format.

3. During Sequence No. 6, and No. 8, pulses are counted by a count-rate meter at the surface and produce Gate 1 and Gate 2 on the log format. The sigma panel takes the count rates at Gate 1 and Gate 2, and computes the sigma curve that is presented on the log format.

4. During Sequence No. 10, Sequence No. 1 is initiated again, the source is turned on, and the cycle is repeated.

Figure 12.4 illustrates the typical cycle for DNLL data.
The actual calculation of sigma is made using the relation

$$\Sigma = (10{,}500/\Delta T)(\log, N_1/N_2),$$

where N_1 and N_2 are count rates from Gates 1 and 2 of the near spacing detector and ΔT is normally 300 μs. The DNLL system does not measure background gamma rays. The detectors are shielded to count only gamma rays with energies greater than 2.2 MeV, which reportedly eliminates background effects.

TDT-M

The TDT-M is a modern version of the TDT-K with 16 gates instead of 3. The 16 gates are fixed, but depending on the value of τ, different gates are used to estimate Σ. Figure 12.5 illustrates two cases for the TDT-M and the resulting use of gates. Table 12.2 details which gates are used to calculate τ depending on the range of the previously determined τ.

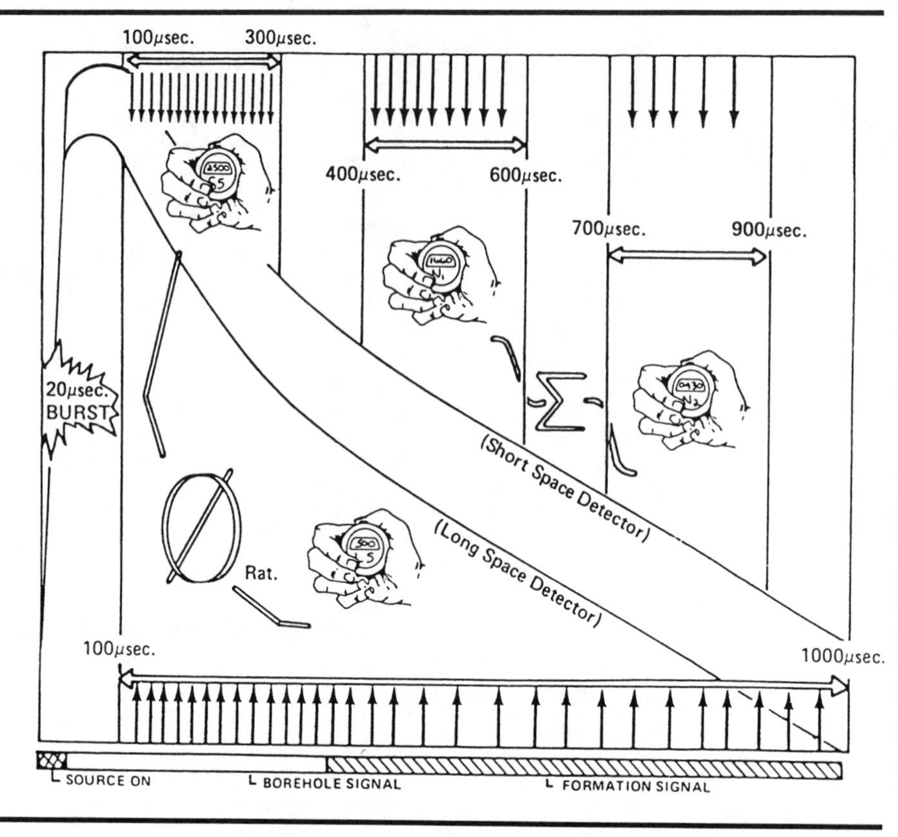

FIGURE 12.4 *Typical Cycle of DNLL Data. Courtesy Dresser Atlas.*

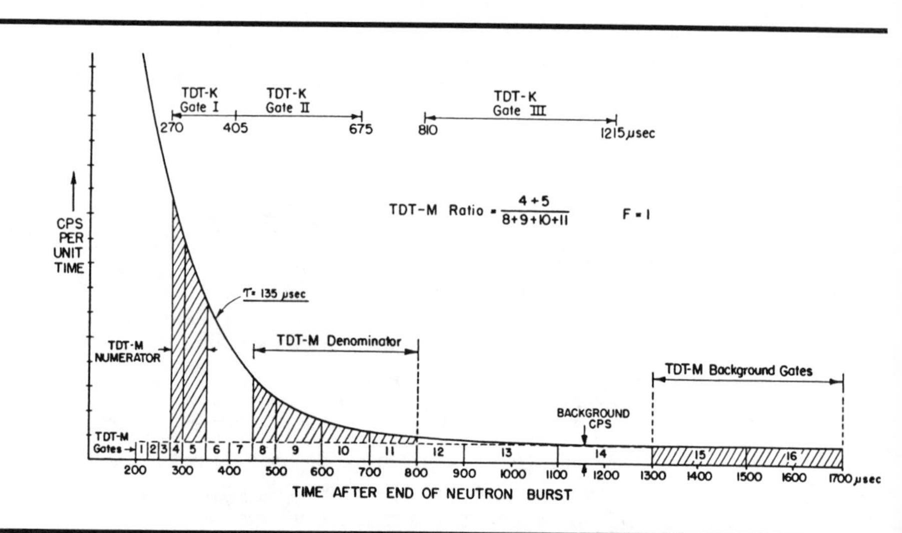

FIGURE 12.5 *TDT-M Gating. Reprinted by permission of SPE-AIME from Hall et al. 1982. © 1982 SPE-AIME.*

TABLE 12.2 *TDT-M Calculations for Measurement Cycle Lasting 1.8 Milliseconds*

τ_{NEAR} Range (μs)	Ratio (R) (Integers are Gate Numbers)	Equation for τ
$\tau < 106.3$	$\frac{1 + 2}{5 + 6 + 7 + 8 + 9}$	$\tau = 38.0 + 69.9R^{-1}$
106.3–118.8	$\frac{2 + 3}{6 + 7 + 8 + 9 + 10}$	$\tau = 48.4 + 73.1R^{-1}$
118.8–131.3	$\frac{3 + 4}{7 + 8 + 9 + 10 + 11}$	$\tau = 58.2 + 76.1R^{-1}$
131.3–143.8	$\frac{4 + 5}{8 + 9 + 10 + 11}$	$\tau = 61.3 + 124.1R^{-1}$
143.8–162.5	$\frac{5 + 6}{9 + 10 + 11 + 12}$	$\tau = 67.6 + 164.2R^{-1}$
162.5–187.5	$\frac{6 + 7}{9 + 10 + 11 + 12 + 13}$	$\tau = 53.6 + 130.3R^{-1}$
$\tau > 187.5$	$\frac{7 + 8}{10 + 11 + 12 + 13}$	$\tau = 63.2 + 164.6R^{-1}$

Note: Reprinted, by permission, from J. E. Hall, C. W. Johnstone, J. L. Baldwin, and L. A. Jacobson: "A New Thermal Neutron Decay Logging System—TDT*-M," *J. Pet. Tech.* (January 1982), table I. © 1982 SPE-AIME.

TMD

The Welex Thermal Multigate Decay logging system is designed to measure Σ, the thermal-neutron capture cross section of the formation. Like prior pulsed neutron systems, the TMD tool uses a 14-MeV pulsed neutron generator source to create a time-dependent thermal neutron, and hence capture gamma ray, distribution in the vicinity of two detectors within the tool. The decay rate of the capture gamma radiation measured by the tool is used to obtain Σ. At this point, however, many similarities between the TMD system and the other commercial systems end.

At each detector, the TMD tool measures count rates in six different time gates between each neutron burst. These gates span the decay from very near the end of one neutron burst until almost the beginning of the next neutron burst. The first two gates are positioned soon after the end of the burst (see fig. 12.6) such that: virtually all neutrons from the burst have thermalized before the first gate opening, and a significant number of counts from neutron

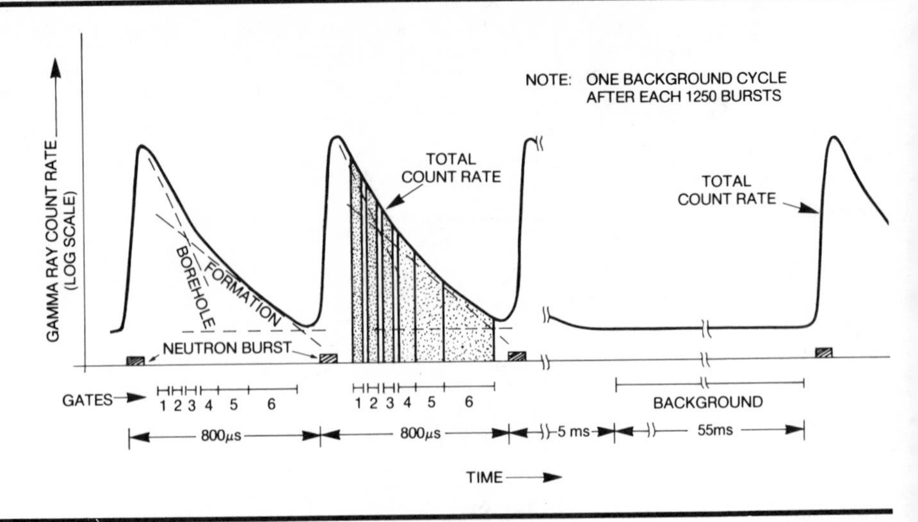

FIGURE 12.6 ***WELEX TMD Tool Pulse-Gate-Time Parameters. Courtesy WELEX, a Halliburton Company.***

capture within the borehole fluid, or the closely surrounding materials, are recorded in each gate. The last four gates are each progressively wider at longer delay times from the neutron burst. In gates progressively later than the burst, the percentage of counts from the formation, relative to the borehole, will increase. The last gate (Gate 6) is sufficiently delayed from the burst so that a negligible number of counts (generally $< 3\%$) in the gate are generated from captures in the borehole fluid.

The TMD system has a fixed repetition rate of 1250 neutron bursts per second, with each neutron burst as narrow as possible (~60 microseconds), consistent with maintaining the full output from the generator. This optimizes the count rates from the borehole in the first two gates.

After one second of operation (1250 bursts), the neutron generator is turned off for approximately 60 milliseconds, with the last 55 milliseconds being used to measure in each detector the background count rate in a seventh (background) gate. This is shown schematically in figure 12.6. Unlike some prior pulsed neutron tools that measure background after each neutron burst, the TMD system measures background during a relatively long time interval after 1250 bursts have been completed. After 1.06 seconds, the operating cycle is repeated. The background, which is due primarily to neutron-induced iodine activation within the NaI (T1) detector scintillation crystal, is filtered and subtracted from the dead-time corrected count rates in each of the six primary data gates. The half-life of the crystal activation I^{128} (produced by the I^{127} (n, γ)I^{128} reaction) is approximately 25 minutes; thus, variation in this background is negligible during the 1.06-second operating cycle.

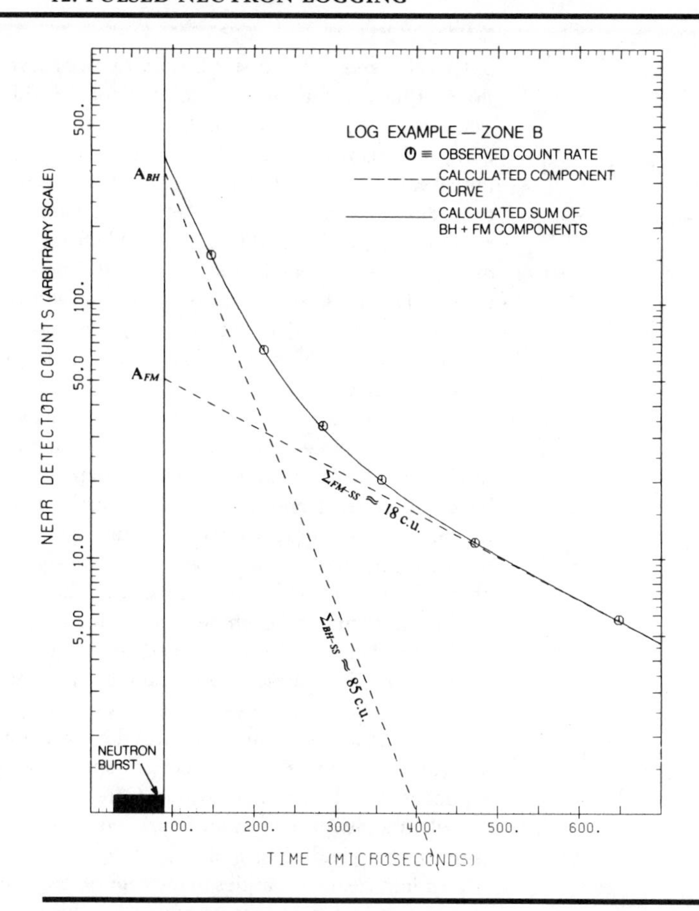

FIGURE 12.7 *TMD Decay Curve. Courtesy WELEX, a Halliburton Company.*

After each burst, the six resulting dead-time and background-corrected count rates in each detector represent points on the composite formation and borehole decay curve. These count rates are then adaptively filtered (i.e., averaged over many bursts) to form a composite decay curve integrated over a short vertical interval in the borehole (one to several feet, depending upon filtering parameters). The circled points in figure 12.7 illustrate the six gate count rates measured in the short-spaced detector in an actual field well interval.

The main field computer program then inputs these six points along the composite formation/borehole decay curve, and uses a sophisticated iterative least-squares technique to separate the composite curve into the borehole and formation decay exponentials. Figure 12.7 also illustrates the formation and borehole components calculated by the program for this data set (dashed curves), as well as the sum of the two calculated components (solid curve). The computer calculates the formation capture cross section, Σ_{FM}, from the slope of the formation decay component. In addition, the computer calculates in real time the borehole capture cross section Σ_{BH} component, and also calculates the intercepts for each component (i.e., initial values A_{FM} and A_{BH} at the end of the neutron bursts). This procedure is completed for decay data from both detectors. The resulting Σ_{FM} data are as free as possible from borehole effects since, during computer calculation of Σ_{FM}, the borehole count rates are essentially subtracted from the total observed count rate. For each six-gate-count–rate data set, the program makes an initial estimate of the formation decay curve using count rates in two gates remote from the burst.

Hence, the starting point of the TMD calculation process is very similar in principle to the final result obtained with the fixed-gate technique. This approximate formation exponential curve is projected back into Gates 1 and 2 to obtain an estimate of formation counts in these gates. These formation counts are then subtracted from the total counts in Gates 1 and 2, respectively, leaving estimates of the borehole counts in these gates. These borehole counts are used to determine a borehole exponential curve, which is then projected into Gates 3 through 6 and subtracted from the total count rates observed in these gates, leaving each gate with an estimate of the formation count rate alone. These four points are then least-squares fitted to obtain an estimate of the formation decay exponential that yields Σ_{FM} and A_{FM}. This better approximation to the formation exponential is then projected into Gates 1 and 2, and processed as described above to obtain an improved borehole decay exponential estimate that yields Σ_{BH} and A_{BH}.

To minimize uncertainties in the capture cross section and intercept estimations, and to reduce computation, this procedure is iteratively repeated a varying number of times depending on an input-specified borehole condition option. Also based on the option selected and/or the results of a series of convergence tests, data about the borehole parameters in one data set can be transferred to the next. This saves computer time by taking advantage of the infrequent changes in borehole fluid salinity. In addition, fewer iterations are required to achieve convergence on the long-spaced detector data, since the relative borehole-to-formation component magnitude is smaller at the increased distance from the source.

In intervals where borehole counts in Gates 1 and 2 are very low (i.e., comparable to statistical fluctuations in the total gate count rates), the computer program automatically reverts to a one exponential decay model to calculate Σ_{FM} using Gates 3 through 6. This condition occurs most often in gas-filled casings. Σ_{BH} is not calculated in this situation and is so indicated on the log by a straight line on the Σ_{BH} curve.

At present, the short- and long-spaced detector Σ_{FM} curves are both output on the log, as is a "composite" Σ_{FM}^{CORR}. The normal TMD field log presentation is a two-magazine display of log values,

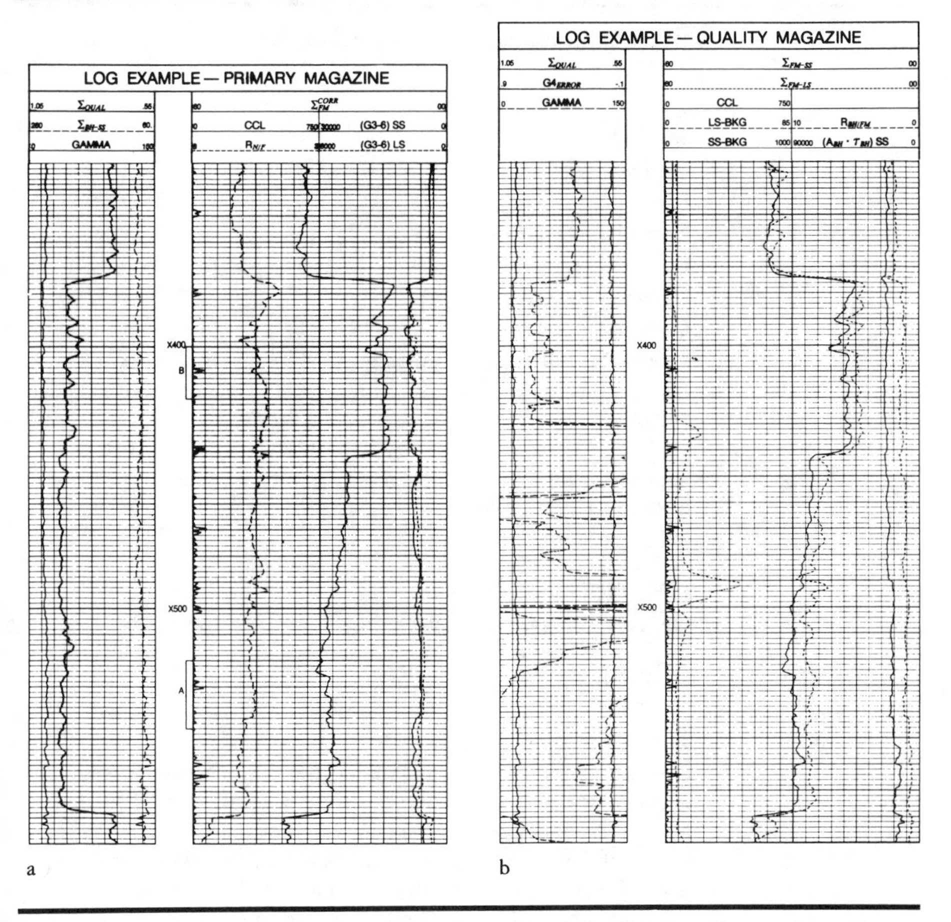

FIGURE 12.8 *TMD Primary and Quality Log Presentations. Courtesy WELEX, a Halliburton Company.*

as shown in figure 12.8 (a). Primary magazine will include: (1) gamma, (2) collars, (3) Σ_{QUAL}, (4) $R_{N/F}$, (5) Σ_{FM}^{CORR}, (6) $\Sigma_{BH\text{-}SS}$, (7) (G3-6)SS, (8) (G3-6)LS. This log will present all information required for a basic formation interpretation.

In addition to the curves' output on the primary magazine, the secondary magazine (fig. 12.8 [b]) will present supplemental information useful in evaluating log quality. Curves included are: (1) gamma, (2) collars, (3) Σ_{QUAL}, (4) $G4_{ERROR}$, (5) SS-BKG, (6) LS-BKG, (7) $\Sigma_{FM\text{-}SS}$, (8) $\Sigma_{FM\text{-}LS}$, (9) $R_{BH/FM}$ and (10) $(A_{BH} * \Sigma_{BH})$SS.

Also produced in the field are digital tapes containing all output parameters described earlier and all unfiltered TMD and gamma ray count rates. They are recorded on a 0.25-foot depth basis. These tapes can be replayed for alternate presentations of the existing output curves (such as scale changes) or to regenerate the output curves

TABLE 12.3 *Curves in a TDT-K Log Presentation*

Curve Name	Units	Log Track	Remarks
Sigma (Σ)	cu	2 & 3	Main curve
Tau (τ)	μs	2 & 3	
Ratio	—	2	Pseudoporosity
Near counts (N)	cps	3	Near detector, Gate 1
Far counts (F)	cps	3	Far detector, Gate 1
Monitor or background	cps	1	Near detector, Gate 3
Quality control	—	1	Check of tau loop
Gamma Ray (GR)	API	1	Natural gamma
Casing-collar log	—	1	Memorized and direct

after changing one or more of the operator-selectable program options described earlier.

LOG PRESENTATIONS

On a typical pulsed neutron log, there may be up to 9 curves displayed. The curves listed in table 12.3 are illustrated in the TDT-K log presentation of figure 12.9.

Sigma Curve

The Σ curve is the principal pulsed neutron measurement and behaves rather like an openhole resistivity curve, that is, it deflects to the left (high values of Σ) in wet zones and to the right (low values of Σ) in hydrocarbon-bearing zones or low-porosity formations. Σ values in shales are quite high, tending to mask the effect of hydrocarbons. Thus shaly pay zones can appear to be water-bearing on the first inspection. Figure 12.10 shows a schematic comparison of Σ with resistivity.

Tau Curve

Tau is just another way of looking at Σ. In fact, τ (the decay-time constant for the thermal-neutron population) is the basic measurement of a pulsed neutron tool. However, all interpretation equations for pulsed neutron logs are linear functions of Σ; and it is much easier to work with Σ rather than with τ. It is recommended that τ be recorded on tape but left off the log presentation, since its scaled reciprocal, Σ, gives exactly the same information in a form that is easier to work with.

Ratio Curve

The ratio curve is a porosity indicator. It is derived by taking the ratio of the gamma ray counts seen during Gate 1 at the near and far detectors respectively. The ratio curve behaves very much like a CNL porosity curve—it deflects to the right (low ratio) in low

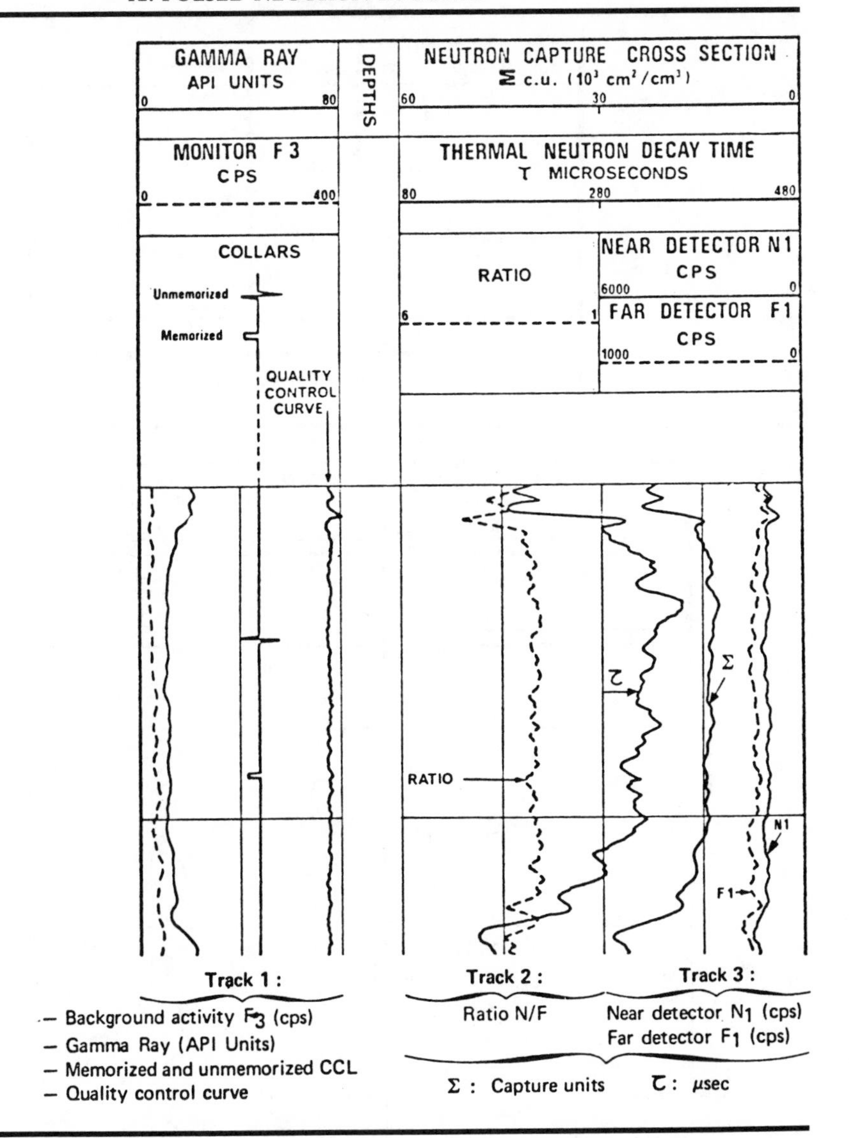

FIGURE 12.9 *TDT-K Log Presentation. Courtesy Schlumberger Well Services.*

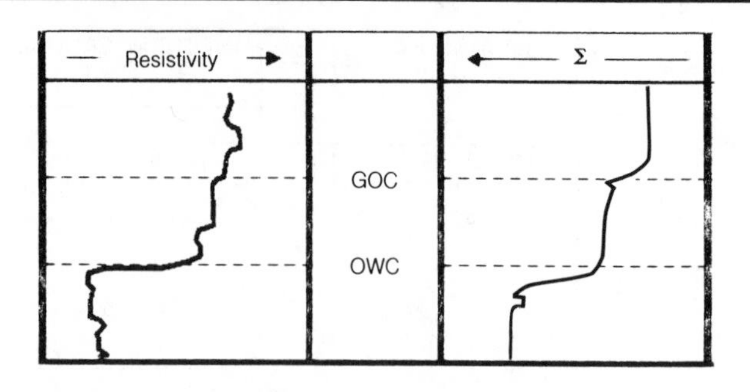

FIGURE 12.10 *Σ–Resistivity Comparison.*

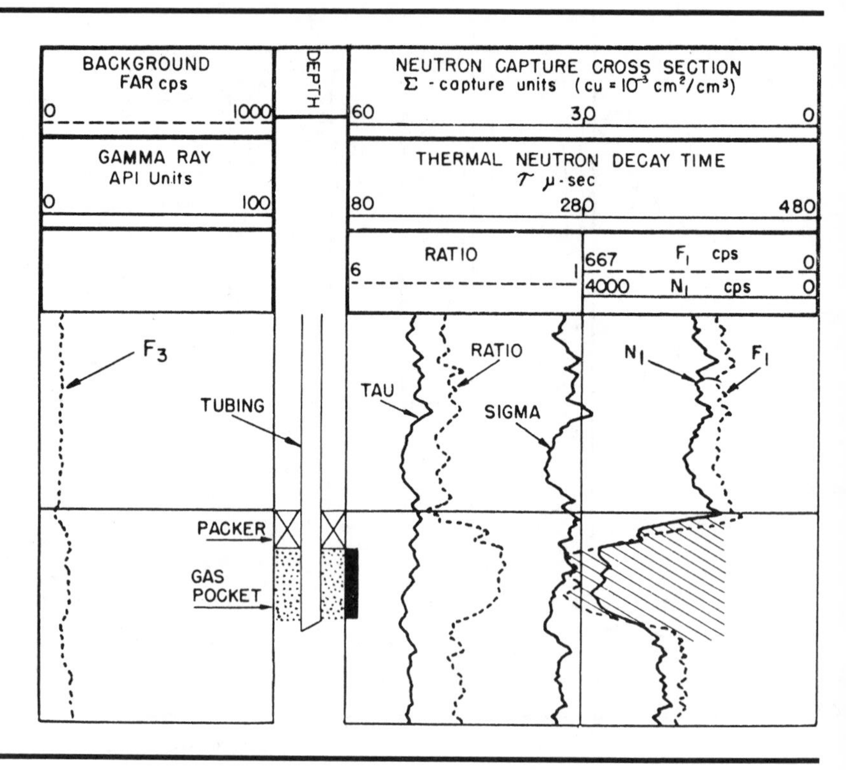

FIGURE 12.11 *Ratio Curve Response to Gas. Courtesy Schlumberger Well Services.*

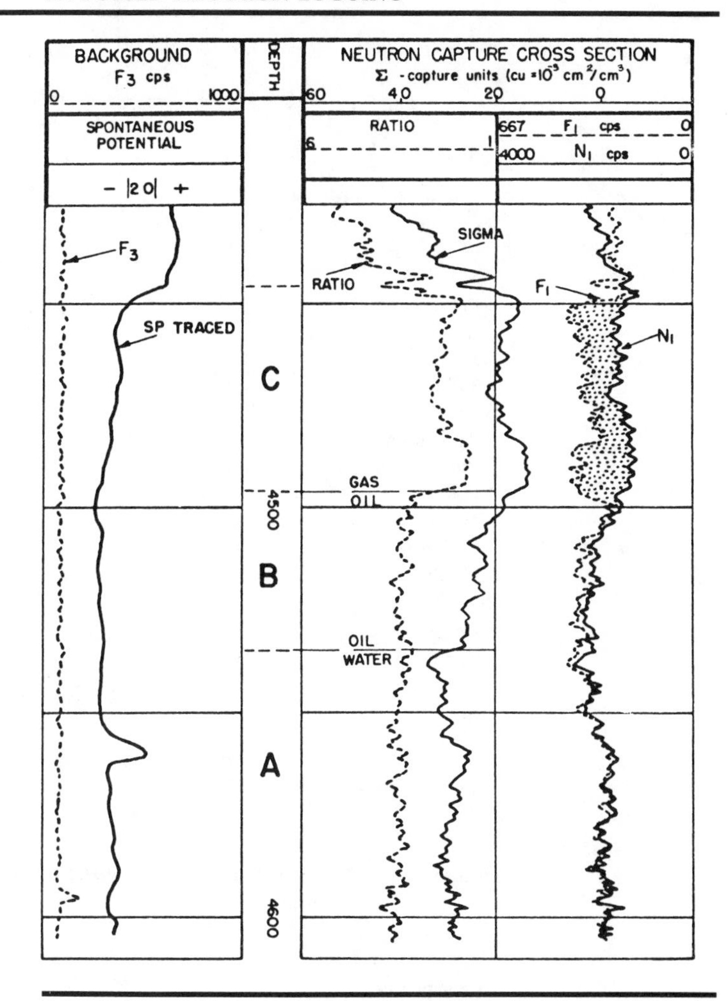

FIGURE 12.12 *Near/Far Count-Rate Display. Courtesy Schlumberger Well Services.*

porosity, or in the presence of gas. Figure 12.11 shows the ratio curve response to a pocket of gas trapped below a packer behind a tubing nipple. In the absence of any openhole porosity logs, the ratio can be used in combination with Σ to find formation porosity.

Near and Far Count-Rate Display

In Track 3, the near and far count rates are displayed as an overlay. Figure 12.12 illustrates this presentation. When the correct scales are chosen for the near (N_1) and far (F_1) count-rate displays, a useful quick-look log results, with the following properties:

in gas, $F_1 > N_1$ (dotted line is left of solid),
in shales, $F_1 < N_1$ (dotted line is right of solid),

TABLE 12.4 *Capture Cross Sections of Elements*

Element	Σ (cu)
Common elements	
Chlorine	570
Hydrogen	200
Nitrogen	83
Potassium	32
Iron	28
Sodium	14
Sulfur	9.8
Calcium	6.6
Aluminum	5.4
Phosphorus	3.9
Silicon	3.4
Magnesium	1.7
Carbon	0.16
Oxygen	0.01
Rare elements	
Boron	45,000
Cadmium	18,000
Lithium	6,200
Mercury	1,100
Manganese	150

and in clean oil- or water-bearing zones, the two curves lie practically on top of one another.

QUESTION #12.1
Refer to figure 12.12.

a. Color code the gas-bearing intervals on the near/far count-rate display using red or pink.
b. Color code the shale zone at the top of the log green.
c. Why do you think the oil–water contact is marked where it is (i.e., at 4535 ft)?
d. Read the average value of Σ in the water-bearing zone.

Background and Quality Curves

The background curve is a very insensitive natural gamma ray curve. There should be little movement on this curve except in "hot" zones, which are very radioactive. This curve is sometimes omitted without any great loss.

The quality curve is a way of monitoring the feedback loop that slides the timing of the gates as directed by the previous duty cycle's measurement of Σ. At a bed boundary with a large abrupt change of Σ, the loop may become unstable. In this case, the monitor curve will appear hashy. In such places, the log should be considered unworthy of a quantitative interpretation. As long as the monitor stays as a smooth straight line, the log may be considered acceptable.

Summary

The most important curves of the pulsed neutron log are:

Σ	for water saturation
Ratio	for porosity
GR	for shale content
Near/far display	for gas indications

CAPTURE CROSS SECTIONS

The capture cross section of a formation depends on the elements that make up the formation and their relative abundance. Σ values vary over a wide range. The capture cross sections for thermal neutrons are listed in table 12.4 for a number of elements and in table 12.5 for various compounds found in rocks.

BASIC INTERPRETATION

Clean Formations

Practical interpretation of pulsed neutron logs is conceptually very simple. The total formation capture cross section, Σ, recorded on the log, is just the sum of the products of the volume fractions found in the formation and their respective capture cross sections. Thus, in its simplest form,

$$\Sigma \log = \Sigma \text{ matrix} \cdot (1 - \phi) + \Sigma \text{ fluid} \cdot \phi.$$

Figure 12.13 should clarify the mathematical relationship. Of course, the "fluid" may, in fact, be a mixture of oil and water, in which case, the log response is described by

$$\Sigma_{Log} = \Sigma_{ma}(1 - \phi) + \Sigma_w \phi S_w + \Sigma_{hy} \phi (1 - S_w).$$

By rearrangement of the equation, we have

$$S_w = \frac{(\Sigma_{Log} - \Sigma_{ma}) - \phi(\Sigma_{hy} - \Sigma_{ma})}{\phi(\Sigma_w - \Sigma_{hy})}.$$

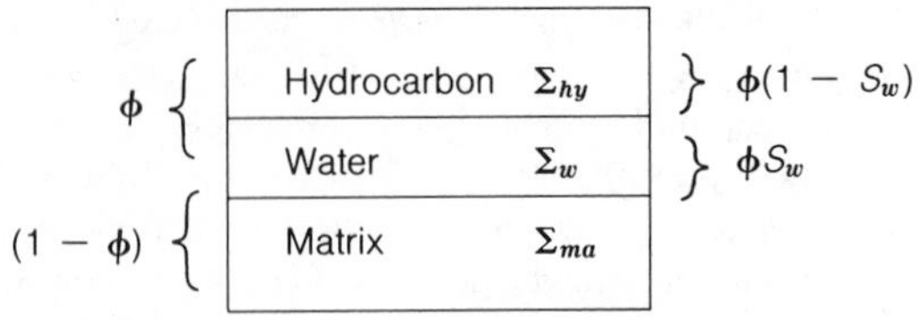

	Layer	Σ	
ϕ	Hydrocarbon	Σ_{hy}	$\phi(1 - S_w)$
	Water	Σ_w	ϕS_w
$(1 - \phi)$	Matrix	Σ_{ma}	

FIGURE 12.13 ***Components of*** Σ_{log}.

TABLE 12.5 *Capture Cross Sections of Compounds*

Compound	Elements	Σ (cu)
Basic minerals		
Quartz	SiO_2	4.2
Calcite	$CaCO_3$	7.3
Dolomite	$CaCO_3 \cdot MgCO_3$	4.8
Feldspars		
Albite	$NaAlSi_3O_8$	7.6
Anorthite	$CaAlSi_2O_8$	7.4
Orthoclase	$KAlSi_3O_8$	15.0
Evaporites		
Anhydrite	$CaSO_4$	13
Gypsum	$CaSO_4 \cdot 2H_2O$	19
Halite	$NaCl$	770
Sylvite	KCl	580
Carnalite	$KCl \cdot MgCl_2 \cdot 6H_2O$	370
Borax	$Na_2B_4O_7 \cdot 10H_2O$	9,000
Kermite	$Na_2B_4O_7 \cdot 4H_2O$	10,500
Iron-bearing minerals		
Geothite	$FeO(OH)$	89
Hematite	Fe_2O_3	104
Magnetite	Fe_3O_4	107
Limonite	$FeO(OH) \cdot 3H_2O$	80
Pyrite	FeS_2	90
Siderite	$FeCO_3$	52
Miscellaneous clays and micas		
Glauconite		22 + 5
Chlorite		25 + 15
Mica/biotite		35 + 10
Pyrolusite	MnO_2	440
Manganite	$MnO(OH)$	400
Cinnabar	HgS	7,800
Shales		35 to 55

TABLE 12.6 *Values for Log Interpretation*

	Material	Sigma (cu)
Σ matrix	Sand	8 to 10
	Limestone	12
	Dolomite	8
Σ hydrocarbon	Oil (function of R_s)	22 (av.)
	Gas (function of γ_g, p, & T)	8 (av.)
Σ water	Fresh	23
	Seawater	34
	Brine	122

QUESTION #12.2
Given:

Σ_{Log} = 25.3 cu.
Σ_{ma} = 10.0 cu.
Σ_{hy} = 22 cu.
Σ_w = 100 cu.
ϕ = 30%.

Find S_w = ____ .

In the previous example, the values for Σ_{ma}, Σ_{hy}, and Σ_w were given. However, in practice, these values may not be known. As a guide, table 12.6 gives values of Σ for commonly found materials. More exact methods for finding Σ_{ma}, Σ_w, and Σ_{hy} will be covered later in the text.

Log interpretation in clean formations is straightforward. The linear equations can also be thought of graphically. If, for example, a crossplot is made of Σ (on the Y-axis) and ϕ (on the X-axis), straight lines represent mixtures of pairs of components. Figure 12.14 shows this method. Note that all water-bearing points fall on the line joining the matrix (porosity = 0, Σ = 10) to the water (porosity 100, Σ = 100). Likewise, all 100% oil-bearing points lie on the line joining the matrix to the oil (porosity = 100, Σ = 22). Thus, all points lie inside a solution triangle covered by the matrix, water, and oil points. By simple constructions, lines of constant S_w can be drawn. The dashed lines on figure 12.14 join at the point given in question #2, (ϕ = 30%, Σ_{Log} = 25.3).

How much confidence can be placed in the value of S_w derived as in figure 12.11? To a large extent, the accuracy depends on the difference between Σ_{hy} and Σ_w. In fresh formation waters (low Σ_w),

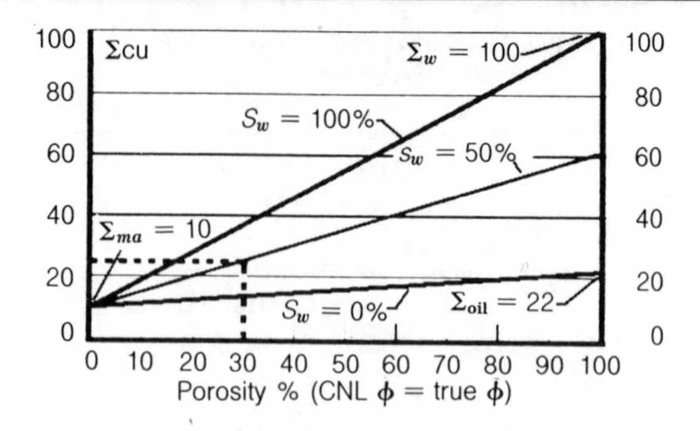

FIGURE 12.14 ***Interpretation in Clean Formations.***

the interpretation will be very questionable; but in very salty formation waters (high Σ_w), it will be reliable. Figure 12.15 illustrates this concept.

EXERCISE
Color the "reliable" area of figure 12.15 with a yellow highlighter Do the conditions in your district fall in the "reliable" area?

Shaly Formations

The addition of shale to the formation can be handled in the same linear fashion as before. That is,

$$\Sigma_{Log} = \Sigma_{ma} \cdot (1 - V_{sh} - \phi) + \Sigma_{sh} V_{sh} + \Sigma_w \phi S_w + \Sigma_{hy} \phi(1 - S_w)$$

which gives

$$S_w = \frac{(\Sigma - \Sigma_{ma}) - \phi(\Sigma_{hy} - \Sigma_{ma}) - V_{sh}(\Sigma_{sh} - \Sigma_{ma})}{\phi(\Sigma_w - \Sigma_{hy})}.$$

Figure 12.16 illustrates the shaly formation. Note that the solution for S_w requires a value for ϕ. In most practical cases, the porosity device used will also be affected by the presence of shale. For example, if a CNL neutron is used as the porosity device, its reading will have to be corrected using

$$\phi = \phi N - V_{sh} \cdot \phi N_{sh},$$

where ϕN is the log reading, V_{sh} is the shale volume % (from GR, etc.), and ϕN_{sh} is the response of the CNL in 100% shale.

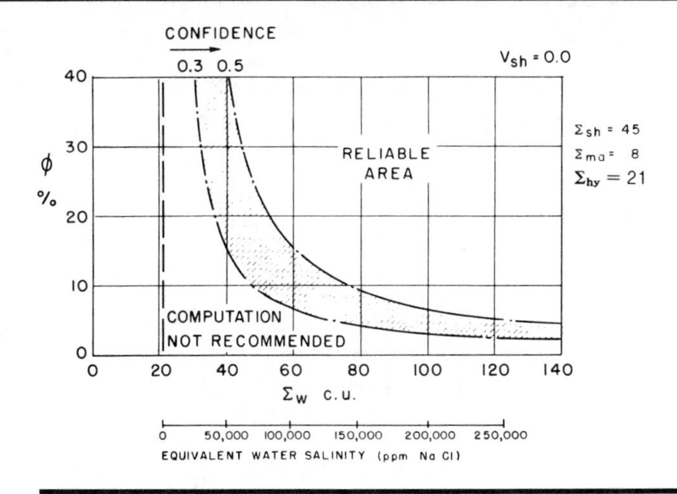

FIGURE 12.15 ***Reliability of*** S_w ***Calculation in Clean Formations. Courtesy Schlumberger Well Services.***

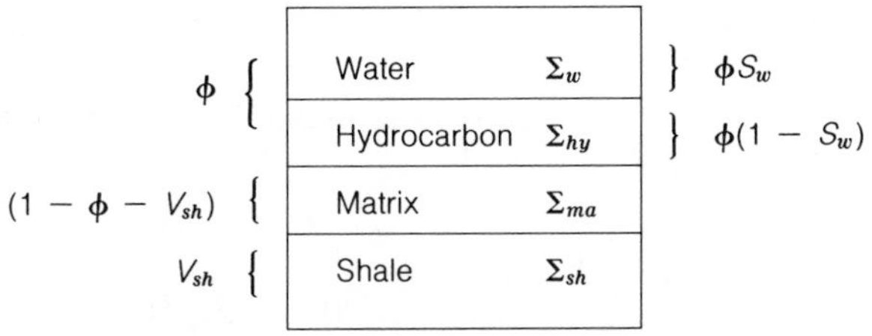

Volume	Component	Σ	Fraction
ϕ	Water	Σ_w	ϕS_w
	Hydrocarbon	Σ_{hy}	$\phi(1 - S_w)$
$(1 - \phi - V_{sh})$	Matrix	Σ_{ma}	
V_{sh}	Shale	Σ_{sh}	

FIGURE 12.16 ***Shaly Formation Schematic.***

QUESTION #12.3 Shaly Formation

Given:

$$\begin{aligned}
\Sigma_{Log} &= 30.3 \text{ cu.}\\
\Sigma_{ma} &= 10 \text{ cu.}\\
\Sigma_{hy} &= 22 \text{ cu.}\\
\Sigma_{w} &= 100 \text{ cu.}\\
\phi N &= 37\%.\\
\phi N_{sh} &= 35\%.\\
V_{sh} &= 20\%.\\
\Sigma_{sh} &= 35 \text{ cu.}
\end{aligned}$$

Find:

a. $\phi =$ ____ .
b. $S_w =$ ____ .
c. $S_w =$ ____ , if you had assumed this was a clean instead of a shaly formation.

Note that the equation for Σ_{Log} in shaly formations could be rearranged to read

$$\Sigma_{Log} = \underbrace{\Sigma_{ma}(1 - \phi) + \Sigma_w \phi S_w + \Sigma_{hy}\phi(1 - S_w)}_{\text{clean component}} + \underbrace{V_{sh}(\Sigma_{sh} - \Sigma_{ma})}_{\text{shale component}}.$$

Thus, an equally valid method of computing S_w would be to correct Σ_{Log} for shale using the relationship

$$\Sigma_{cor} = \Sigma_{Log} - V_{sh}(\Sigma_{sh} - \Sigma_{ma}).$$

Figure 12.17 illustrates the shaly case pictorially by crossplotting Σ with CNL porosity. Note that the net effect of adding shale to a formation is to move a plotted point to the northeast of the plot and set it on a line of constant S_w that is erroneously high. Correction for shale thus moves plotted points to the southwest and lower values of S_w.

Needless to say, the accuracy of S_w results in shaly formations will be reduced due to uncertainties about the exact values of ϕ, V_{sh}, and Σ_{sh}. Figure 12.18 shows the reliable areas as a function of ϕ and Σ_w for a V_{sh} of 20%.

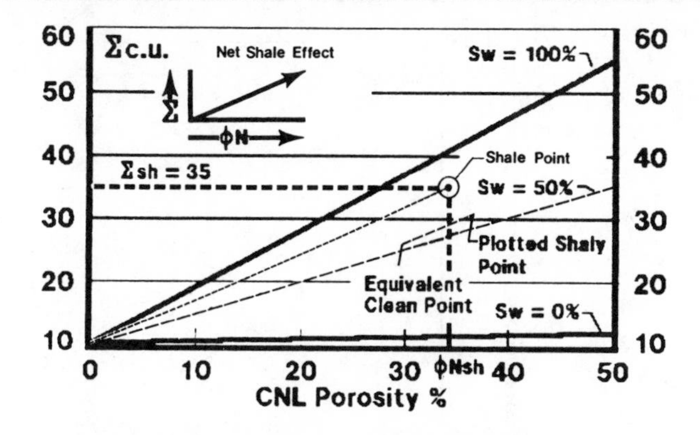

FIGURE 12.17 *Interpretation in Shaly Formations.*

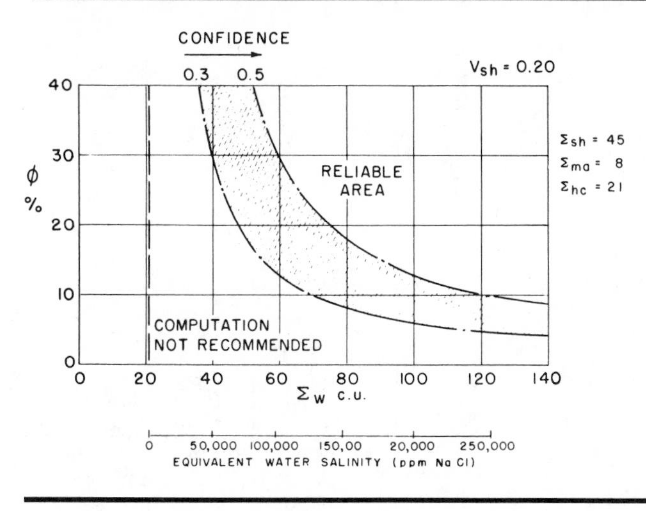

FIGURE 12.18 *Reliable Area for Interpretation in Shaly Formations. Courtesy Schlumberger Well Services.*

Finding Interpretation Parameters

In order to perform a quantitative interpretation of a pulsed neutron log, certain parameters need to be known. These are:

Σ matrix
Σ water
Σ hydrocarbon (oil and/or gas)
Σ shale

We will now explore various methods of finding these parameters.

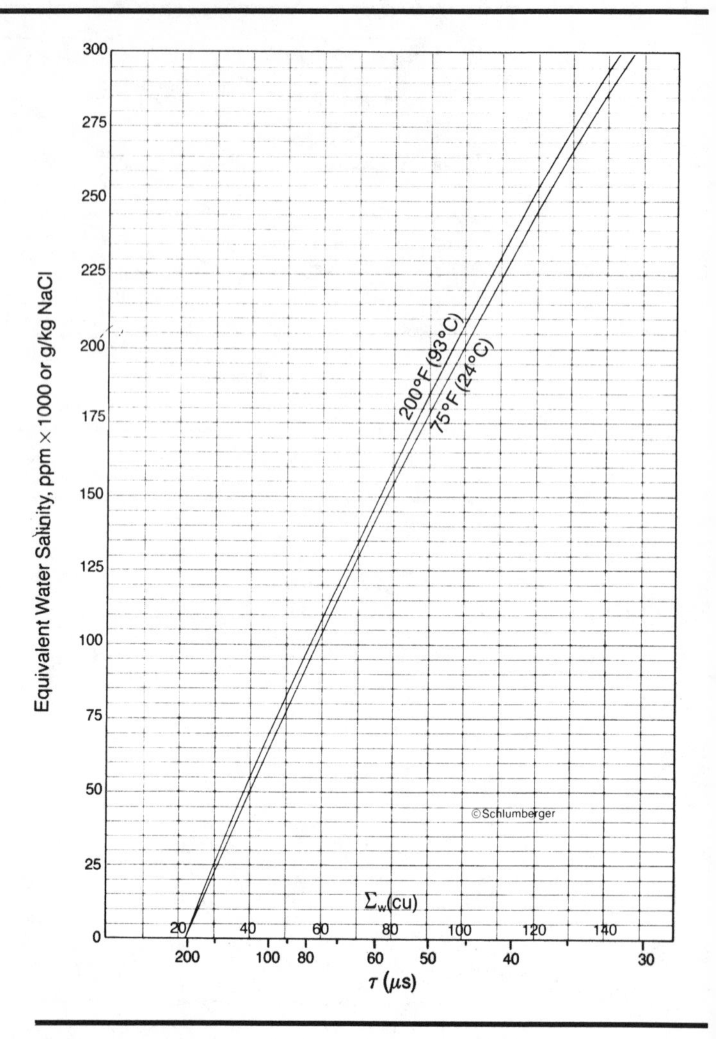

FIGURE 12.19 Σ_w *as a Function of Salinity and Temperature. Courtesy Schlumberger Well Services.*

SIGMA WATER. Σ_w is a simple function of the water salinity (ppm NaCl) and temperature (see fig. 12.19),

QUESTION #12.4 Σ_w
Salinity = 230,000 ppm NaCl.
Temperature = 200°F.

Find Σ_w = _______ .

If the salinity is not known, then a 100% water-bearing section can give us the required data. Note that if $S_w = 100$, the basic equation reduces to

$$\Sigma_{Log} = \Sigma_{ma}(1 - \phi) + \Sigma_w \phi,$$

which can be rewritten to give

$$\Sigma_w = \frac{\Sigma_{Log} - \Sigma_{ma}(1 - \phi)}{\phi}.$$

Thus, provided Σ_{ma} and ϕ are known, Σ_w can be back calculated directly. Note that this method is similar to the R_{wa} technique used with openhole logs. For that reason, the derived value of Σ_w is referred to as Σ_{wa} or *sigma water apparent.* If an extensive water-bearing interval has been logged, a graphical method can provide a "double whammy," both Σ_{ma} and Σ_w from one plot. If pairs of values of Σ and ϕ are plotted on a graph such as the one given in figure 12.20, all points at $S_w = 100\%$ will fall on a straight line connecting Σ_{ma} (at $\phi = 0\%$) and Σ_w (at $\phi = 100\%$).

QUESTION #12.5 Σ_w

Plot Σ vs. ϕ for the following log readings:

Level	Σ (cu)	ϕ (%)
1	17	28
2	21	32
3	23	26
4	27	28
5	21	13
6	24	20
7	34	34
8	28	26
9	26	23
10	29	27

Now draw in the $S_w = 100\%$ line.

Find:

a. Σ_{ma} = _____ .
b. Σ_w = _____ .

SIGMA OIL. Σ_o is a function of the solution GOR (R_s) of the liquid hydrocarbon in question. Light, gassy oils tend to have lower values of Σ_o. Dead, heavy oils have a maximum Σ_o of about 22 cu. Figure 12.21 shows Σ_o as a function of R_s.

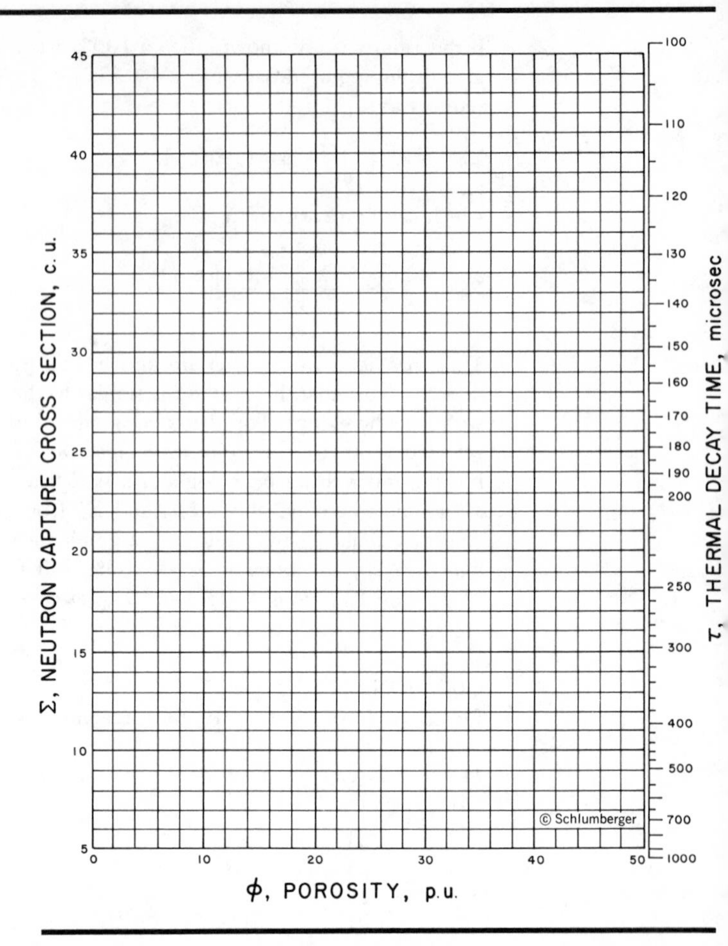

FIGURE 12.20 Σ *vs.* ϕ *Crossplot for* Σ_{ma} *and* Σ_w. *Courtesy Schlumberger Well Services.*

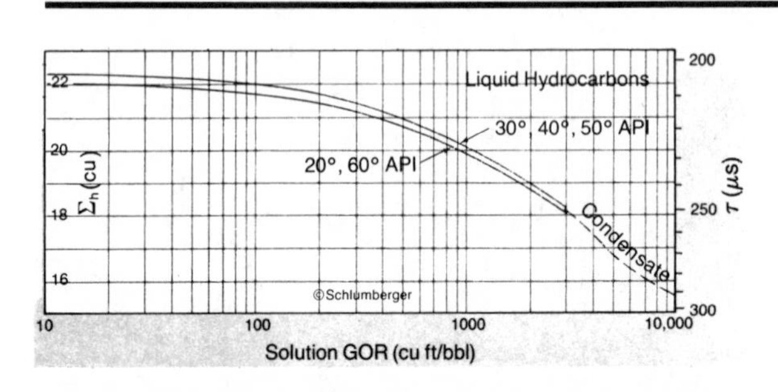

FIGURE 12.21 *Finding* Σ_o. *Courtesy Schlumberger Well Services.*

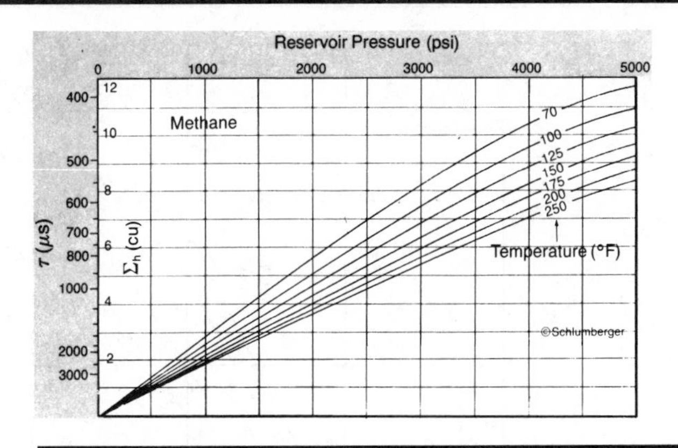

FIGURE 12.22 *Finding Σ Methane. Courtesy Schlumberger Well Services.*

QUESTION #12.6 Σ_o

An oil has a specific gravity of 40° API and an R_s of 400 cu ft/bbl.

Find Σ_o = ______.

SIGMA GAS. Σ_g is a function of pressure, temperature, and gas gravity. Figures 12.22 and 12.23 allow calculation of Σ_g for a variety of formation conditions. Figure 12.22 finds Σ methane. If the gas in question is methane, then no further work is required. However, if the gas in question is heavier than methane, figure 12.23 will convert Σ CH_4 into the appropriate value for Σ gas.

QUESTION #12.7 Σ_g

Given:

p = 3000 psi.
T = 150 °F.
γ_g = 0.65.

Find Σ_g = ______.

FINDING SIGMA SHALE. Σ_{sh} may be found by inspection of the log. Look for places where:

Gamma ray reads high.
Near/far display indicates shale.
Σ_{Log} is relatively high.

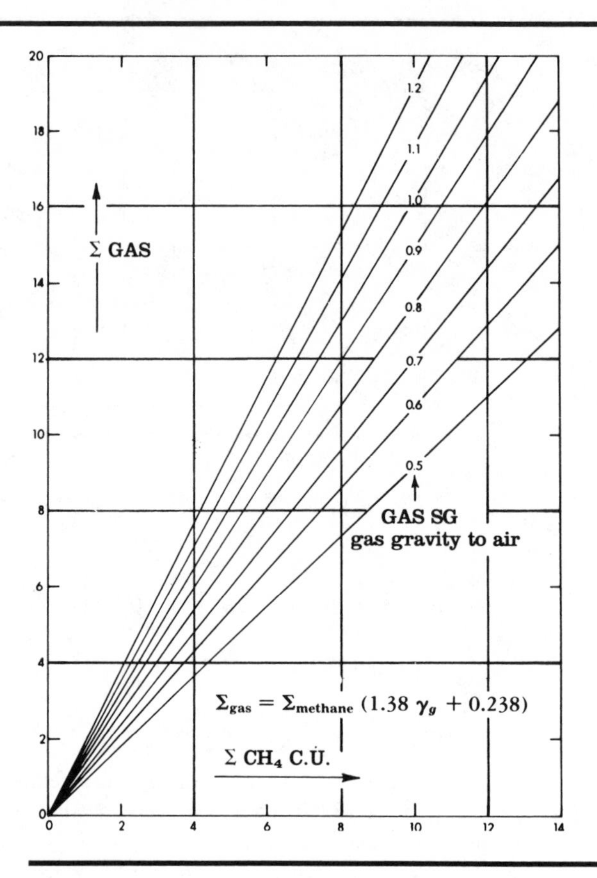

FIGURE 12.23 *Conversion of Σ Methane to Σ Gas.*

QUESTION #12.8 Σ_{sh}

Inspect the log shown in figure 12.24 and pick a value for Σ_{sh}.

Σ_{sh} = ______.

Sigma–Ratio Crossplot

If no openhole logs are available, one essential piece of data will be missing, namely, the porosity. Fortunately, there is a method for finding porosity from pulsed neutron logs. It requires only the values of sigma and ratio, read directly from the log. These two readings are then crossplotted to give porosity. A side benefit is that the plot also gives values of Σ_{wa}, the apparent capture cross section of the water. Five combinations of casing size and borehole-fluid salinity are covered in figures 12.25 through 12.29. To find

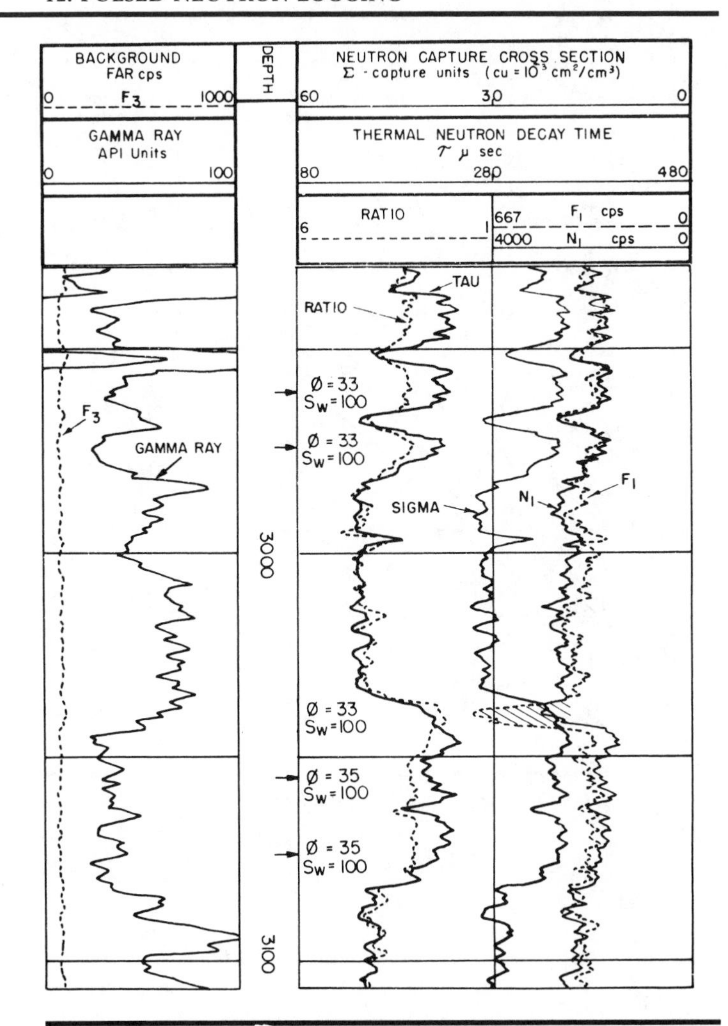

FIGURE 12.24 *Picking Σ Shale. Courtesy Schlumberger Well Services.*

porosity and sigma water apparent, pick the chart that most closely fits the conditions found in the well.

QUESTION #12.9 Σ–Ratio Porosity

A log is run in 5½-in. casing with an 8⅝-in. openhole. Borehole fluid salinity is 80,000 ppm NaCl; Σ_{Log} = 20 cu; and ratio = 2.8.

a. Find ϕ = ______ %.
b. Find Σ_{wa} = ______ cu.

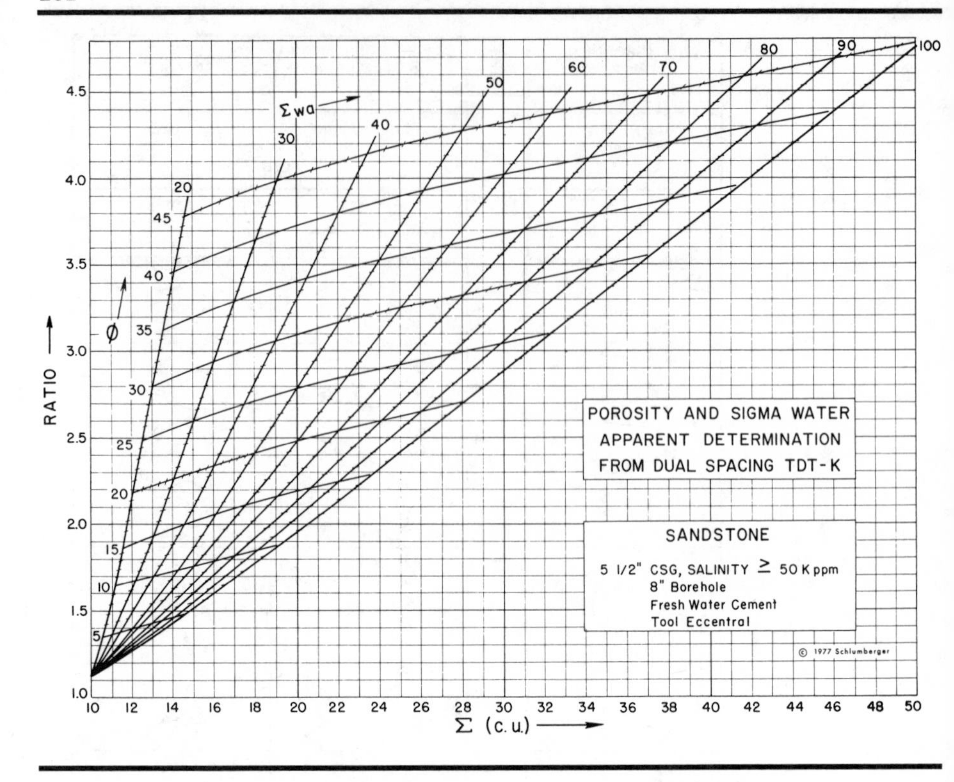

FIGURE 12.25 *ϕ and Σ_{wa}, 5½-in. Casing, High Salinity.*

This kind of plot is particularly useful for practical log analysis, but it has one drawback. The ratio curve, which is derived rather like a CNL porosity curve, is affected by gas; and, just as ϕ CNL reads too low in gas, so does the ratio curve. This means that porosity from the sigma–ratio plot will be too low in gas. The best practical solution to this problem is to guess a formation porosity on the basis of surrounding beds that are oil or water bearing and thus are not adversely affecting the ratio curve.

PRACTICAL LOG ANALYSIS

All the required data is now at hand for the practical log analyst to interpret pulsed neutron logs in a logical manner. The process can be defined as follows:

1. Determine all required parameters— The parameters required are: Σ_{ma}, Σ_g, Σ_w, Σ_{sh}, Σ_o, GR_{cl}, GR_{sh}, casing size, and borehole salinity.

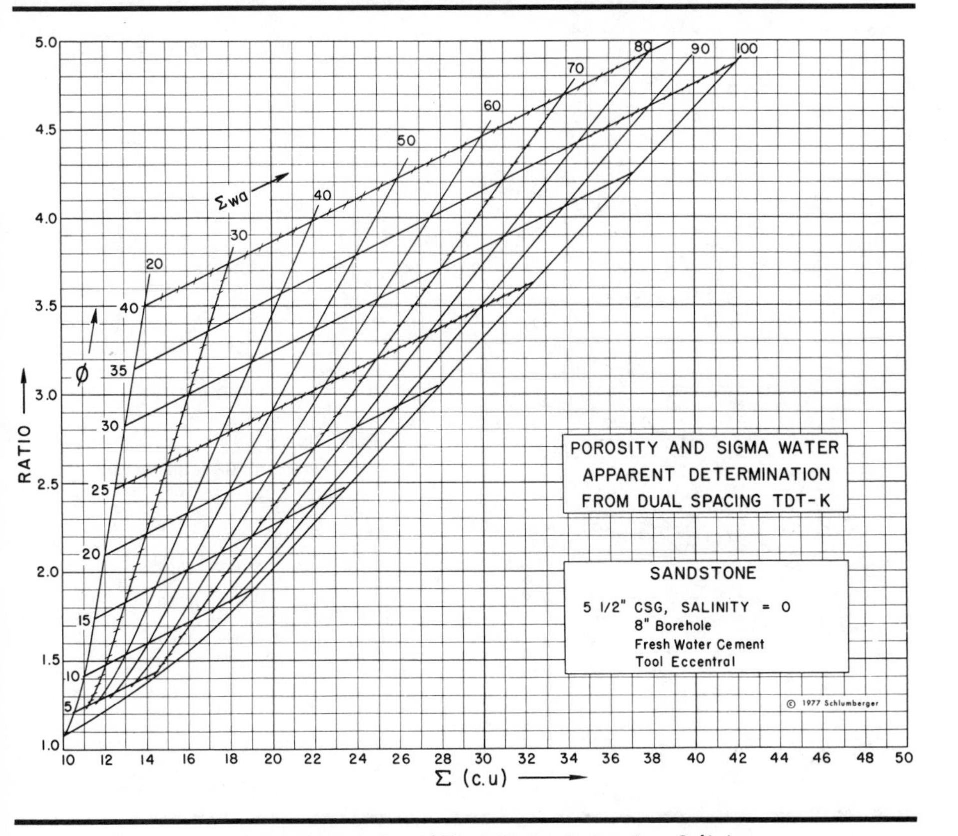

FIGURE 12.26 ϕ *and* Σ_{wa}, *5½-in. Casing, Low Salinity.*

2. Read logs— The log readings required are sigma, ratio and/or openhole porosity, near/far display, and GR and/or openhole shale indicator.

3. Make calculations— The calculations made will give: ϕ, V_{sh}, and S_w.

In order to place the process in perspective, the flow chart given in figure 12.30 should be followed to find all the necessary parameters. Gamma ray parameters may be found by inspection of the gamma ray log as discussed in chapter 11, where calculation of V_{sh} is also discussed. Σ_{sh} may also be found by inspection. A flow chart for final calculation of formation parameters is provided in figure 12.31. A handy format for these calculations is the table shown in figure 12.32.

Note that there is a column for ϕ, derived from the Σ-ratio plot, and a column for ϕ_e, the effective porosity, calculated using the equation

$$\phi_e = \phi - V_{sh}\phi_{sh}$$

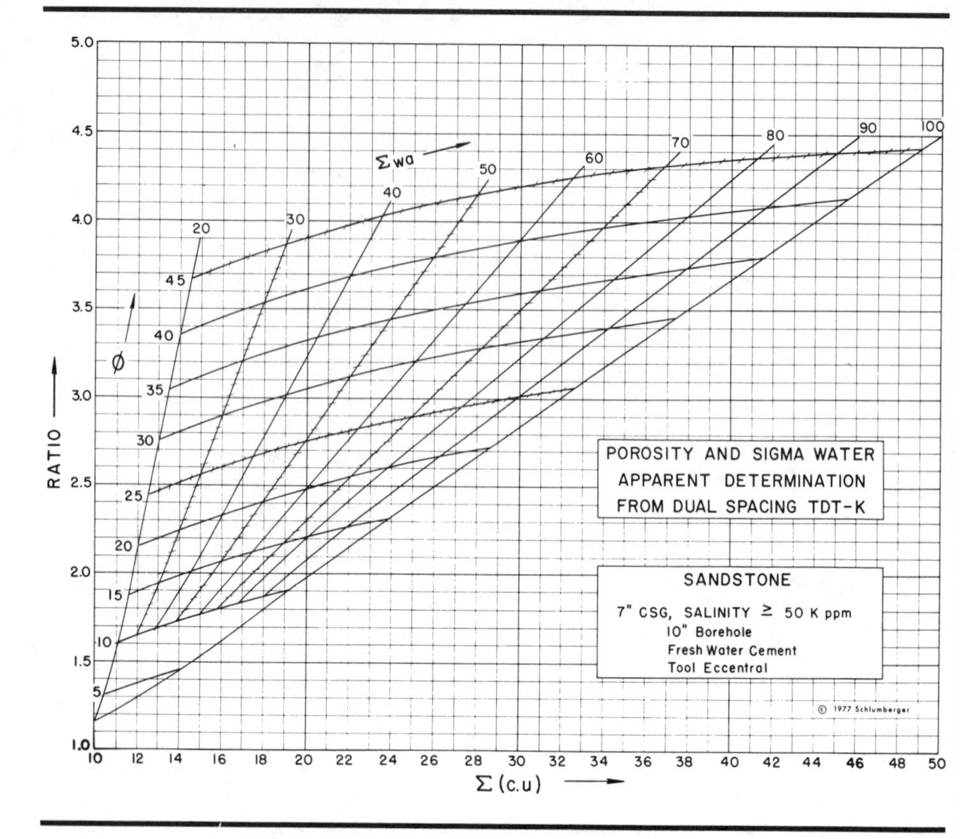

FIGURE 12.27 ϕ *and* Σ_{wa}, *7-in. Casing, High Salinity.*

where ϕ_{sh} is the apparent (or total) porosity picked from a Σ-ratio plot by inputting the values for Σ and ratio read from the log in a shale section.

RESERVOIR MONITORING—TIME-LAPSE TECHNIQUE

Pulsed neutron logs are useful for monitoring the depletion of a reservoir. The *time-lapse* method is used. A base log is run in the well shortly after initial completion but before substantial depletion of the producing horizons. A few days, weeks, or even months of production are required to clean up near wellbore effects of the drilling operation, such as mud-filtrate invasion, etc. Once a base log is obtained, the well may be relogged at time intervals over the life of the field. Typically, a log will be run every six months or once a year, depending on production rate.

Successive logs may be overlaid so that changes in saturation can be easily spotted by changes in sigma. A good example of this is

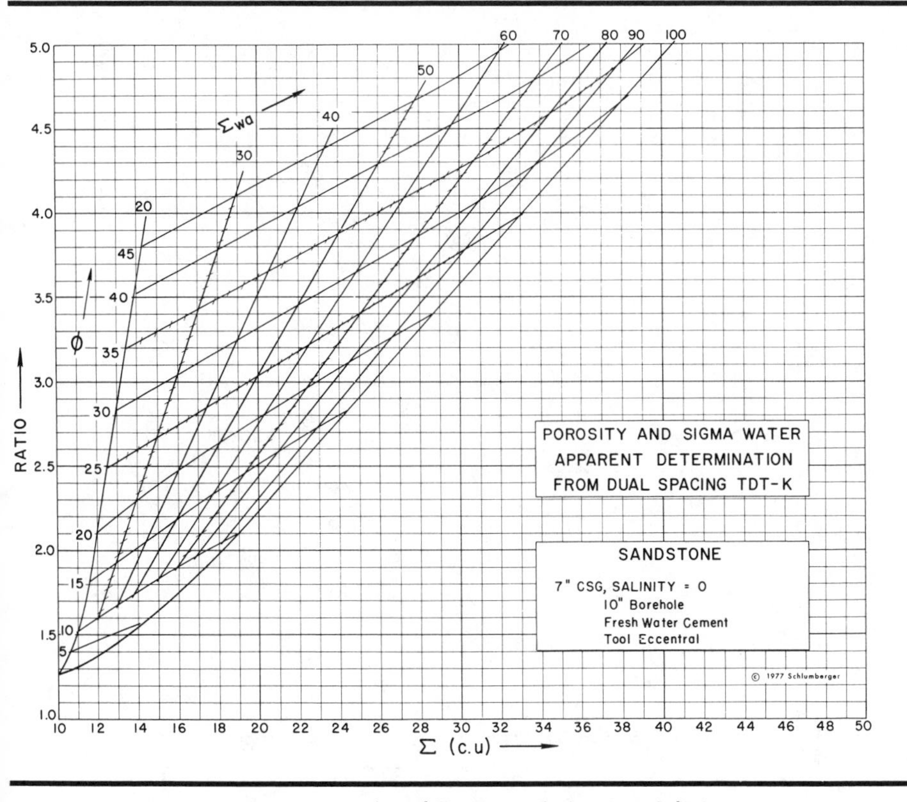

FIGURE 12.28 ϕ *and* Σ_{wa}, *7-in. Casing, Low Salinity.*

given by figure 12.33, which shows a base log and three additional logs at roughly six-month intervals. Note the rapid rise of the oil–water contact(s) with passage of time.

It is simple to calculate changes in S_w. Consider the state of affairs at time 1:

$$S_{w1} = \frac{(\Sigma_1 - \Sigma_{ma}) - \phi(\Sigma_{hy} - \Sigma_{ma})}{\phi(\Sigma_w - \Sigma_{hy})},$$

and some time later at time 2:

$$S_{w2} = \frac{(\Sigma_2 - \Sigma_{ma}) - \phi(\Sigma_{hy} - \Sigma_{ma})}{\phi(\Sigma_w - \Sigma_{hy})}.$$

The change in S_w is, therefore,

$$\Delta S_w = S_{w1} - S_{w2} = \frac{(\Sigma_1 - \Sigma_2)}{\phi(\Sigma_w - \Sigma_{hy})} = \frac{\Delta\Sigma}{\phi\Delta\Sigma \text{ fluids}}.$$

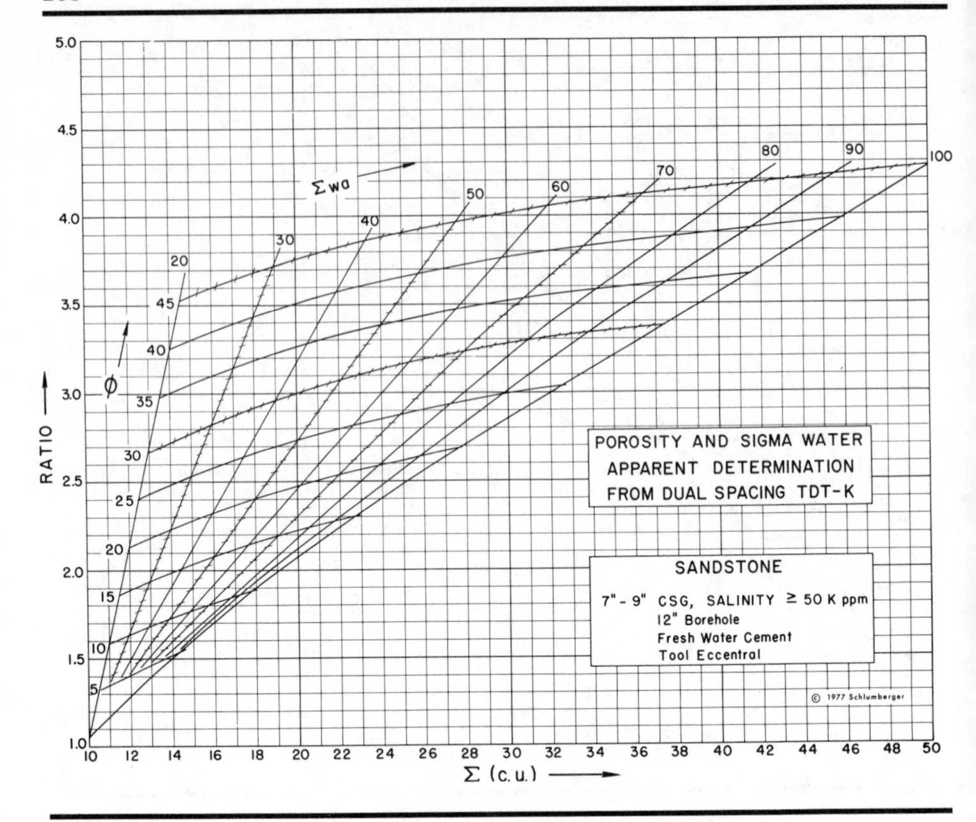

FIGURE 12.29 ϕ *and* Σ_{wa}, *7- to 9-in. Casing, Low Salinity.*

LOG-INJECT-LOG

The log-inject-log technique is used to find residual oil saturations. A base log is run and then the formation is injected with brine and logged again. Finally, the formation is injected with fresh water and logged a third time (see fig. 12.34). Provided the capture cross section of the fresh and brine flushes are known, all the unknown quantities may be normalized out and the residual oil saturation found using

$$S_o = 1 - \frac{\Sigma_{Log}\ \text{brine} - \Sigma_{Log}\ \text{fresh}}{\phi(\Sigma\ \text{brine} - \Sigma\ \text{fresh})}.$$

Note that it is not necessary to know either Σ_{ma} or Σ_o. The technique has many variations, some of which use specially chlorinated oil that has a high capture cross section.

DEPARTURE CURVES

Ideally, pulsed neutron logs should be usable for quantitative interpretation without having to make any corrections to the values read from the log. However, in some cases (e.g., when a base log is run with a fresh completion fluid and a subsequent log is run with a salty completion fluid in the borehole, or if the base log is run without a liner and a subsequent log with a liner), corrections will be required to the raw log measurement of sigma before quantitative interpretation can be made. The required corrections are a function of three variables: casing size, hole size, and salinity of the borehole fluid. Many sets of departure curves are published for the TDT-K tool as functions of openhole size and casing size (see fig. 12.35).

Considerable controversy exists in the literature regarding the need for departure curves. One school of thought holds that the diffusion of neutrons from the borehole to the formation necessitates the use of departure curves. Others maintain that the gating systems used are the base cause for the need of corrections. The NLL and the TDT-M are supposedly free of diffusion effects and the need for departure curves.

DEPTH OF INVESTIGATION

Another item of interest is the depth of investigation of the pulsed neutron tool. As with most radioactivity measurements, there is no fixed depth of investigation. Rather, a geometric factor describes what percentage of the total signal comes from what radial distance from the borehole wall. Figure 12.36 shows the response of the TDT-K in 5½-in. casing with a 1-in. cement sheath. Note that "depth of investigation" is somewhat deeper if salt water has invaded the formation. At all events, the majority of the signal comes from within one foot of the borehole wall.

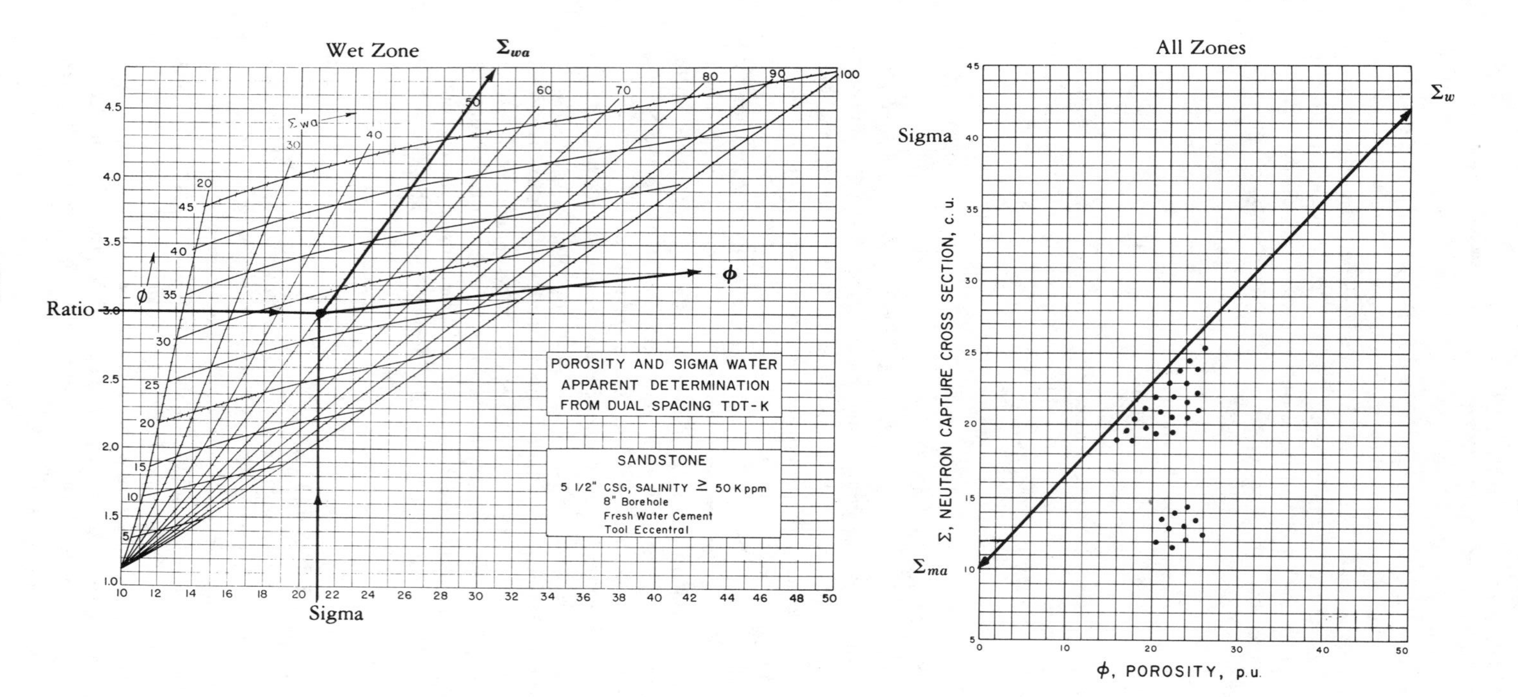

FIGURE 12.30 ***Pulsed Neutron Interpretation Parameters (for Clean Formations Only).***

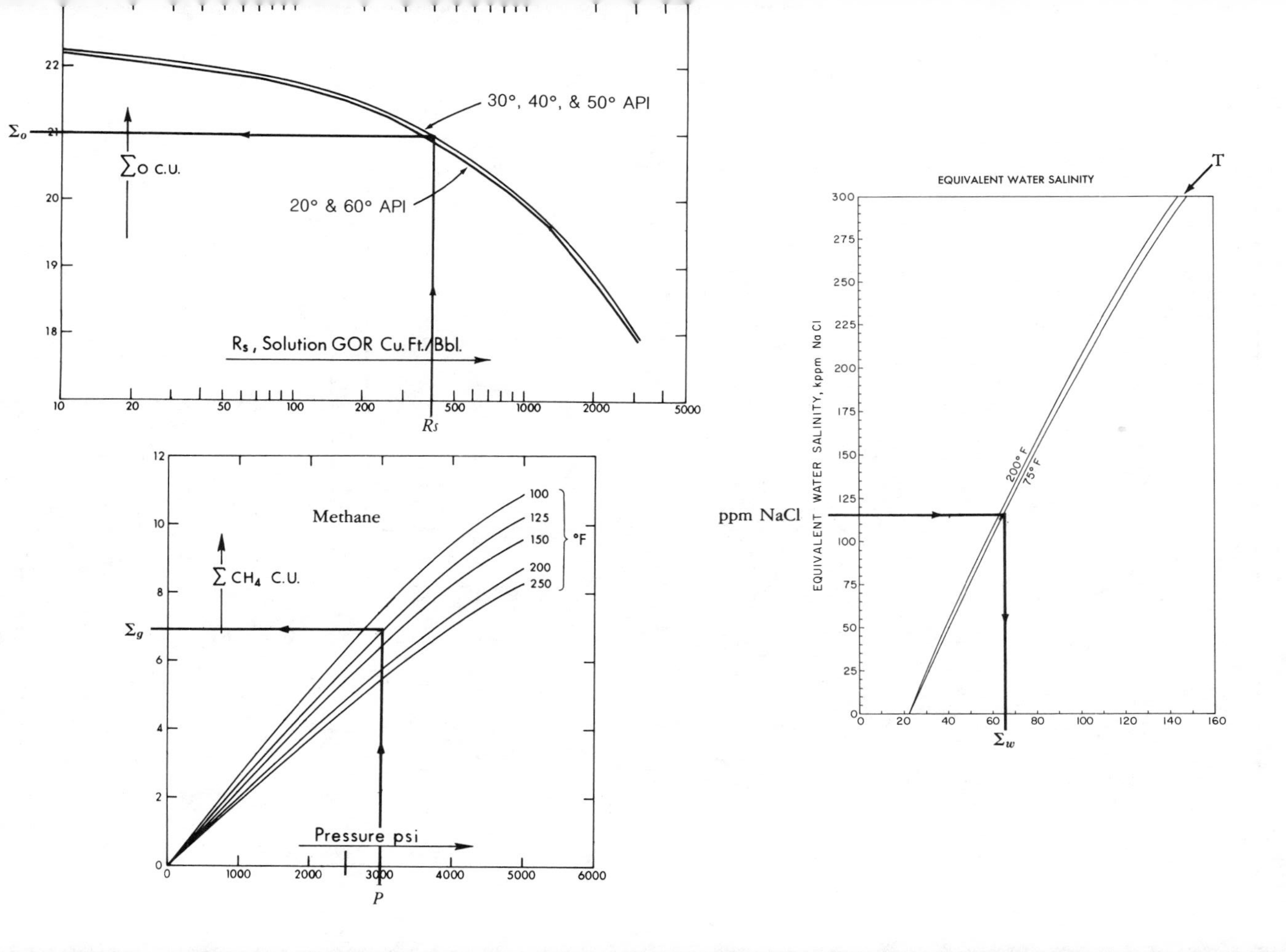

Σ_o
$\sum$o C.U.
30°, 40°, & 50° API
20° & 60° API
R_s, Solution GOR Cu. Ft./Bbl.
R_s
Methane
100
125
150
200
250
°F
$\sum CH_4$ C.U.
Σ_g
Pressure psi
P
EQUIVALENT WATER SALINITY
EQUIVALENT WATER SALINITY, kppm Na Cl
200° F
75° F
T
ppm NaCl
Σ_w

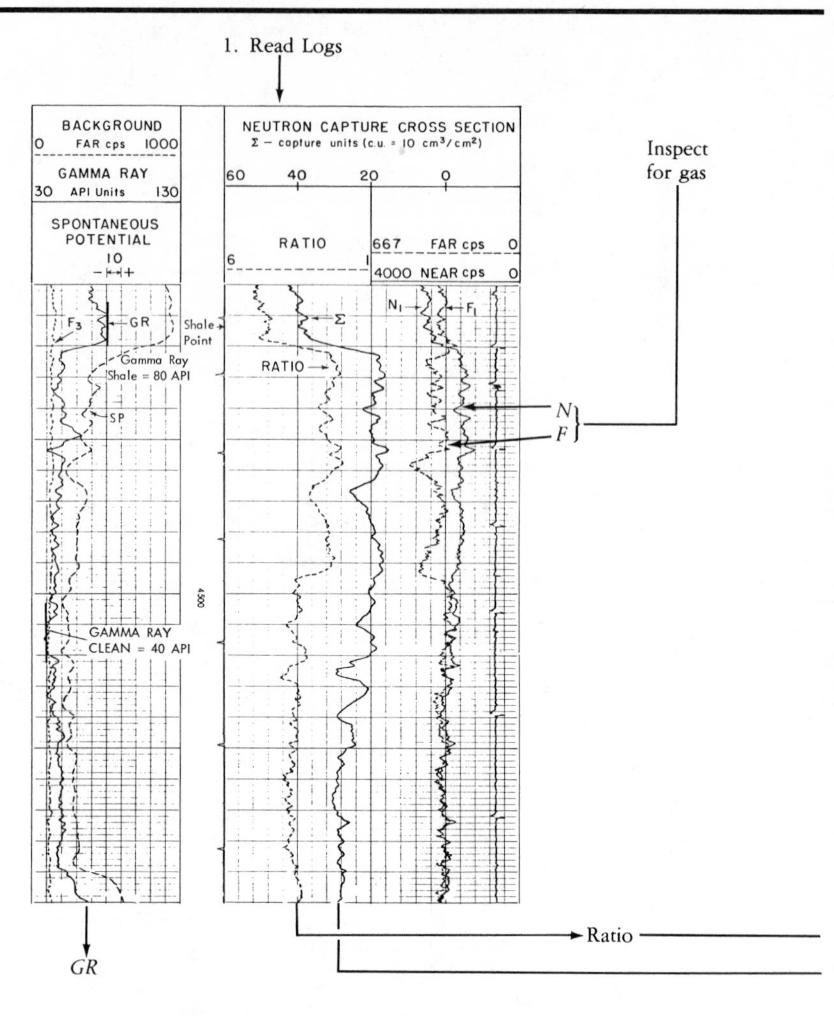
1. Read Logs
BACKGROUND
0 FAR cps 1000
GAMMA RAY
30 API Units 130
SPONTANEOUS POTENTIAL
10
NEUTRON CAPTURE CROSS SECTION
Σ — capture units (c.u. = 10 cm³/cm²)
60 40 20 0
RATIO
6 1
667 FAR cps 0
4000 NEAR cps 0
F3
GR
Shale Point
Gamma Ray Shale = 80 API
SP
GAMMA RAY CLEAN = 40 API
Σ
RATIO
N1
F1
4500
Inspect for gas
N
F
Ratio
GR

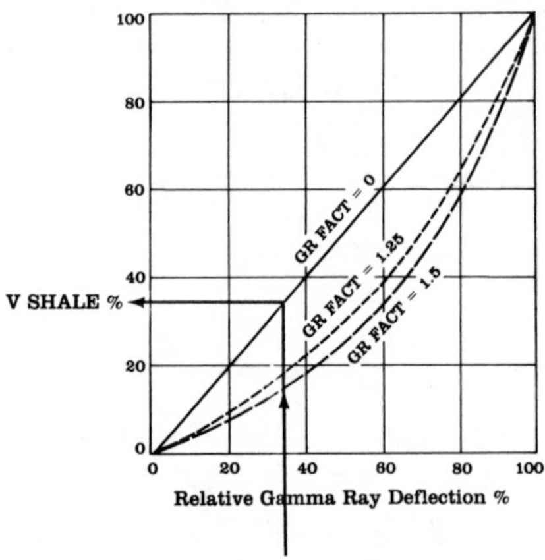
V SHALE %
GR FACT = 0
GR FACT = 1.25
GR FACT = 1.5
Relative Gamma Ray Deflection %
0
20
40
60
80
100
3. Find V-Shale

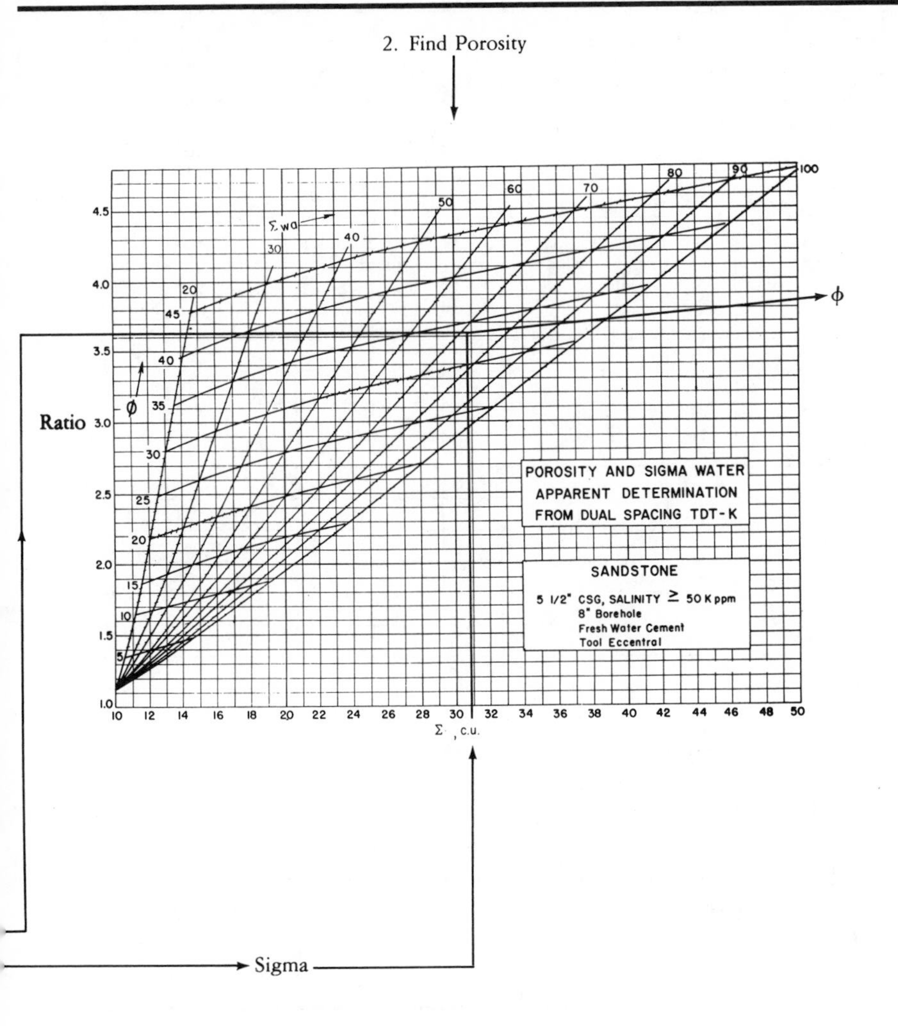

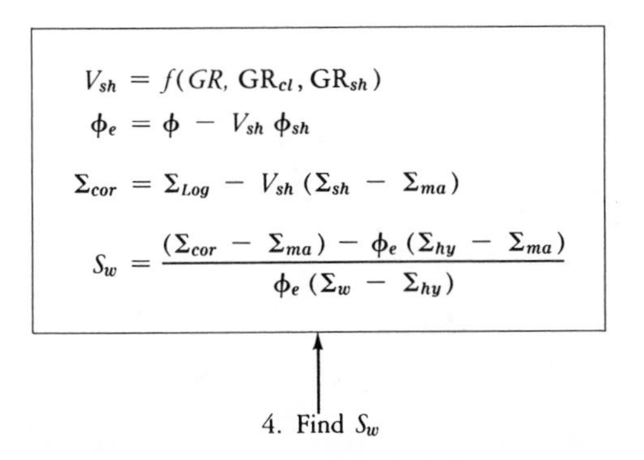

$$V_{sh} = f(GR, \mathrm{GR}_{cl}, \mathrm{GR}_{sh})$$

$$\phi_e = \phi - V_{sh}\,\phi_{sh}$$

$$\Sigma_{cor} = \Sigma_{Log} - V_{sh}\,(\Sigma_{sh} - \Sigma_{ma})$$

$$S_w = \frac{(\Sigma_{cor} - \Sigma_{ma}) - \phi_e\,(\Sigma_{hy} - \Sigma_{ma})}{\phi_e\,(\Sigma_w - \Sigma_{hy})}$$

FIGURE 12.31 ϕ, V_{sh}, *and* S_w *from Pulsed Neutron Logs.*

Well: . . . . Date: . . . . Analyst: . . . .

Parameters

Σ_{ma} . . . . Σ_o . . . . GR_{cl} . . . . Casing . . . .

Σ_w . . . . Σ_g . . . . GR_{sh} . . . . Salinity . . . .

Σ_{sh} . . . . Σ_{hy} . . . . GR factor . . . .

Data		Log Readings			Calculations					Comments
#	Depth	Σ	Ratio	GR	ϕ	V_{sh}	Σ_{cor}	ϕ_e	S_w	

FIGURE 12.32 ***Tabular Method for Pulsed Neutron Log Interpretation.***

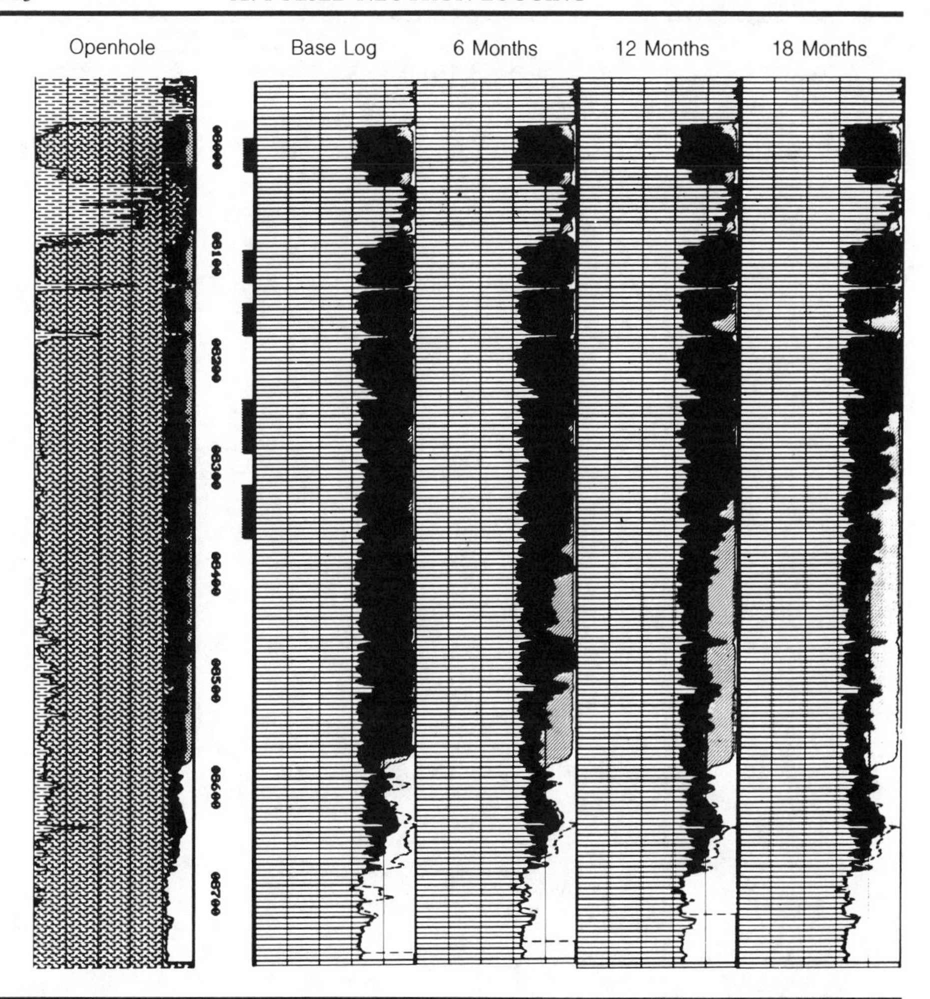

FIGURE 12.33 *Time-Lapse Logging.*

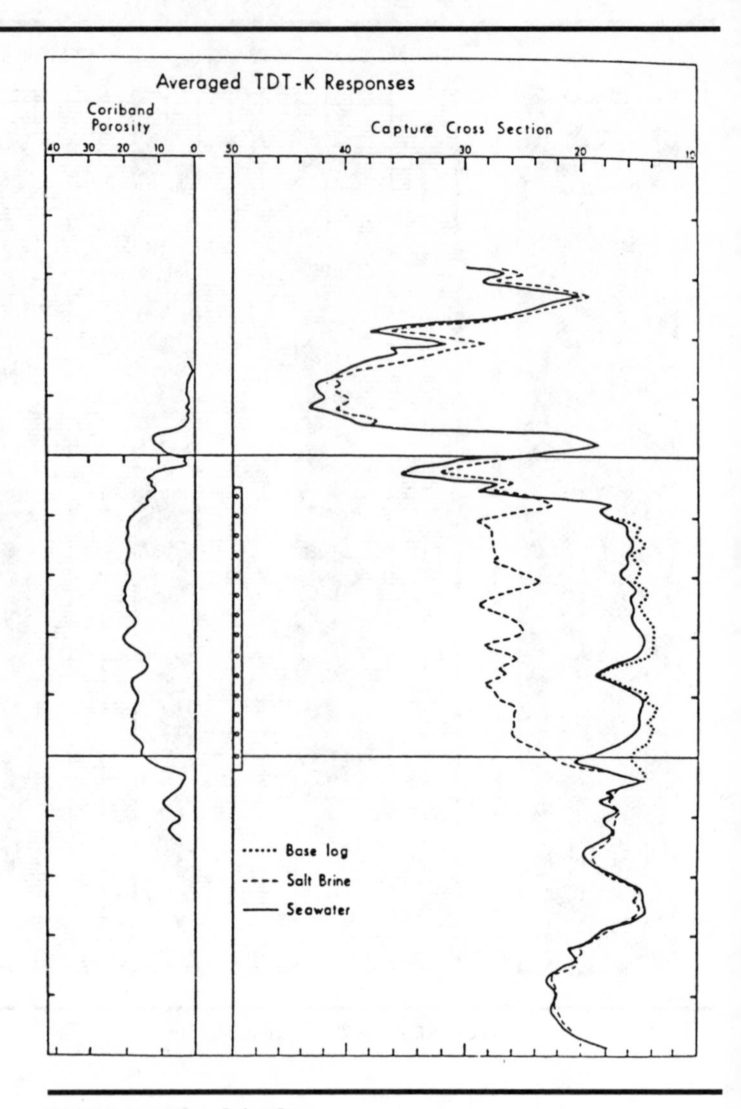

FIGURE 12.34 *Log-Inject-Log.*

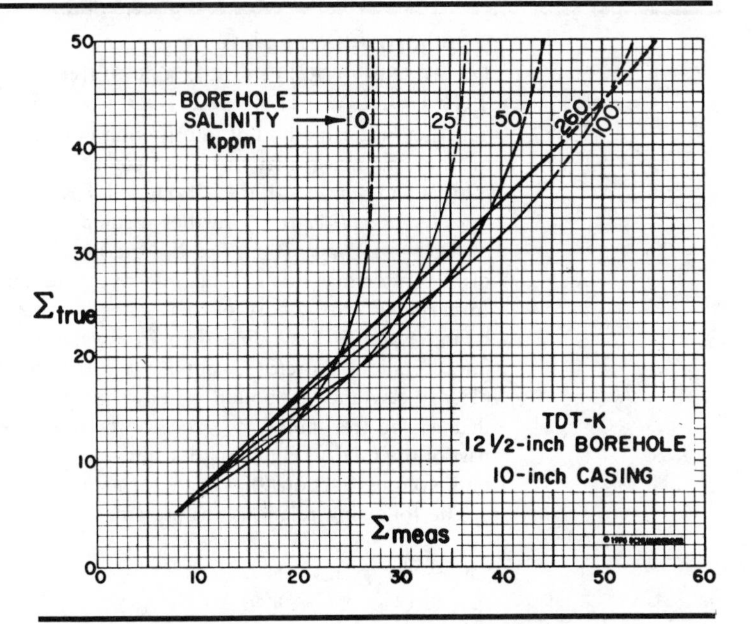

FIGURE 12.35 *TDT Departure Curve. Courtesy Schlumberger Well Services.*

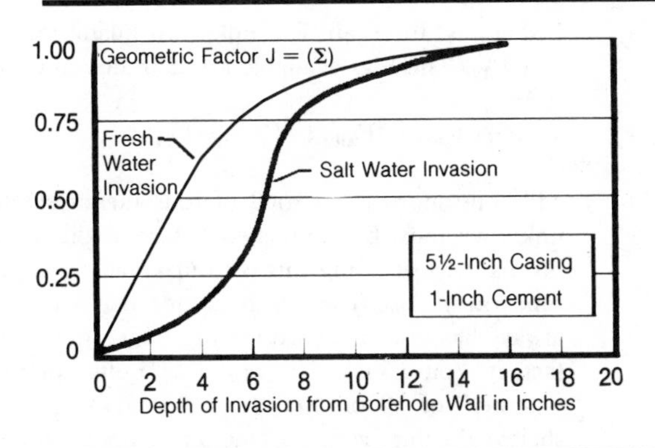

FIGURE 12.36 *Depth of Investigation of TDT-K.*

APPENDIX: INTERPRETATION OF PULSED NEUTRON LOGS USING THE DUAL-WATER METHOD

The dual-water method of interpreting pulsed neutron logs is based on the assumption that shales are composed of dry-clay, crystalline minerals to whose surface is bounded a layer of water. This water is called *bound water.* A further assumption is that the properties of the bound water (e.g., R_w, Σ_w, etc.) may be different from those of *free* water that exists in the effective, interconnected pore space. In particular, the theory of dual-water interpretation proposes that bound water is less saline than free water in most cases. Correct interpretation, therefore, calls for a means to find the amount of (1) dry clay and (2) bound water. The concept of *total porosity* ϕ_T, that is, the free fluids, ϕ_e, and the bound water, is an important part of the theory. Figure 12A.1 illustrates the concepts by mapping bulk volume fractions of a shaly formation.

The following relationships pertain:

$$\phi_e = \phi_T - V_{wb}.$$
$$S_{wT} = (V_{wf} + V_{wb})/\phi_T.$$
$$S_{we} = V_{wf}/\phi_e.$$
$$V_{sh} = V_{wb} + V_{dc}.$$

Essentially, there are five unknown quantities: V_{ma}, V_{dc}, V_{wb}, V_{wf}, and V_{hy}. The logs available are Σ, ratio, and GR. The identity

$$V_{ma} + V_{dc} + V_{wb} + V_{wf} + V_{hy} = 1$$

adds one more for a total of four measurements. Therefore, one unknown must be eliminated before a solution can be found. The normal way of doing this is to make an assumption about V_{wb} as a function of V_{dc}. That is, to assume that a unit volume of dry clay always has associated with it the same amount of bound water. In fact, in "pure shale," it would be quite common to find a "total porosity" of 30 or 40% (as reflected by neutron log readings in shales). In this case, the amount of bound water associated with a dry clay can be back-calculated. For example, if a 100% shale has a total porosity of 35%, it follows that

$$V_{wb} = 35\% \quad \text{and} \quad V_{dc} = 65\%$$

and hence that

$$V_{wb} = \alpha \cdot V_{dc},$$

where α is some constant which, in this example, is numerically equal to $35/65 = 0.538$.

Having reduced the unknowns to four (V_{ma}, V_{dc}, V_{wf}, and V_{hy}), since V_{wb} can now be assumed equal to αV_{dc}, the solution to the

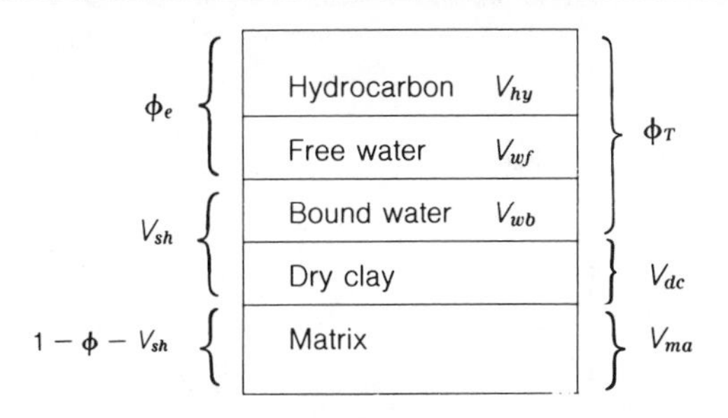

FIGURE 12A.1 *Dual-Water Shaly Formation.*

dual-water problem becomes straightforward. The following steps are required:

1. Find all necessary parameters Σ_{ma}, Σ_{dc}, Σ_{wb}, Σ_{wf}, Σ_{hy}, GR_{ma}, GR_{dc}.
2. Find ϕ_T and V_{dc}.
3. Solve for ϕ_e and S_{we}.

Finding Parameters

Crossplot techniques are particularly useful for finding the required parameters. The log data points should be divided into two groups: The 100% shales and the clean-formation intervals.

In clean formations, a plot of Σ vs. ϕ will define Σ_{ma} and Σ_{wf}, provided there is sufficient variation in porosity and enough points at 100% water saturation. Figure 12A.2 shows the procedure schematically.

A similar plot for finding Σ_{dc} and Σ_{wb} is shown in figure 12A.3 (all points must come from the shale sections). Note that, on both plots, ϕ_T, derived from the Σ vs. ratio crossplot, is used. This entails an assumption that porosity measured in this way is, in fact, equal to total porosity.

Σ_{hy} can be found by conventional means. Thus, only the gamma ray response to dry clay and response to the matrix remain to be found. It is assumed that neither formation water nor hydrocarbon contribute to the gamma ray response, so it can be written

$$GR = (1 - \phi_T)GR_{ma} + V_{dc}GR_{dc}.$$

From which it follows that, in shales,

$$GR_{dc} = GR/(1 - \phi_T),$$

and, in clean intervals,

$$GR_{ma} = GR/(1 - \phi_T).$$

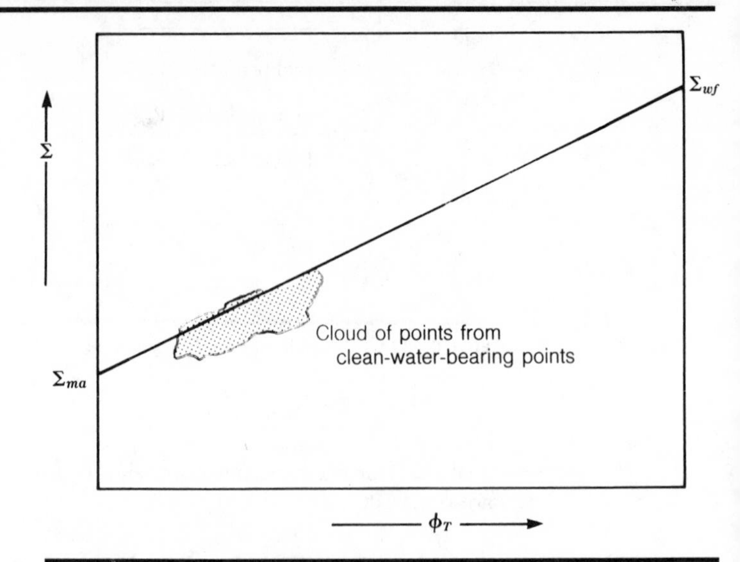

FIGURE 12A.2 *Finding* Σ_{ma} *and* Σ_{wf}.

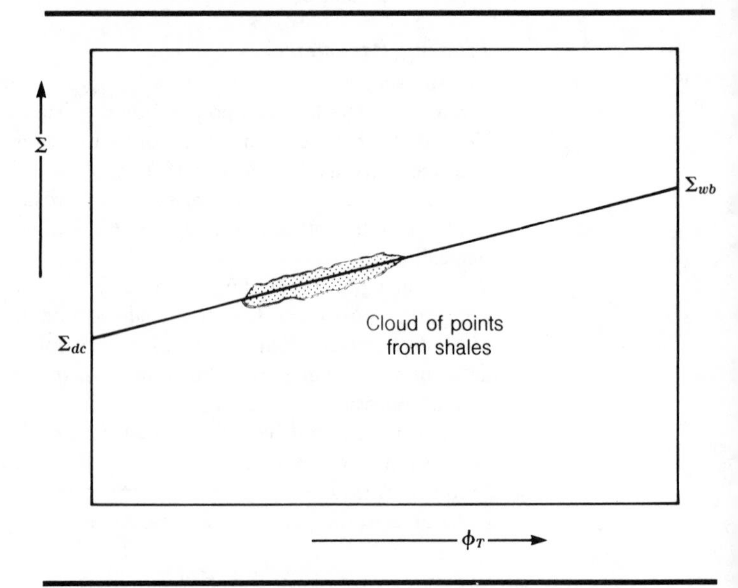

FIGURE 12A.3 *Finding* Σ_{dc} *and* Σ_{wb}.

For example, in a shale, $GR = 110$ and $\phi_T = 33\%$; but, in a clean section, $GR = 25$ and $\phi_T = 25\%$, so it follows that

$$GR_{dc} = 110/(1 - 0.33) = 149.25, \text{ and}$$
$$GR_{ma} = 25/(1 - 0.25) = 33.3.$$

Finding ϕ_T *and* V_{dc}

As already stated, ϕ_T is found from the Σ vs. ratio crossplot. V_{dc} can be found from the GR using

$$V_{dc} = \frac{GR - GR_{ma}(1 - \phi_T)}{GR_{dc} - GR_{ma}}.$$

Solving for ϕ_e *and* S_{we}

Once V_{dc} and ϕ_T are established, the following relationships hold:

$$\phi_e = \phi_T - V_{wb} \quad \text{(which also} = V_{hy} + V_{wf}\text{)},$$
$$V_{ma} = 1 - \phi_T - V_{dc}, \text{ and}$$
$$V_{wb} = \alpha V_{dc},$$

where α has been established in the shales as

$$\phi_{Tsh}/(1 - \phi_{Tsh}).$$

The response of the pulsed neutron log itself can be written as

$$\Sigma = \Sigma_{ma} V_{ma} + \Sigma_{dc} V_{dc} + \Sigma_{wb} \alpha V_{dc} + \Sigma_{wf} V_{wf} + \Sigma_{hy} V_{hy},$$

whence

$$\Sigma_{hy} V_{hy} + \Sigma_{wf} V_{wf} = \Sigma - \Sigma_{ma} V_{ma} - V_{dc}(\Sigma_{dc} + \alpha\Sigma_{wb}).$$

The right side of the equation can be evaluated since all the parameters and variables have now been defined. If this quantity is, in fact, Σ^*, then

$$V_{wf} = \frac{\Sigma^* - \phi_e \Sigma_{hy}}{\Sigma_{wf} - \Sigma_{hy}}.$$

By definition,

$$S_{we} = V_{wf}/\phi_e.$$

Practical Applications

Table 12A.1 gives a flow chart to guide the user through the chain of calculations. Note that the process does not lend itself to graphical solutions using nomographs. The use of a programmable calculator is recommended. Most logging service companies offer this or similar computations using on-board computers.

TABLE 12A.1 *Flow Chart for Dual-Water Interpretation of Pulsed Neutron Logs*

1. Find ϕ_T from Σ and ratio.
2. Find V_{dc} from $V_{dc} = \dfrac{GR - GR_{ma}(1 - \phi_T)}{GR_{dc} - GR_{ma}}$.
3. Find V_{ma} from $V_{ma} = 1 - \phi_T V_{dc}$.
4. Find V_{wb} from $V_{wb} = \alpha V_{dc}$.
5. *Find* ϕ_e from $\phi_e = \phi_T - V_{wb}$.
6. Find V_{wf} from $V_{wf} = \dfrac{\Sigma^* - \phi_e \Sigma_{hy}}{\Sigma_{wf} - \Sigma_{hy}}$,

   where

$$\Sigma^* = \Sigma_{Log} - V_{ma}\Sigma_{ma} - V_{dc}\Sigma_{dc} - V_{wb}\Sigma_{wb}.$$

7. Find S_{we} from $S_{we} = V_{wf}/\phi_e$.
8. Recompute V_{sh} from $V_{sh} = V_{dc} + V_{wb}$.

BIBLIOGRAPHY

Antkiw, S.: "Depth of Investigation of the Dual-Spacing Thermal Neutron Decay Time Logging Tool," *SPWLA Trans.* 17th Logging Symposium, Denver, Colo., June 1976.

Blackburn, J. S., and Brimage, R. C.: "Estimation of Formation Pressures in Clean Gas Sands from the Dual-Spacing TDT Log," *SPWLA Trans.* 19th Logging Symposium, El Paso, Tex., June 1978.

Clavier, C., Hoyle, W., and Meunier, D.: "Quantitative Interpretation of Thermal Neutron Decay Time Logs: Part I. Fundamentals and Techniques, Part II. Interpretation Examples, Interpretation Accuracy, and Time-Lapse Technique," *J. Pet. Tech.* (June 1971) 743–763.

"Departure Curves for the Thermal Decay Time Log," Schlumberger, publication C-11989 (1976).

Dewan, J. T., Johnstone, C. W., Jacobson, L. A., Wall, W. B., and Alger, R. P.: "Thermal Neutron Decay Time Logging Using Dual Detection," *SPWLA Trans.* 14th Logging Symposium, Lafayette, La., May 1973.

Hall, J. E., Johnstone, C. W., Baldwin, J. L., and Jacobson, L. A.: "A New Thermal Neutron Decay Logging System-TDT-M," paper SPE 9462 presented at the SPE 55th Annual Technical Conference and Exhibition, Dallas, Sept. 21–24, 1980; *J. Pet. Tech.* (January 1982).

McGhee, B. F., McGuire, J. A., and Vacca, H. L.: "Examples of Dual Spacing Thermal Neutron Decay Time Logs in Texas Coast Oil & Gas Reservoirs," *SPWLA Trans.* 15th Annual Logging Symposium, June 1974.

Murphy, R. P. and Owens, W. W.: "The Use of Special Coring and Logging Procedures for Defining Reservoir Residual Oil Saturations," *J. Pet. Tech.* (July 1973) 841–850.

Murphy, R. P., Foster, G. T., and Owens, W. W.: "Evaluation of Waterflood Residual Oil Saturations Using Log-Inject-Log Procedures," *J Pet. Tech.* (February 1977) 178–186.

Wahl, J. S., Nelligan, W. B., Frentrop, A. H., Johnstone, C. W., and Schwartz, R. J.: "The Thermal Neutron Decay Time Log," *Soc. Pet. Eng. J.* (December 1970) 365–379.

Wiese, Harry C.: "TDT Log Applications in California," *J. Pet. Tech.* (February 1983) 429–444.

Answers to Text Questions

QUESTION #12.1
(c) Σ increase
(d) 30 cu

QUESTION #12.2
S_w = 50%

QUESTION #12.3
(a) ϕ = 30%
(b) S_w = 50%
(c) S_w = 55%

QUESTION #12.4
Σ_w = 110 cu

QUESTION #12.5
(a) Σ_{ma} = 13 cu
(b) Σ_w = 75 cu

QUESTION #12.6
Σ_o = 21.2 cu

QUESTION #12.7
Σ methane = 6.5 cu
Σ_g = 8.0 cu

QUESTION #12.8
Σ_{sh} = 33 cu

QUESTION #12.9
(a) ϕ = 25%
(b) Σ_{wa} = 50 cu

CHAPTER THIRTEEN

CEMENT BOND LOGGING

Cement bond logging (CBL) is an important part of a well-completion program and is also recommended for most workover programs. Most of the cementing-related problems encountered can be diagnosed by use of the CBL. However, successful interpretation of a CBL depends on certain minimum requirements:

a. The tool must be run strictly centralized.
b. A transit-time curve must be recorded.
c. A wave-train or VDL display must be available.

Before explaining these requirements in detail, however, it is worthwhile to review the principles of oilwell cementing, since many problems can be traced back to the primary cement job and the way it was conducted.

PRINCIPLES OF OILWELL CEMENTING

Oilwell cementing is the process of mixing a slurry of cement, water, and additives and pumping it down through steel casing to the annular space between the wellbore and the outside of the casing. Figure 13.1 illustrates the conventional cementing process. Sometimes it is advisable to cement a well in two or more stages; in which case a more complex completion string is required, as shown in figure 13.2. This technique is employed when two or more zones have different cement requirements.

Cement has three principal functions in a well:

1. to restrict fluid movement between formations,
2. to bond the casing to the formation, and
3. to provide support for the casing.

As long as an effective hydraulic seal is provided between porous and permeable zones in the well, a cement job may be considered as good. Important examples include isolation of gas-bearing formations and wet zones from primary oil-producing horizons and isolation of shallow drinking-water aquifers from deeper saltwater formations.

Cements themselves are classified by API classification letters A through J. Table 13.1 details the well depths, temperatures, and slurry weights for these cements. For further details refer to service-company handbooks.

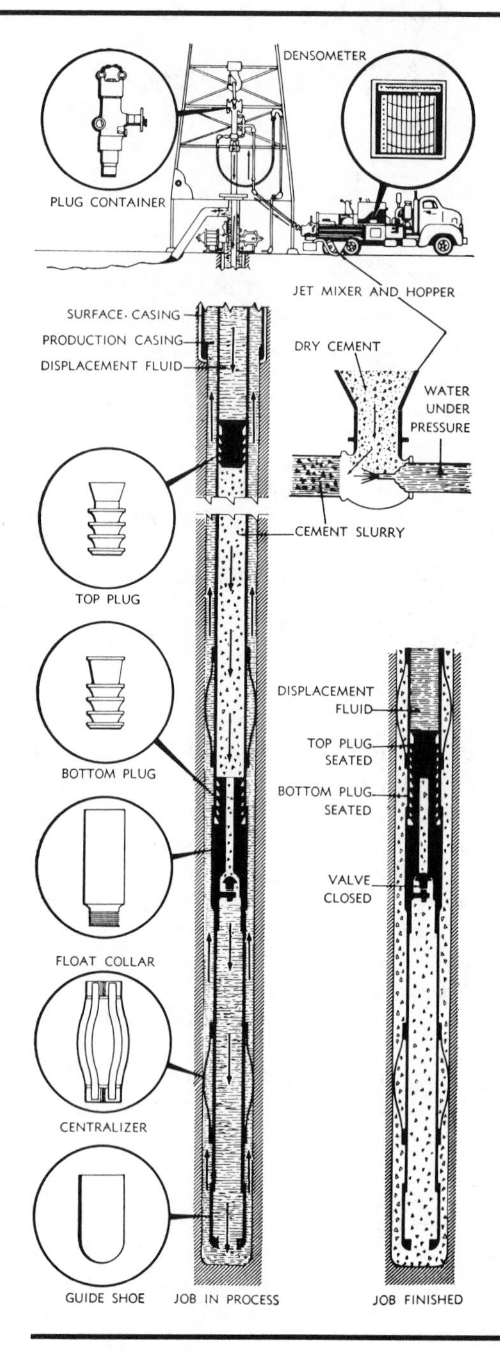

FIGURE 13.1 *Primary Cementing Procedures. Courtesy Halliburton Services.*

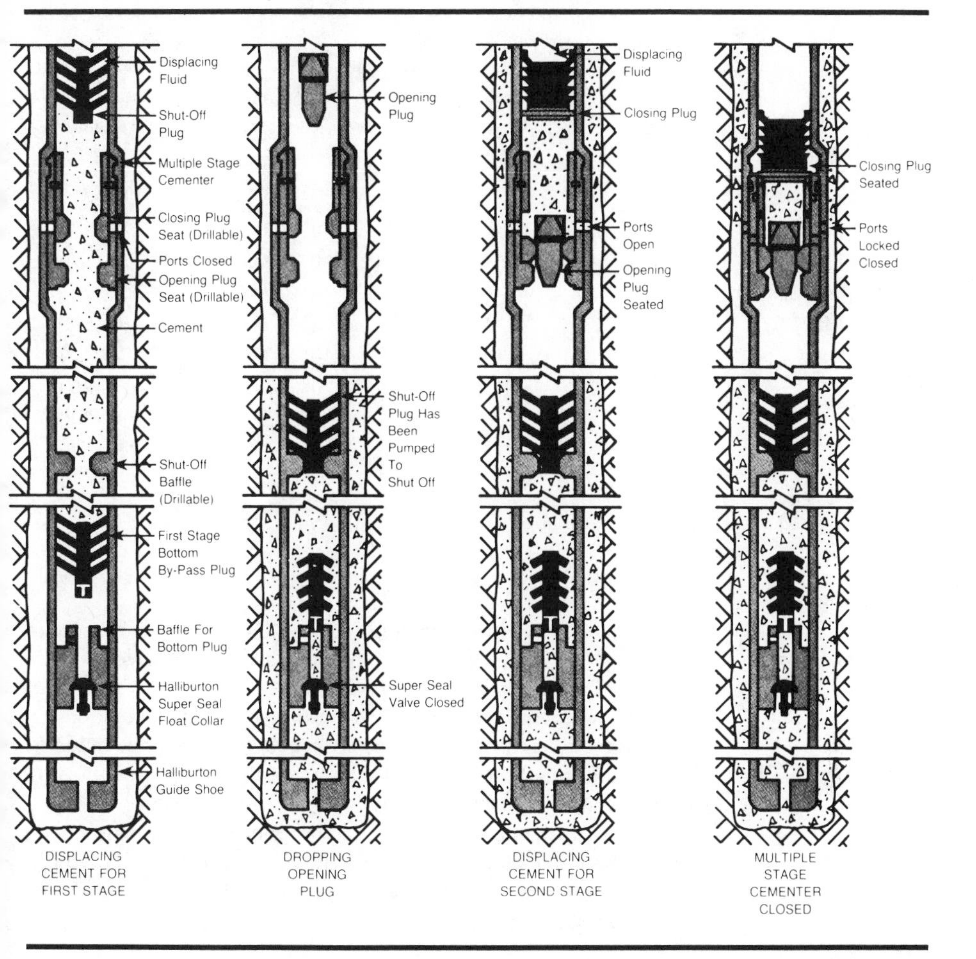

FIGURE 13.2 *Successive Steps for Stage Cementing. Courtesy Halliburton Services.*

TABLE 13.1 *API Classification of Cements*

API Classification	Mixing Water (gal/sack)	Slurry Weight (lb/gal)	Well Depth (ft)	Static Temperature (°F)
A (Portland)	5.2	15.6	0– 6,000	80–170
B (Portland)	5.2	15.6	0– 6,000	80–170
C (High Early)	6.3	14.8	0– 6,000	80–170
D (Retarded)	4.3	16.4	6–12,000	170–260
E (Retarded)	4.3	16.4	6–14,000	170–290
F (Retarded)	4.5	16.2	10–16,000	230–320
G (Basic)*	5.0	15.8	0– 8,000	80–200
H (Basic)*	4.3	16.4	0– 8,000	80–200
J	4.9	15.4	12–16,000	260–320

* Can be accelerated or retarded for most well conditions.
Note: Reprinted by permission of Halliburton Services.

Various additives are commonly used with cements to control such variables as density, viscosity, setting time, and compressive strength. Table 13.2 details the effects of additives on cements.

Other practices commonly used during cementing operations include the use of casing centralizers, scratchers (to remove mudcake from the borehole wall), and turbolizers (to induce turbulent flow in the casing/formation annulus).

The use of cement additives and/or mechanical accoutrements can dramatically affect the quality of the final cement job. Anything, within reason, that assists in obtaining a good hydraulic seal should not be overlooked.

PRINCIPLES OF CEMENT BOND LOGGING

Conventional CBL tools rely for their operation on the fact that a compressional (acoustic) wave transmitted along the wall of a steel pipe becomes attenuated if the pipe has cement bonded to it. The relationship between the compressive strength of the cement and the attenuation rate (measured in db/ft) is shown in figure 13.3. Note that the type of cement is relatively unimportant and that, given the attenuation rate, a cement compressive strength can be deduced.

The data for figure 13.3 refers to steel pipes completely surrounded by cement. In the case of partial cementation, it is of interest to note the data of figure 13.4, where the percentage of the pipe's circumference is plotted against the attenuation rate. This relationship is linear (i.e., the smaller the percentage of the circumference cemented, the less the attenuation rate). It is also worth noting that at least a ¾-in. thickness of cement is required in order for these

TABLE 13.2 *Effects of Some Additives on the Physical Properties of Cement*

		Bentonite	Diatomaceous Earth	Pozzolan	Sand	Heavy Weight	Accelerator	Sodium Chloride*	Retarder	Friction Reducer	Low-Water-Loss Materials	Lost-Circulation Materials
Density	Decreased	⊗	⊗	⊗								
	Increased				⊗	⊗		X				
Water required	Less											
	More	⊗	⊗	X	X					X		X
Viscosity	Decreased						X	X	⊗	⊗		
	Increased	X	X	X	X	X						X
Thickening time	Accelerated						⊗	⊗				
	Retarded	X	X					X	⊗	X	X	
Early strength	Decreased	X	X	X					⊗		X	X
	Increased						⊗	⊗		X		
Final strength	Decreased	⊗	⊗	X		X					X	X
	Increased								X	X		
Durability	Decreased	X	X									X
	Increased			⊗								
Water loss	Decreased	⊗							X	X	⊗	X
	Increased		X									

Note: Reprinted by permission of Halliburton Services.
X denotes minor effect.
⊗ denotes major effect and/or principal purpose for which used.
* small percentages of sodium chloride accelerate thickening. Large percentages may retard API class A cement.

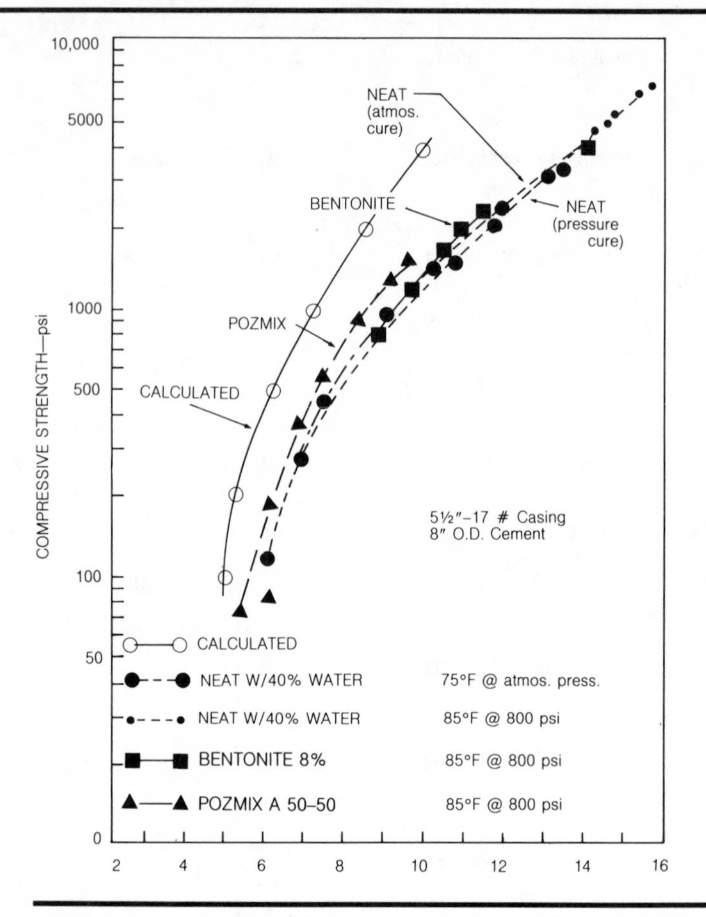

FIGURE 13.3 *Compressive Strength vs. Attenuation Rate for Various Cements. Reprinted by permission of the SPE-AIME from Pardue et al. 1963, fig. 7, p. 548. © 1963 SPE-AIME.*

relationships to hold. Figure 13.5 shows that after a thickness of ¾ in. is exceeded, attenuation rate is constant. Thus, measurement of the attenuation rate of a compressional wave propagated along the casing gives information regarding the bonding of the cement to the casing. Information regarding the bonding of the cement to the formation has to come from a separate source such as the VDL (variable-density display) or wave-train recording. In order to understand this more fully the tools used must be studied.

TOOLS AVAILABLE

Table 13.3 details the tools available for cement bond logging. In general, there are through-tubing tools for workover jobs and full-

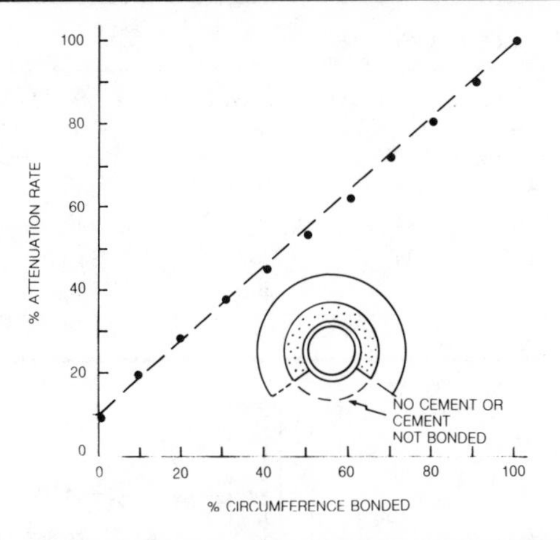

FIGURE 13.4 *Percent Attenuation vs. Percent Circumference Bonded. Reprinted by permission of the SPE-AIME from Pardue et al. 1963, fig. 7, p. 548. © 1963 SPE-AIME.*

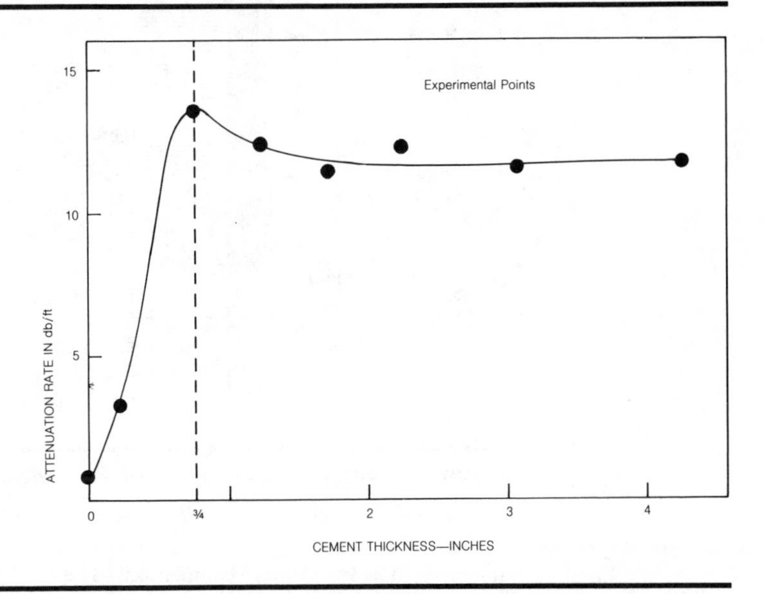

FIGURE 13.5 *Attenuation Rate vs. Cement Thickness. Reprinted by permission of the SPE-AIME from Pardue et al. 1963, fig. 20, p. 552. © 1963 SPE-AIME.*

TABLE 13.3 *CBL Tools*

Type	Tool OD (inches)	Max. Press. (psi)	Max. Temp. (°F)	Min. Pipe Size (inches)
CBL	1$^{11}/_{16}$	16,500	300	2
CBL	3⅝	20,000	350	4½
CET	3⅜,	20,000	350	4½
	5	20,000	350	5½

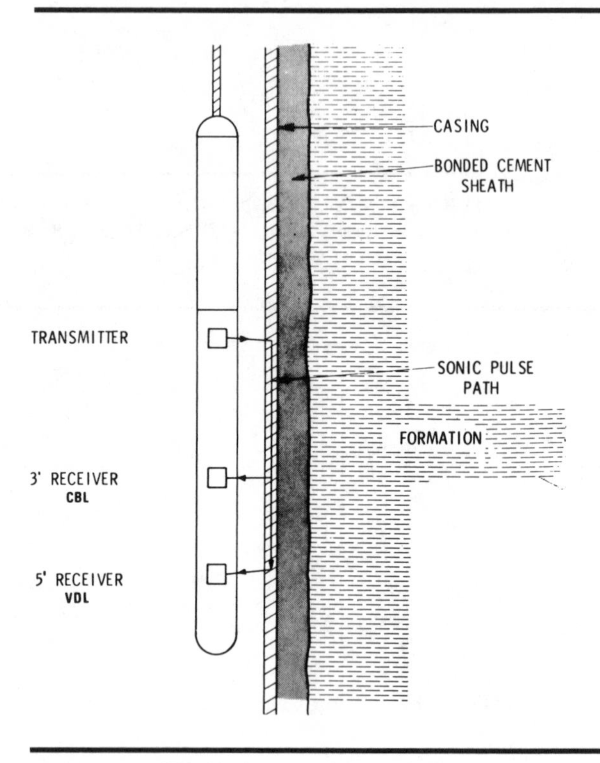

FIGURE 13.6 *CBL Tool. Courtesy Schlumberger Well Services.*

diameter tools for primary completions. The CET will be dealt with later in a separate section.

Figure 13.6 illustrates a conventional CBL tool. It consists of an acoustic transmitter and two receivers. In actual practice, the tool may be identical with a BHC sonic tool but with only one transmitter and two receivers being used. The near receiver is placed 3 ft from the transmitter and is used for amplitude measurements. The far receiver is placed 5 ft from the transmitter and is used for wave-train recordings.

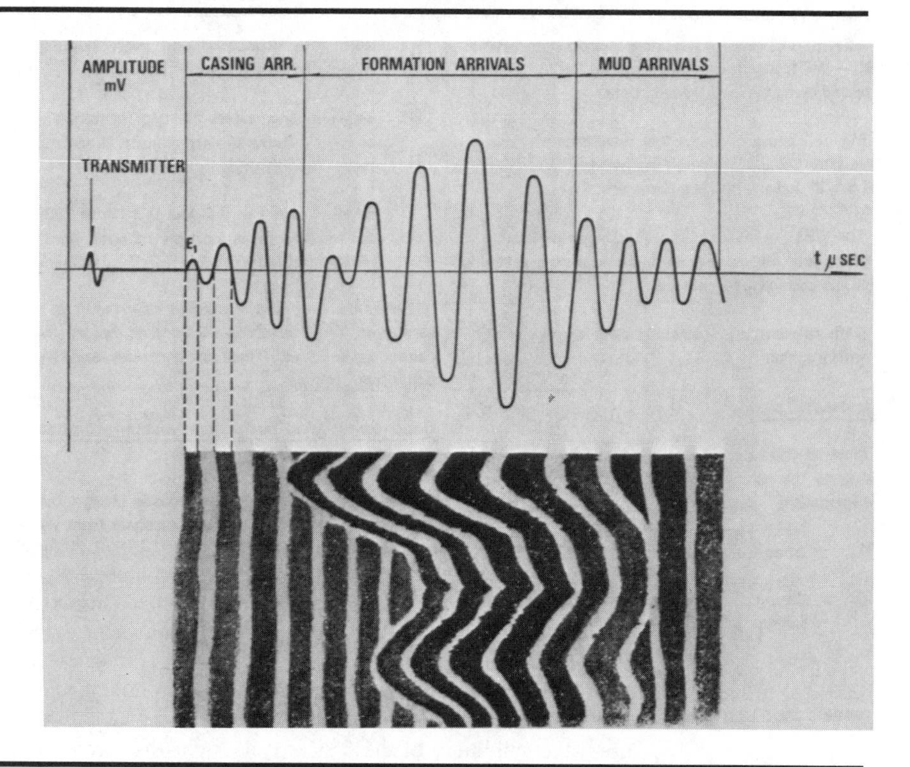

FIGURE 13.7 *CBL Wave Train. Courtesy Schlumberger Well Services.*

OPERATING PRINCIPLES

Three paths are available for a compressional wave emanating from the transmitter:

1. through the casing,
2. through the formation, and
3. through the borehole fluid.

Typical travel times for these three media are 56 μs/ft for casing, 60 to 100 μs/ft for formation, and 170 to 200 μs/ft for the borehole fluid. The shortest time path for the acoustic energy to travel is through the casing and the longest is through the borehole fluid. Thus, at the receiver, the signal recorded will have three major components. The initial transmitted pulsed will be spread out into a wave train. Figure 13.7 illustrates a typical wave train seen at the receiver.

Amplitude Measurement

Since the amplitude of the casing wave is required for the attenuation measurement it is sufficient to measure the amplitude of the first

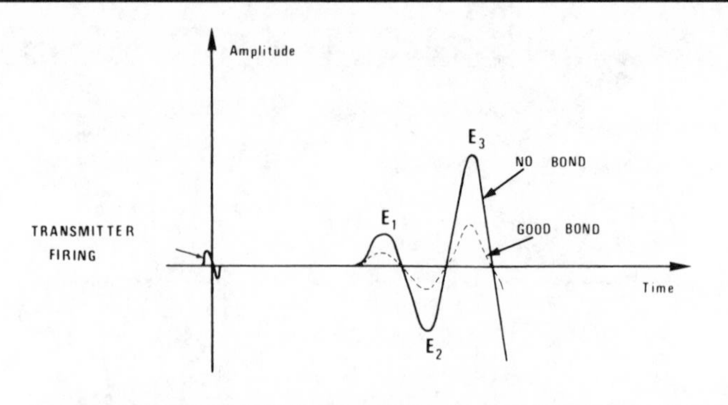

FIGURE 13.8 ***Schematic Receiver Output Signal with Bonded and Unbonded Casing. Courtesy Schlumberger Well Services.***

arrival since this will be the one that has traveled through the casing. For convenience, the various arrival peaks are named E_1, E_2, E_3, etc.—E_1 being the first arrival. The form of the transmitted wave and the first arrivals at the 3-ft receiver are shown in figure 13.8. Where there is no cement, the amplitude of E_1 is large and attenuation is small. Where cement is present, the amplitude of E_1 is small and attenuation is large. Due to the nature of the measuring system used, the amplitude of the first arrival, E_1, is displayed on the log in millivolts. A low value of millivolts means a high attenuation rate and good cement. A high value of millivolts means a low attenuation rate and poor cement. Typically, amplitudes are recorded on a scale of 0 to 50 mV across one track.

Travel-Time Measurement

At the same time the amplitude is measured, the one-way travel time from the transmitter to the receiver is also measured and displayed on the log. In casing of constant size, this travel time should be constant and a function solely of the tool and casing size. As we shall see later, this single-receiver Δt measurement is extremely valuable in diagnosing problems such as cycle skipping and tool eccentricity. Figure 13.9 shows the expected Δt value for $1^{11}/_{16}$- and $3^{5}/_{8}$-in. OD tools. This information is usually displayed in Track 1 of the log, with Δt increasing to the left.

Wave-Train Display

The entire wave train can be conveniently displayed in a number of ways, the most common of which is the *variable density display* (VDL), where bands of alternate dark and light shading reveal the peaks and valleys of the wave train. A common presentation is as shown in figure 13.10, where the VDL display is in Track 3 scaled from 200 to 1200 μs. As examples of alternative ways to display

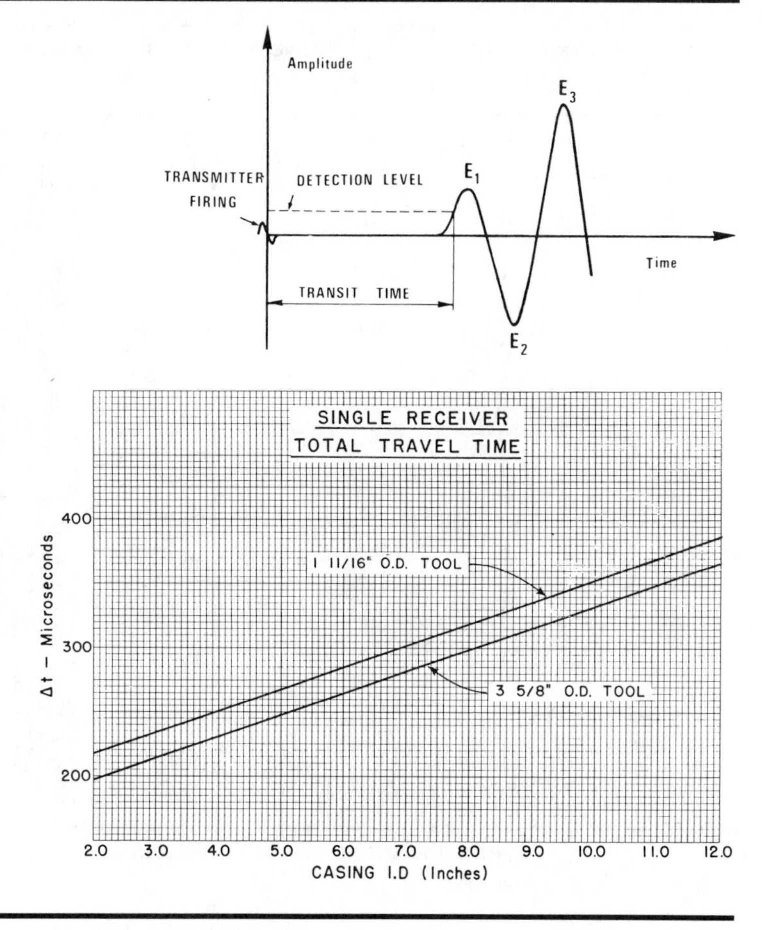

FIGURE 13.9 *Single Receiver (3-ft)* Δt *Values for CBL Tools. Courtesy Schlumberger Well Services.*

the same information, figure 13.11 shows a VDL and a full waveform display over the same section and figure 13.12 shows a half-wave display.

Δ_t *Stretching*

At the time the transmitter pulse is initiated, an electronic clock starts counting. A bias level is adjusted by the logging engineer to detect the first arrival at the 3-ft receiver, at which time the clock stops. Δt is the elapsed time recorded by the clock. The detection level must be set high enough so that the background noise does not trigger the system but not so high as to miss E_1 altogether and skip over to E_3, or even later arrivals. This method of measuring Δt exhibits two distinctive characteristics when the amplitude of E_1

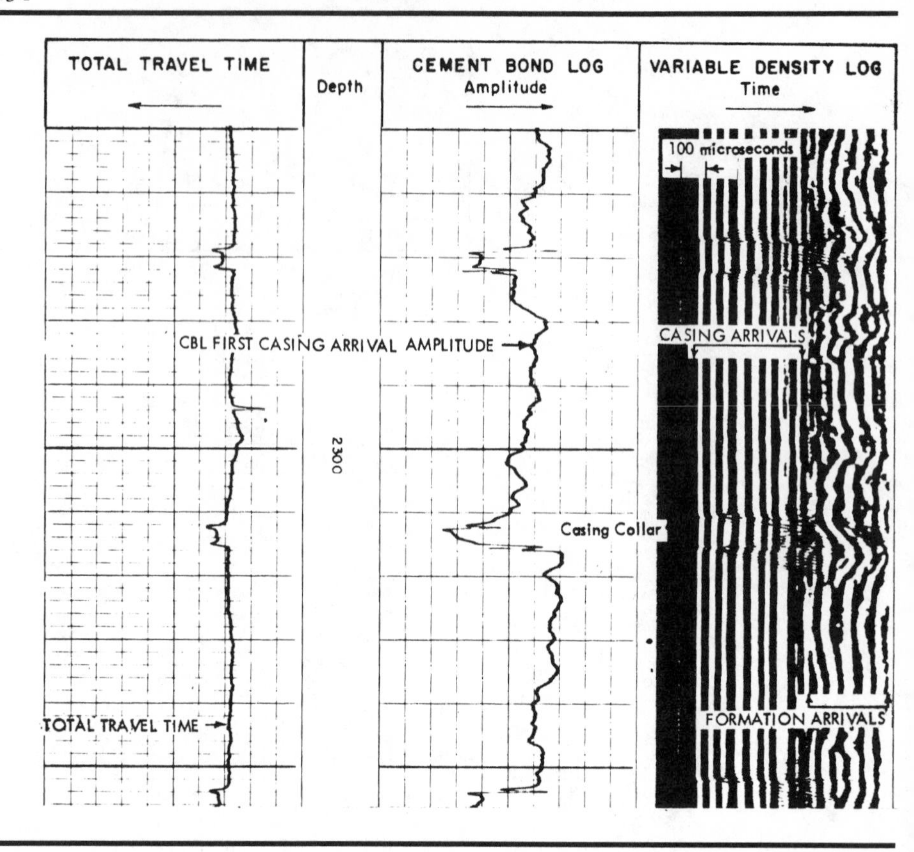

FIGURE 13.10 *CBL Presentation with VDL. Courtesy Schlumberger Well Services.*

is decreased due to good cement bonding. These are Δt stretching and cycle skipping.

Δt stretching is shown in figure 13.13. Note that the amplitude of E_1 is decreased and, since the bias level remains in the same position, the measured transit time, Δt, is somewhat lengthened. This will be evident on the log and is a normal and acceptable phenomenon.

Cycle Skipping

If a further drop in amplitude causes the peak of E_1 to fall below the preselected bias level, detection will take place on E_3 and Δt will increase significantly. This is known as cycle skipping (see fig. 13.14). Cycle skipping may be unavoidable in very good bonding and although it gives a hashy appearance to the Δt curve, it is not all bad; at least there is a good bond! Figure 13.15 gives an example of cycle skipping in a very well-bonded section.

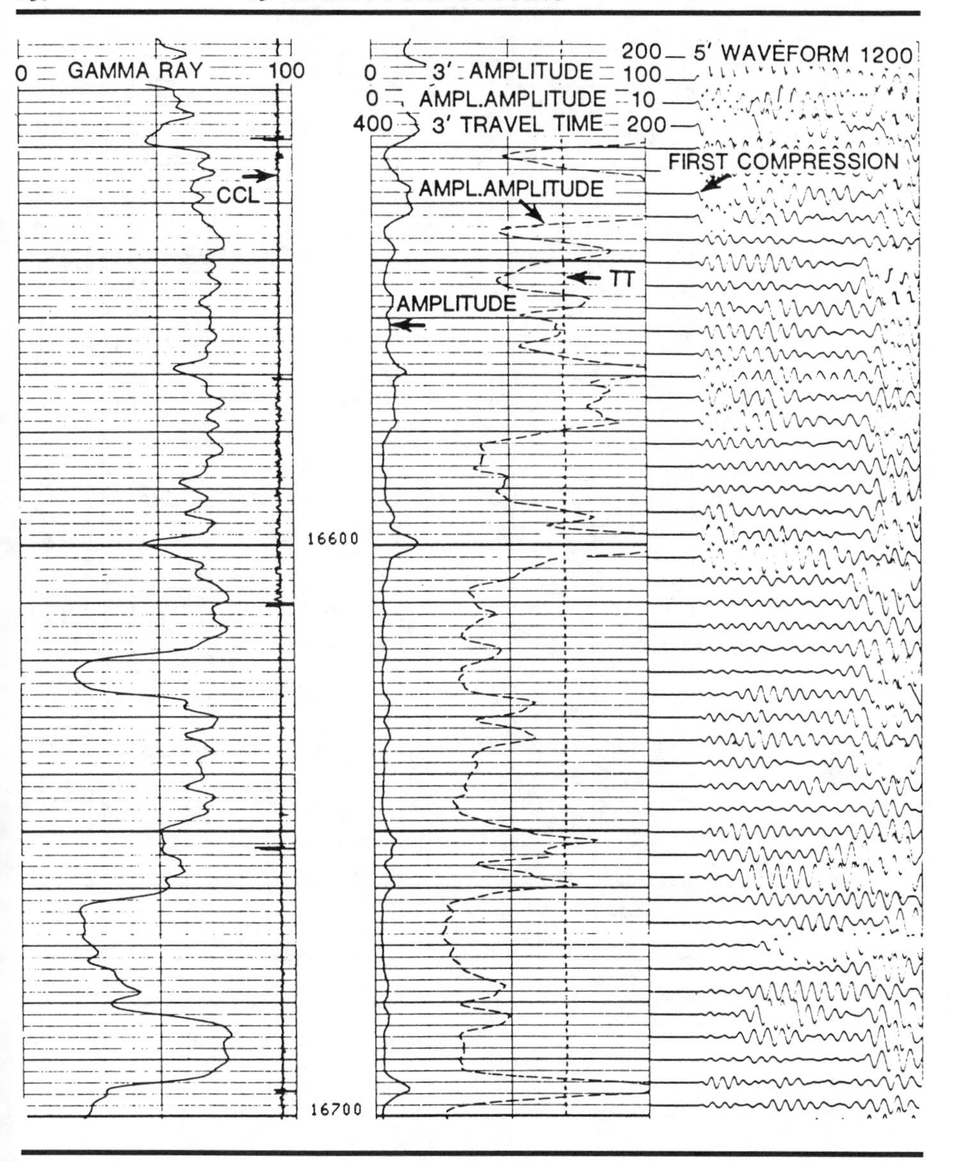

FIGURE 13.11 *CBL and Full-Wave Display.*

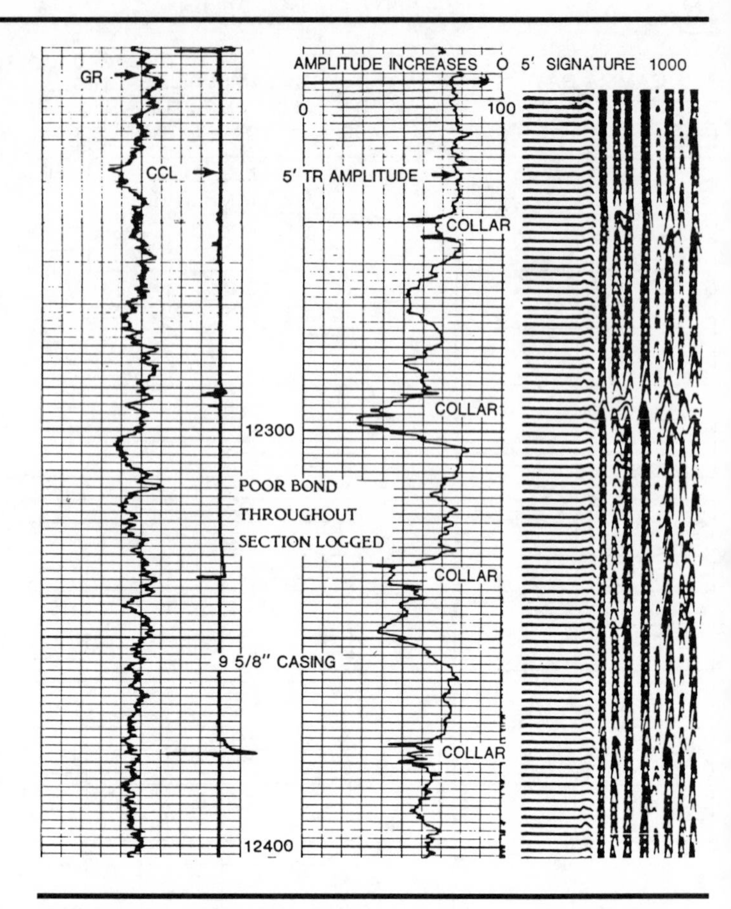

FIGURE 13.12 *Half-Wave Display.*

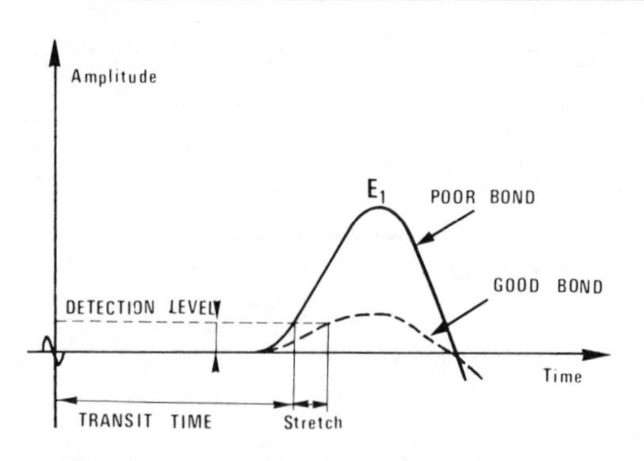

FIGURE 13.13 Δt *Stretching Due to Good Cement Bond. Courtesy Schlumberger Well Services.*

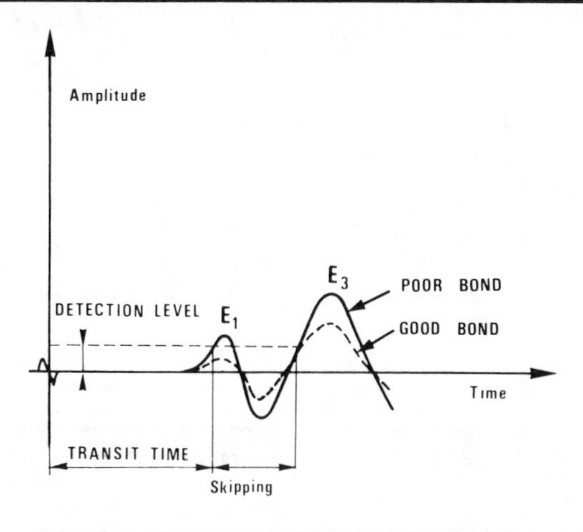

FIGURE 13.14 *Cycle Skipping. Courtesy Schlumberger Well Services.*

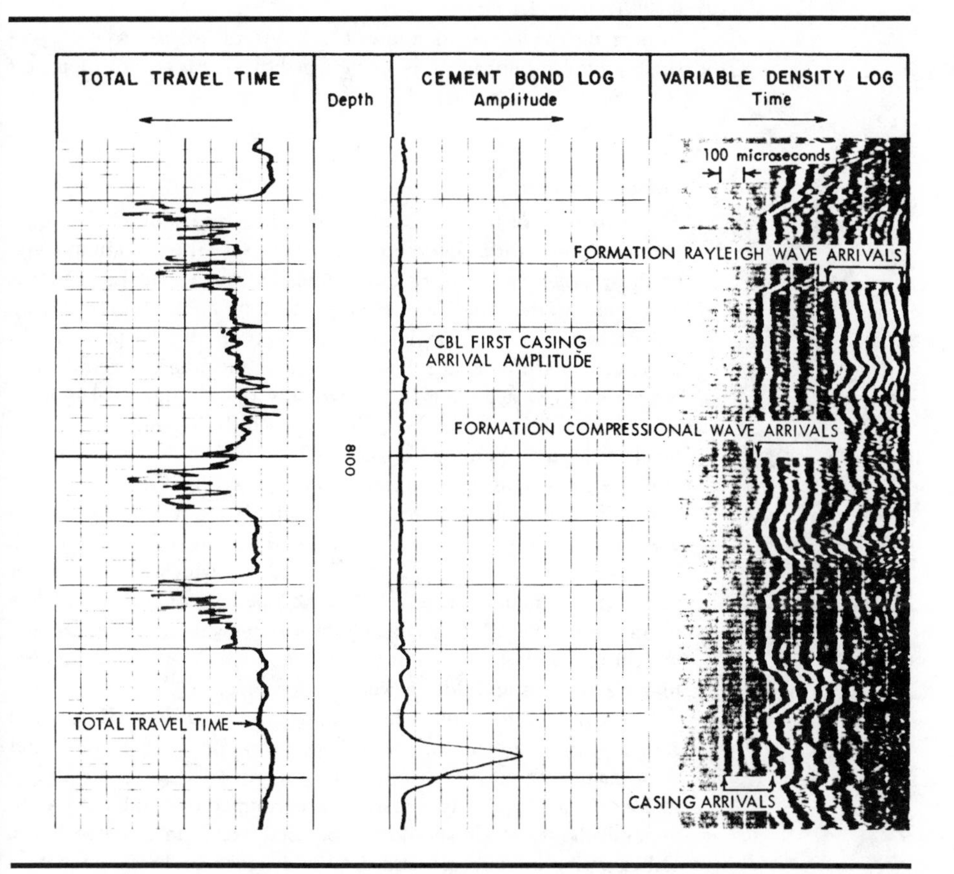

FIGURE 13.15 *Cycle Skipping in Well-Bonded Casing. Courtesy Schlumberger Well Services.*

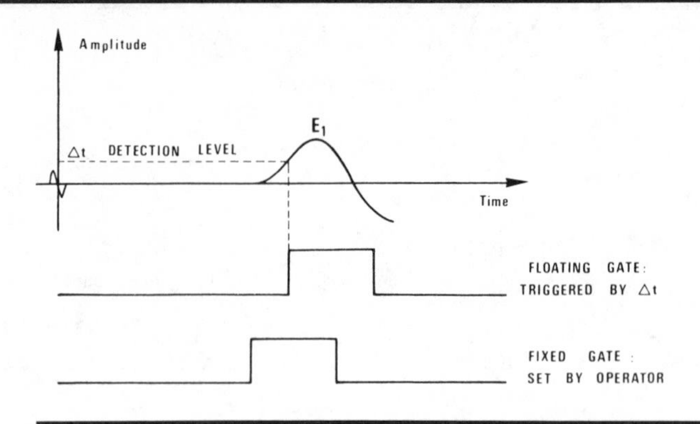

FIGURE 13.16 *Fixed and Floating Gates. Courtesy Schlumberger Well Services.*

QUESTION #13.1

Inspect the Δt curve in figure 13.15. At the places where cycle skipping is most apparent, estimate which peak was actually detected (i.e., E_3, E_5, . . . etc.)

Gating Systems

The amplitude of the E_1 peak is detected by means of an electronic gate that opens for a predetermined time, takes the maximum value of the received wave during the time it is open, and records that amplitude on the log. There are two common methods of achieving the correct gate position in order to key on the E_1 peak: (1) a fixed gate that opens at a fixed time after the transmitter fires (T_o mode) and (2) a floating gate that only opens after E_1 has been detected by the Δt-measuring circuitry, which depends on the bias level set by the operator (T_x mode).

If the transit-time measurement is triggered by E_1, both systems give the same result. If there is cycle skipping, the two gating systems give different results since the fixed gate will read E_1, which will be small, and the floating gate will read E_3, which may be larger than E_1. The floating gate relieves the operator of the task of fine tuning the position of the fixed gate for changes in casing size and weight or borehole fluid, but at the same time it causes erroneously high values of amplitude in well-bonded pipe.

It is essential to know which gating system is used. If it is not stated on the log heading, ask the operator. Figure 13.16 illustrates these gating systems; their effects on recorded CBL logs are shown in figure 13.17. The two systems are compared in table 13.4. In general, the T_x mode should not be used except in deviated holes. Other gating systems are also used and figure 13.18 summarizes a number of CBL systems from various service companies.

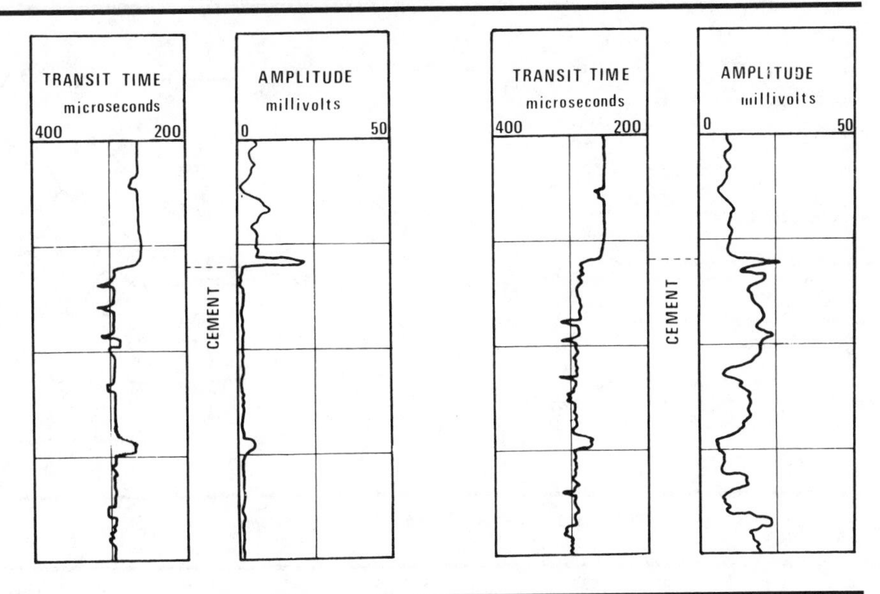

FIGURE 13.17 *Same Log Run* (a) *with Fixed Gate and* (b) *with Floating Gate. Courtesy Schlumberger Well Services.*

TABLE 13.4 *Comparison of Fixed and Floating Gates*

Situation	Effect on the Transit Time (Δt)	Effect on Amplitude Reading	
		Fixed Gate	Floating Gate
Free pipe or poor bond	Δt constant	High-reading E_1	High-reading E_1
Moderate to good bond	Δt constant	Low-reading E_1	Low-reading E_1
Very good bond	Risk of stretching	Very low-reading E_1	Very low-reading E_1
	Cycle skipping	Very low-reading E_1	High-reading E_3

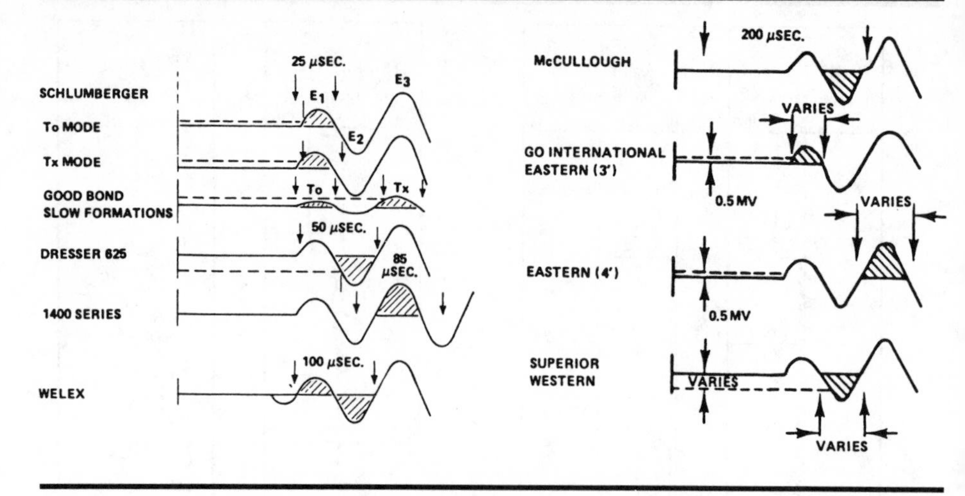

FIGURE 13.18 *Comparison of Gating Systems.*

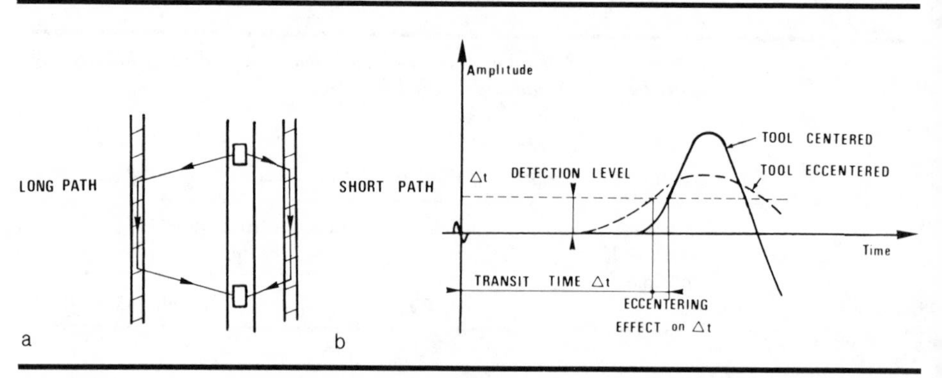

FIGURE 13.19 (a) *Eccentered Tool and* (b) *Waveforms Centered and Eccentered. Courtesy Schlumberger Well Services.*

Deviated Holes and Eccentered Tools

The centering of the tool is very critical for the response of the CBL. Once the tool is off center, essentially two paths become available for the transmitted compressional wave (see fig. 13.19a). As a result, the waveform at the receiver becomes "smeared" and this results in two effects on the log: (1) the measured Δt will be shortened and (2) the measured amplitude will be too low, giving a false indication of good bonding (see fig. 13.19b).

Figure 13.20 shows a section of log in a deviated hole where the tool was not properly centralized. Note that where Δt decreases so does the amplitude. The cure is to insist that the tool be centralized top, middle, and bottom with either rubber fins or spring centralizers.

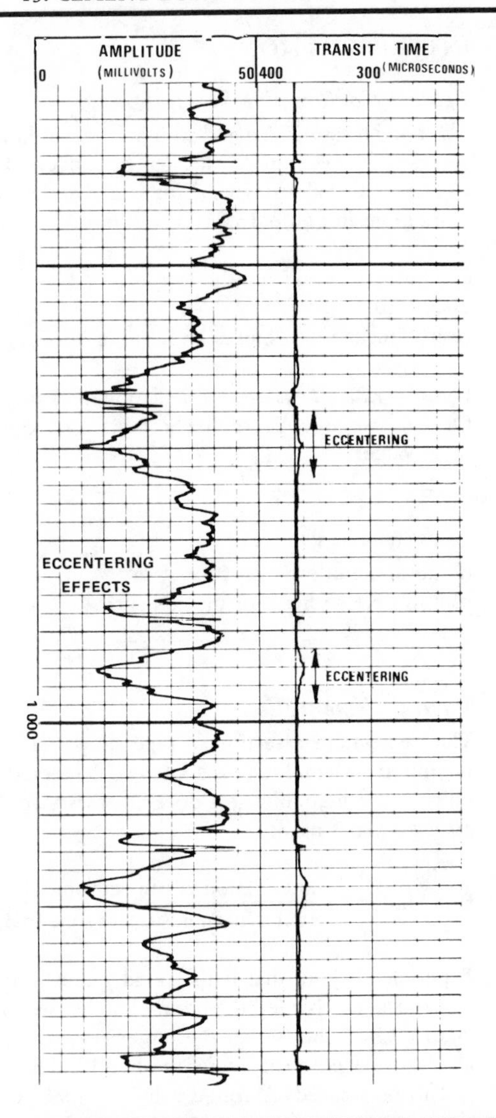

FIGURE 13.20 *CBL Run in Deviated Hole. Courtesy Schlumberger Well Services.*

INTERPRETATION

Cement Compressive Strength

For a given casing size and weight, the amplitude of the transmitted wave, measured in millivolts, may be converted to a cement compressive-strength value in psi. This is conveniently achieved using the nomogram in figure 13.21. The required inputs are:

1. amplitude in mV from the log
2. casing size
3. casing thickness (requires knowledge of casing weight)

Figure 13.22 allows quick calculation of casing thickness from casing OD and weight. Alternatively, table 13.5 can be used to find the same values.

QUESTION #13.2

If the amplitude reads 10 mV in 9⅝-in. 40-lb casing, what is the compressive strength of the cement?

Partial Cementation

The case where a section of pipe shows an amplitude higher than the minimum amplitude read in well-cemented pipe raises the possibility that a hydraulic seal does not exist. To quantify the analysis, a *bond index* is used:

$$\text{Bond index} = \frac{\text{attenuation in suspect zone (dB/ft)}}{\text{attentuation in suspect zone (dB/ft)}}.$$

Experience shows that if the bond index is at least 0.8 a seal can be expected, provided the poorly bonded section is sufficiently long. Figure 13.23 shows that the larger the casing size, the longer the interval needed at or above the bond index = 0.8 condition for a seal to be assured. A rough rule of thumb is to double the casing size in inches and subtract 5 to give the number of feet of section required.

Wave-Train Signatures

Figure 13.24 neatly summarizes the various patterns that may appear on a wave-train display such as the VDL. Typically, the VDL Track will be scaled 0 to 1000 or 200 to 1200 μs. The first 200 or 250 μs represent the time after the transmitter fires but before the first arrival at the 5-ft receiver. The next few hundred microseconds represent casing arrivals closely followed by formation arrivals, with mud arrivals coming last. The clue to reading VDL displays lies in observing the relative strength of the signal in these time intervals. For

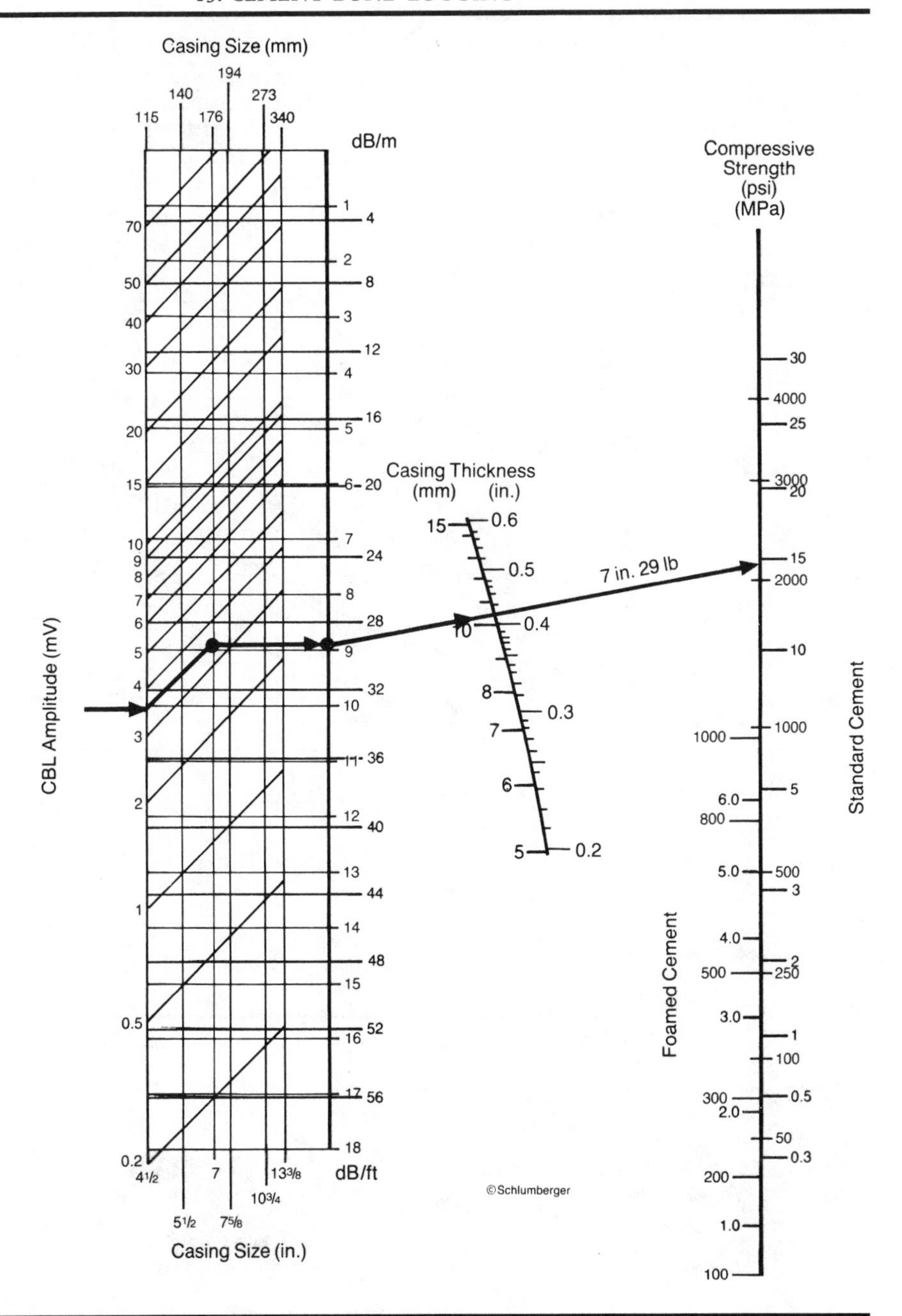

FIGURE 13.21 *CBL Amplitude Interpretation. Courtesy Schlumberger Well Services.*

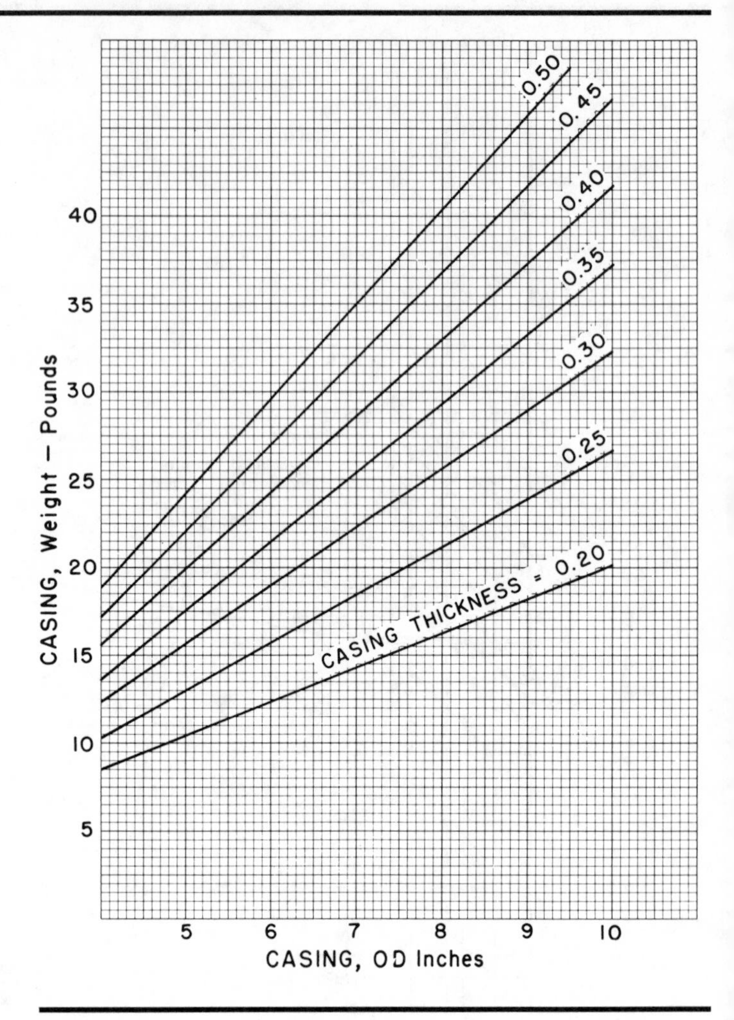

FIGURE 13.22 *Casing OD Weight and Thickness. Courtesy Schlumberger Well Services.*

TABLE 13.5 *Frequently Used Casing Dimensions*

OD (inches)	Weight* per foot	Nominal ID	Drift Diam.†
4	11.60	3.428	3.303
4½	9.50	4.090	3.965
	11.60	4.000	3.875
	13.50	3.920	3.795
4¾	16.00	4.082	3.957
5	11.50	4.560	4.435
	13.00	4.494	4.369
	15.00	4.408	4.283
	17.70	4.300	4.175
	18.00	4.276	4.151
	21.00	4.154	4.029
5½	13.00	5.044	4.919
	14.00	5.012	4.887
	15.00	4.974	4.849
	15.50	4.950	4.825
	17.00	4.892	4.767
	20.00	4.778	4.653
	23.00	4.670	4.545
5¾	14.00	5.290	5.165
	17.00	5.190	5.065
	19.50	5.090	4.965
	22.50	4.990	4.865
6	15.00	5.524	5.399
	16.00	5.500	5.375
	18.00	5.424	5.299
	20.00	5.352	5.227
	23.00	5.240	5.115
6⅝	17.00	6.135	6.010
	20.00	6.049	5.924
	22.00	5.989	5.864
	24.00	5.921	5.796
	26.00	5.855	5.730
	26.80	5.837	5.712
	28.00	5.791	5.666
	29.00	5.761	5.636
	32.00	5.675	5.550
7	17.00	6.538	6.413
	20.00	6.456	6.331
	22.00	6.398	6.273
	23.00	6.366	6.241
	24.00	6.336	6.211
	26.00	6.276	6.151
	28.00	6.214	6.089
	29.00	6.184	6.059
	30.00	6.154	6.029
	32.00	6.094	5.969
	35.00	6.004	5.879

TABLE 13.5 (*Continued*)

OD (inches)	Weight* per foot	Nominal ID	Drift Diam.†
	38.00	5.920	5.795
	40.00	5.836	5.711
7⅝	20.00	7.125	7.000
	24.00	7.025	6.900
	26.40	6.969	6.844
	29.70	6.875	6.750
	33.70	6.765	6.640
	39.00	6.625	6.500
8⅝	24.00	8.097	7.972
	28.00	8.017	7.892
	32.00	7.921	7.796
	36.00	7.825	7.700
	38.00	7.775	7.650
	40.00	7.725	7.600
	43.00	7.651	7.526
	44.00	7.625	7.500
	49.00	7.511	7.386
9	34.00	8.290	8.165
	38.00	8.196	8.071
	40.00	8.150	8.025
	45.00	8.032	7.907
	55.00	7.812	7.687
9⅝	29.30	9.063	8.907
	32.30	9.001	8.845
	36.00	8.921	8.765
	40.00	8.835	8.679
	43.50	8.755	8.599
	47.00	8.681	8.525
	53.50	8.535	8.379
10	33.00	9.384	9.228
10¾	32.75	10.192	10.036
	40.00	10.054	9.898
	40.50	10.050	9.894
	45.00	9.960	9.804
	45.50	9.950	9.794
	48.00	9.902	9.746
	51.00	9.850	9.694
	54.00	9.784	9.628
	55.50	9.760	9.604
11¾	38.00	11.150	10.994
	42.00	11.084	10.928
	47.00	11.000	10.844
	54.00	10.880	10.724
	60.00	10.772	10.616
12	40.00	11.384	11.228
13	40.00	12.438	12.282
13⅜	48.00	12.715	12.559

TABLE 13.5 (*Continued*)

OD (inches)	Weight* per foot	Nominal ID	Drift Diam.†
16	55.00	15.375	15.187
18⅝	78.00	17.855	17.667
20	90.00	19.190	19.002
21½	92.50	20.710	20.522
	103.00	20.610	20.422
	114.00	20.510	20.322
24½	100.50	23.750	23.562
	113.00	23.650	23.462

* Weight per foot (in pounds) is given for plain pipe (no threads or coupling).
† Drift diameter is the guaranteed minimum internal diameter of any part of the casing. Use drift diameter to determine the largest-diameter equipment that can be safely run inside the casing. Use internal diameter (ID) for calculations of volume capacity.

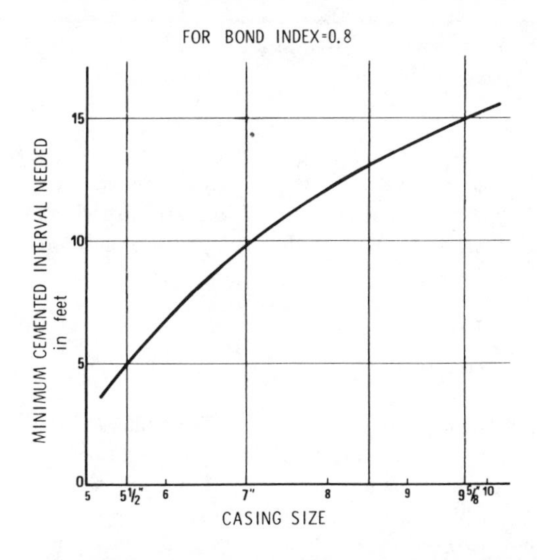

FIGURE 13.23 ***Minimum Requirements for Hydraulic Seal. Courtesy Schlumberger Well Services.***

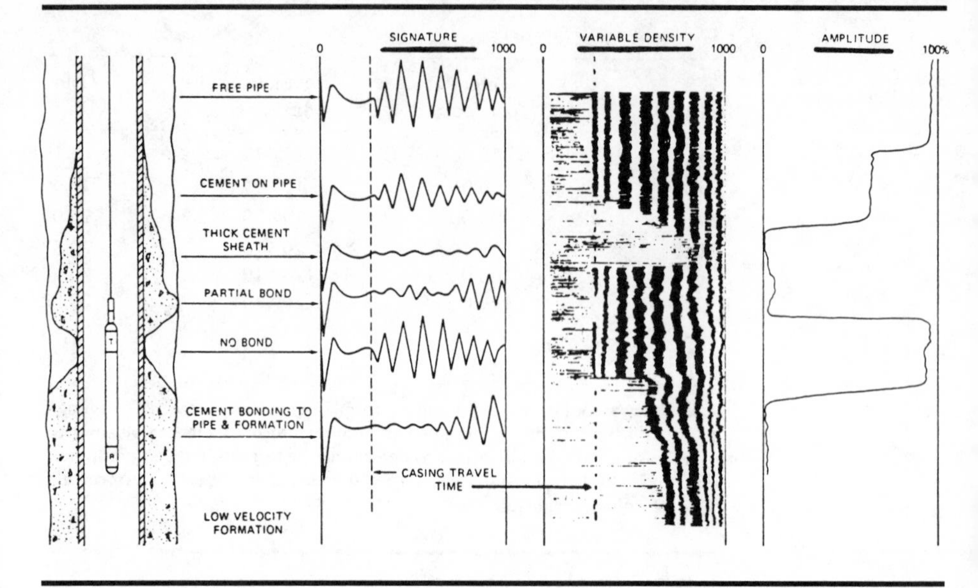

FIGURE 13.24 *CBL Schematic Showing VDL Displays. Courtesy Dresser Atlas.*

example, in free pipe, casing arrivals will be strong and formation arrivals weak. In well-bonded pipe, casing arrivals will be weak and formation arrivals strong, etc. By observing the amplitude, the transit time, and the VDL display, any condition of cementation can be diagnosed.

Free Pipe

Free pipe will exhibit:

1. steady travel time, with casing collars visible
2. high amplitude (~50 mV), with casing collars giving a 10- to 15-mV reduction over a 3-ft depth interval
3. a VDL display, with parallel black and white bands showing chevron patterns at casing collars over a 5-ft depth interval

No formation arrivals will be visible. Figure 13.25 illustrates free-pipe logs.

Free Pipe in Deviated Hole

Where the pipe rests on the low side of a deviated hole, acoustic coupling can take place between the pipe and the formation. In that case, the VDL may show formation arrivals. Do not be fooled. Diagnostics are:

1. transit time slightly reduced (tool eccentered)
2. amplitude high

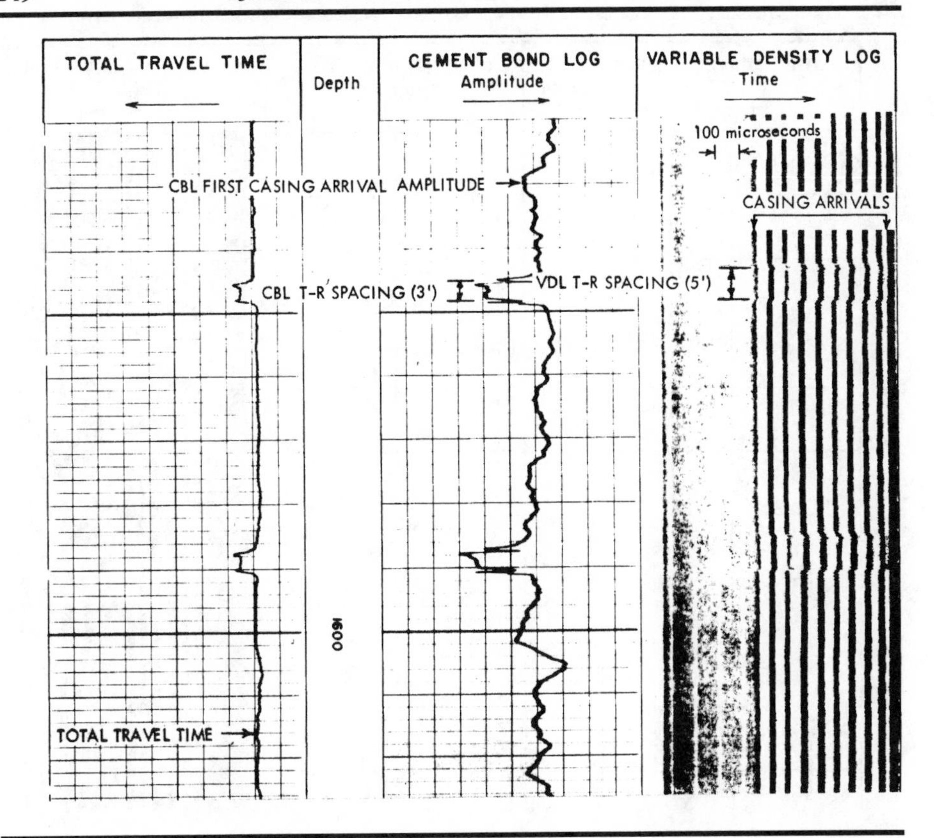

FIGURE 13.25 *Free-Pipe CBL/VDL. Courtesy Schlumberger Well Services.*

3. VDL shows weak casing arrivals and strong to moderate formation arrivals

Figure 13.26 illustrates this case.

Well-Cemented Pipe

In the ideal case of well-cemented pipe, the following diagnostics can be expected:

1. low amplitude (a few millivolts only)
2. transit time increasing due to stretch or cycle skips
3. strong formation arrivals

Figure 13.27 illustrates logs for a well-cemented section.

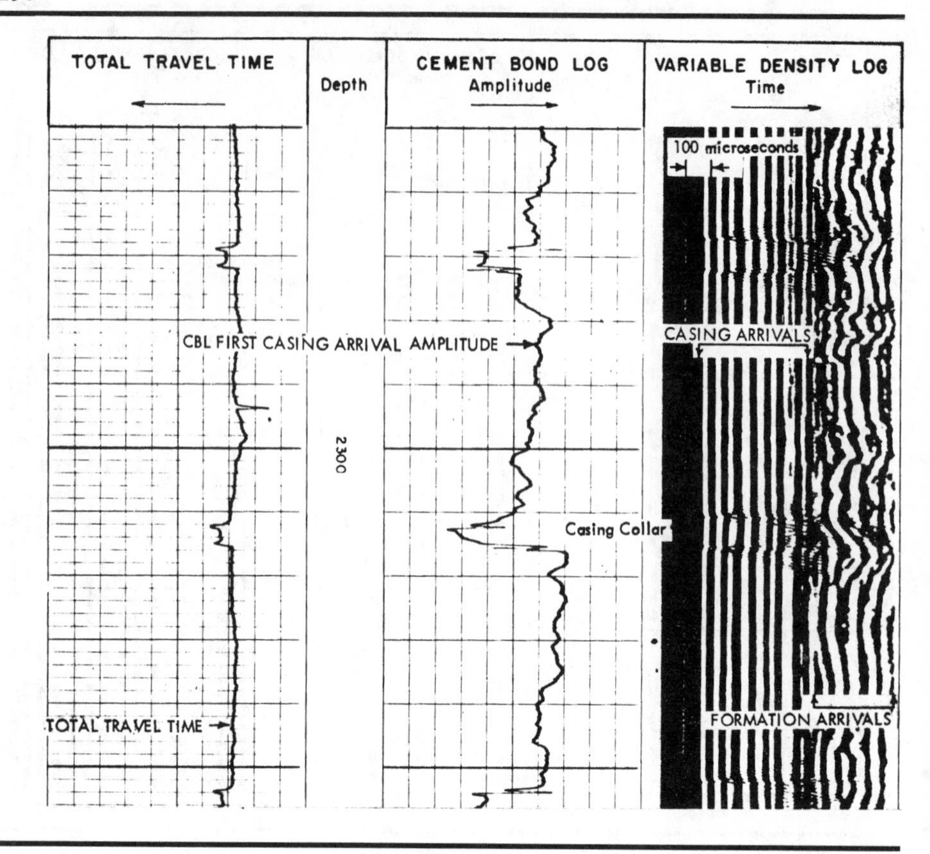

FIGURE 13.26 *CBL/VDL in Deviated Hole. Courtesy Schlumberger Well Services.*

Microannulus/Channeling

It is not uncommon to find cases where it is difficult to assess whether partial bonding is due to channels or due to what is known as a microannulus. Both cases may exhibit:

Medium amplitude,
Moderate casing arrivals, and
Moderate formation arrivals.

One way to determine the cause of partial bonding is to rerun the log with pressure on the casing. The microannulus is a microscopic gap between the cement and the casing that forms very poor acoustic coupling. Thus, although cement is present, the log suggests otherwise. The root cause of the microannulus may be due to one or other of the following:

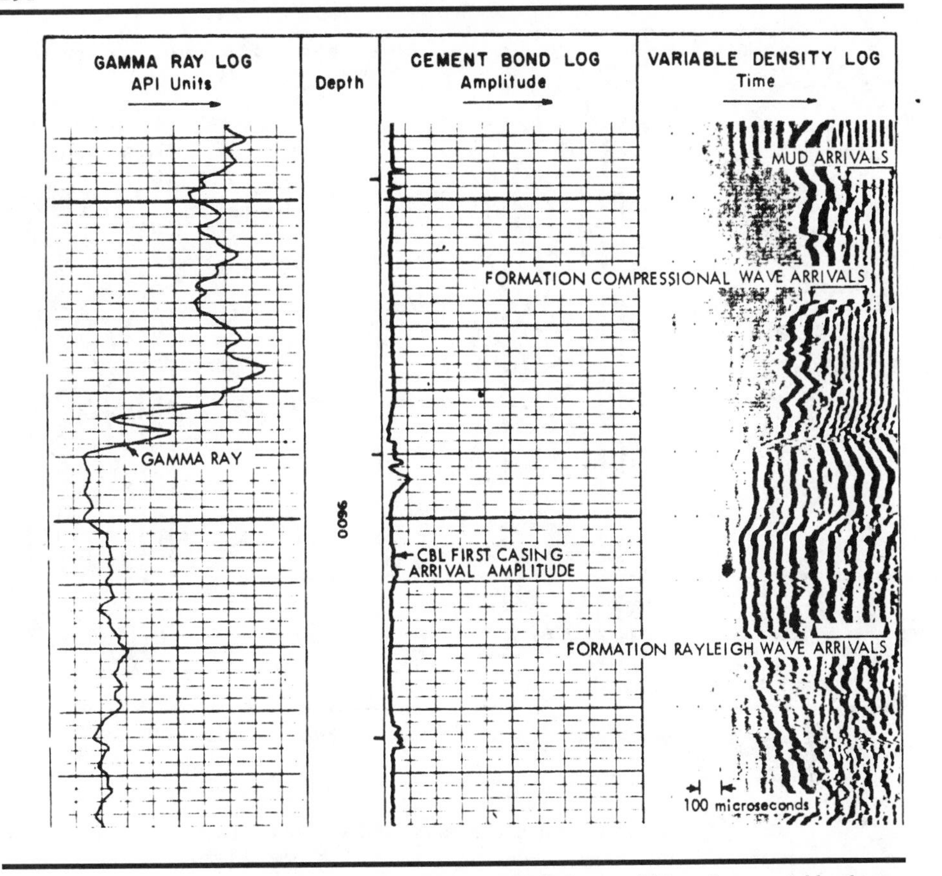

FIGURE 13.27 *CBL/VDL in a Well-Cemented Pipe. Courtesy Schlumberger Well Services.*

a. Pressure was held on the casing while the cement cured. Subsequent release of the casing pressure allowed the microannulus to form.
b. The casing expanded due to the heat generated by the curing process and subsequently cooled and shrank back to normal size.
c. Cold completion fluid was circulated just prior to running the CBL, causing thermal contraction of the casing.
d. A completion fluid of lesser density was used to circulate out a heavy mud used to displace cement. The resulting reduction in pressure in the casing allowed the microannulus to form.

If a microannulus is suspected in the zone of interest, rerun the log with 500 to 1000 psi at the casing head. The service company should come equipped with the appropriate pressure-control equipment to do the job safely. If the CBL amplitude is lower on the rerun under pressure, then there is no need to squeeze. No case

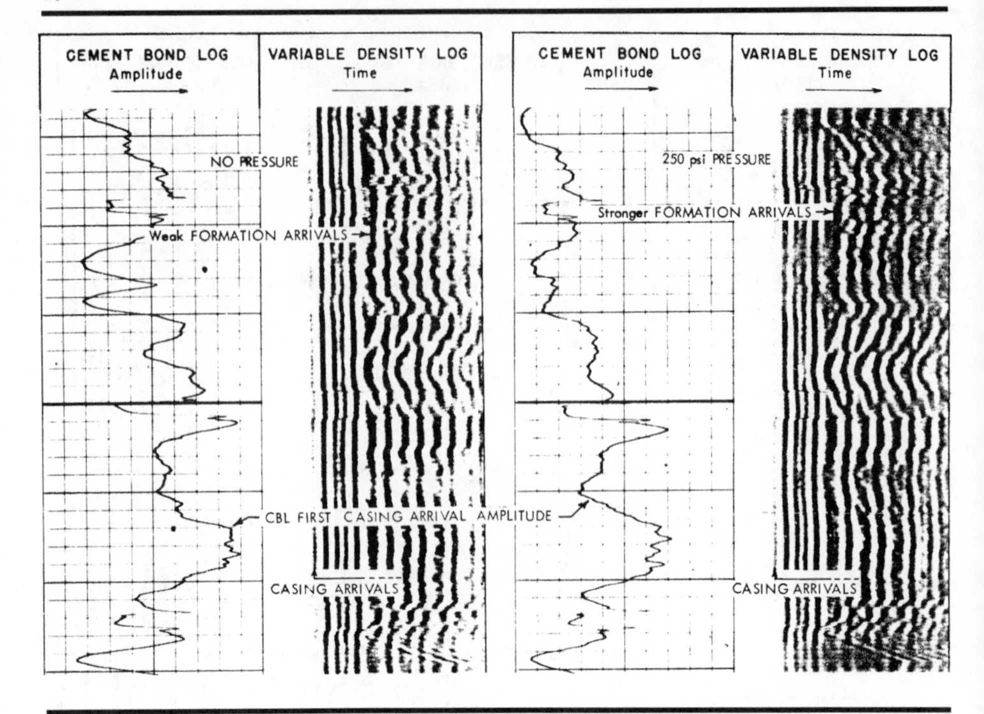

FIGURE 13.28 *Microannulus Case. Courtesy Schlumberger Well Services.*

has ever been documented where communication existed across a microannulus zone. If the CBL does not improve, channeling may be suspected and a squeeze should be attempted.

Figure 13.28 documents a microannulus case. Figure 13.29 is a useful reference for determining the pressure required to expand a given casing by a given amount. Normal microannuli are between 0.005 and 0.01 inches.

LOG QUALITY CONTROL

The two most important quality control aspects of CBL logging are:

1. Good centralization. Insist on at least three fin centralizers near the transmitter/receiver array and spring centralizers top and bottom of the entire tool string.
2. Calibration tails should conform to the published service-company specifications.

CEMENT EVALUATION TOOL (CET)

The CET is a relatively new tool that uses a rather different approach to evaluation of cement bonding problems. The tool comes in two

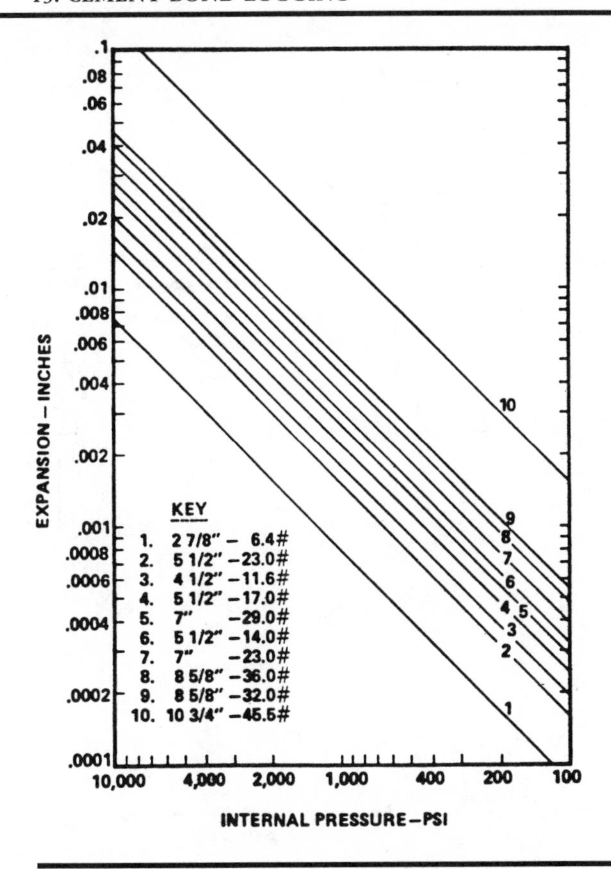

FIGURE 13.29 *Pressure vs. Expansion for Common Casings. Reprinted by permission of the SPE-AIME from Carter and Evans 1964. © 1964 SPE-AIME.*

versions a 4-in. OD sonde for 5½- to 9⅝-in. casings and a 3⅜-in. OD sonde for 4½- to 5½-in. casings. Both versions are rated for 350°F and 20,000 psi. Each carries eight ultrasonic transducers arrayed along the sonde, each pointing to a different 45° segment of the compass. Figure 13.30 illustrates the tool.

Each ultrasonic transducer emits a beam of ultrasonic energy in a 300- to 600-kHz band, which covers the resonant frequency range of most oilfield-casing thicknesses. This energy pulse causes the casing to ring or resonate in its thickness dimensions. The vibrations die out quickly or slowly depending on the material behind the casing. Most of the energy is reflected back to the transducer where it is measured; the remainder passes into the casing wall and echoes back and forth until it is totally attenuated. A ninth transducer continuously measures the acoustic travel time of the fluid column in the casing, so that the other eight transducer travel times can be converted to distance measurements.

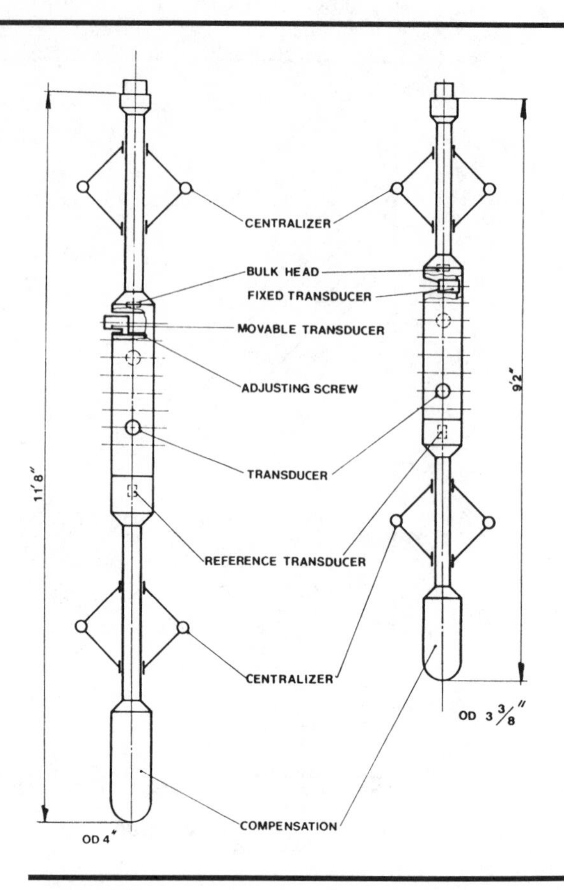

FIGURE 13.30 *CET Tool. Reprinted by permission of the SPE-AIME from Froelich et al. 1982, fig. 8, p. 1839. © 1981 SPE-AIME.*

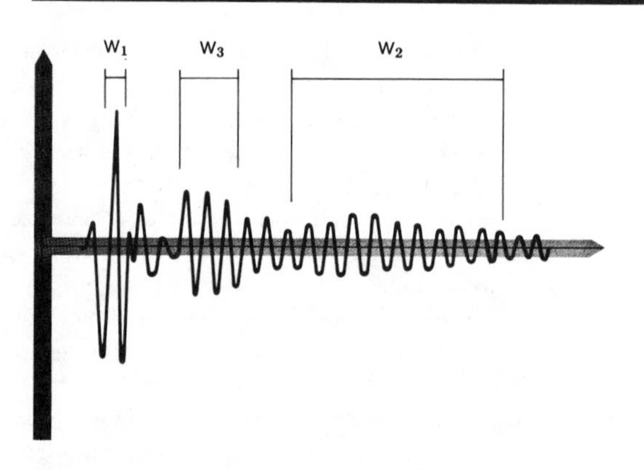

FIGURE 13.31 *CET Wave Train. Courtesy Schlumberger Well Services.*

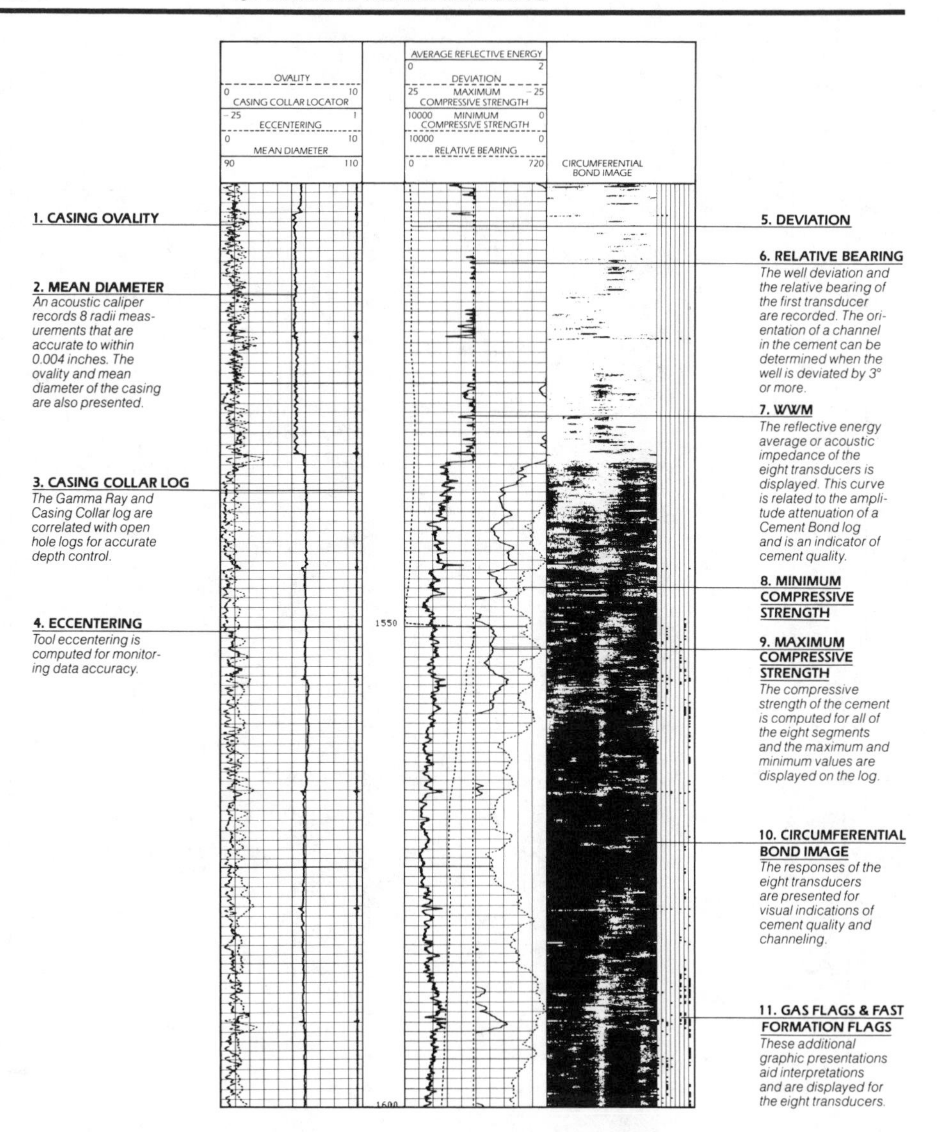

FIGURE 13.32 *CET Presentation. Courtesy Schlumberger Well Services.*

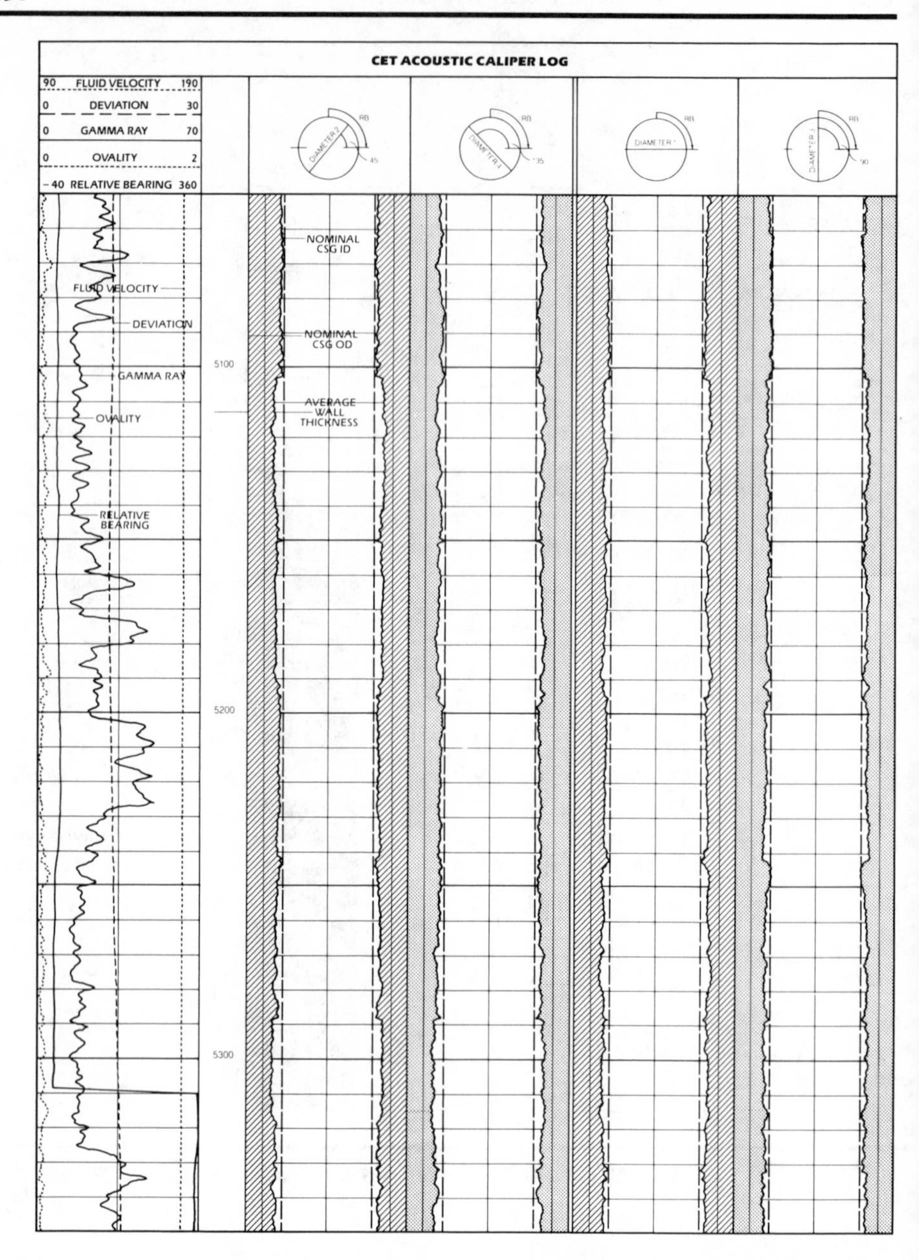

FIGURE 13.33 *Casing-ID Presentation. Courtesy Schlumberger Well Services.*

If the reflected wave train is displayed as a function of time, it appears as shown in figure 13.31. The windows labelled W_1, W_2, and W_3 are used to calculate the following:

1. the first arrival time, used for distance calculations (W_1)
2. compressive strength (W_1 and W_2)
3. detection of hard formations (W_3)

The presentations possible with this tool allow for a radial picture of the cement around the casing, which is helpful in detecting channels. However, the tool is practically useless for determining if the cement is bonded to the formation. A typical CET presentation for cement evaluation is shown in figure 13.32 and a casing-ID inspection presentation is shown in figure 13.33.

BIBLIOGRAPHY

Brown, H. D., Grijalva, V. E., and Raymer, L. L.: "New Developments in Sonic Wave Train Display and Analysis in Cased Holes," *Log Analyst* (January–February 1971) 27–40.

Chang, S. K. and Everhart, A. H.: "A Study of Sonic Logging in Cased Borehole," paper SPE 11034 presented at the 57th Annual Technical Conference and Exhibition, New Orleans, Sept. 26–29, 1982.

"The Essentials of Cement Evaluation," Schlumberger, March 1976.

Froelich, B., Pittman, D., and Seeman, B.: "Cement Evaluation Tool—A New Approach to Cement Evaluation," paper SPE 10207 presented at the 56th Annual Technical Conference and Exhibition, San Antonio, Oct. 5–7, 1981; *J. Pet. Tech.* (August 1982).

Grosmangin, M., Kokesh, F. P., and Majani, P.: "A Sonic Method for Analyzing the Quality of Cementation of Borehole Casings," *J. Pet. Tech.* (Feb. 1961) 165–171; *Trans.*, AIME.

McGhee, B. F., and Vacca, H. L.: "Guidelines for Improved Monitoring of Cementing Operations," SPWLA Trans. 21st Logging Symposium, July 1980.

Pardue, G. H., Morris, R. L., Gollwitzer, L. H., and Moran, J. H.: "Cement Bond Log—A Study of Cement and Casing Variables," *J. Pet. Tech.* (May 1963) 545–555; *Trans.*, AIME.

"Schlumberger's Cement Evaluation Tool," Schlumberger, SMP/5040.

Answers to Text Questions

QUESTION #13.1
E_9 or E_{11}

QUESTION #13.2
400 psi

CHAPTER FOURTEEN

CASING INSPECTION

Inspection of the mechanical state of the completion string is an important aspect of production logging. Many production (or injection) problems can be traced back to mechanical damage to, or corrosion of, the completion string. A number of inspection methods are available, including:

Multifingered caliper logs
Electrical-potential logs
Electromagnetic inspection devices
Borehole televiewers or borehole TV

Of these, the majority measure the extent to which corrosion has taken place. Only the electrical-potential log may indicate where corrosion is currently taking place. With the exception of the caliper logs, all the devices require that the tubing be pulled before running the survey, since (1) most are designed to inspect casing rather than tubing and (2) most are large-diameter tools.

CALIPER LOGS

Various arrangements of caliper mechanisms are available to gauge the internal shape of a casing or tubing string. Figure 14.1 illustrates three such tools. Table 14.1 lists the various sizes available (Dia-Log), their respective number of feelers, and the appropriate casing size.

Tubing Profiles

Tubing-profile calipers will determine the extent of wear and corrosion, and will detect holes in the tubing string—all in a single run into the well. The large number of feelers on each size of caliper insures detection of even very small irregularities in the tubing wall.

In pumping wells, the tubing caliper log may be run by one man and there is no need for a pulling unit crew to be present. A "pull sheet" showing the maximum percentage of wall loss of every joint of tubing in the well may be prepared. Before the well is pulled, a program for rearranging the tubing string can be provided. Moving partially worn joints nearer the surface and discarding thin-wall joints substantially prolongs the effective life of tubing strings and reduces pulling costs in pumping wells. In flowing or gas-lift wells, the tubing-profile caliper provides an economical method to periodically check for corrosion damage, monitor the effectiveness of a corrosion inhib-

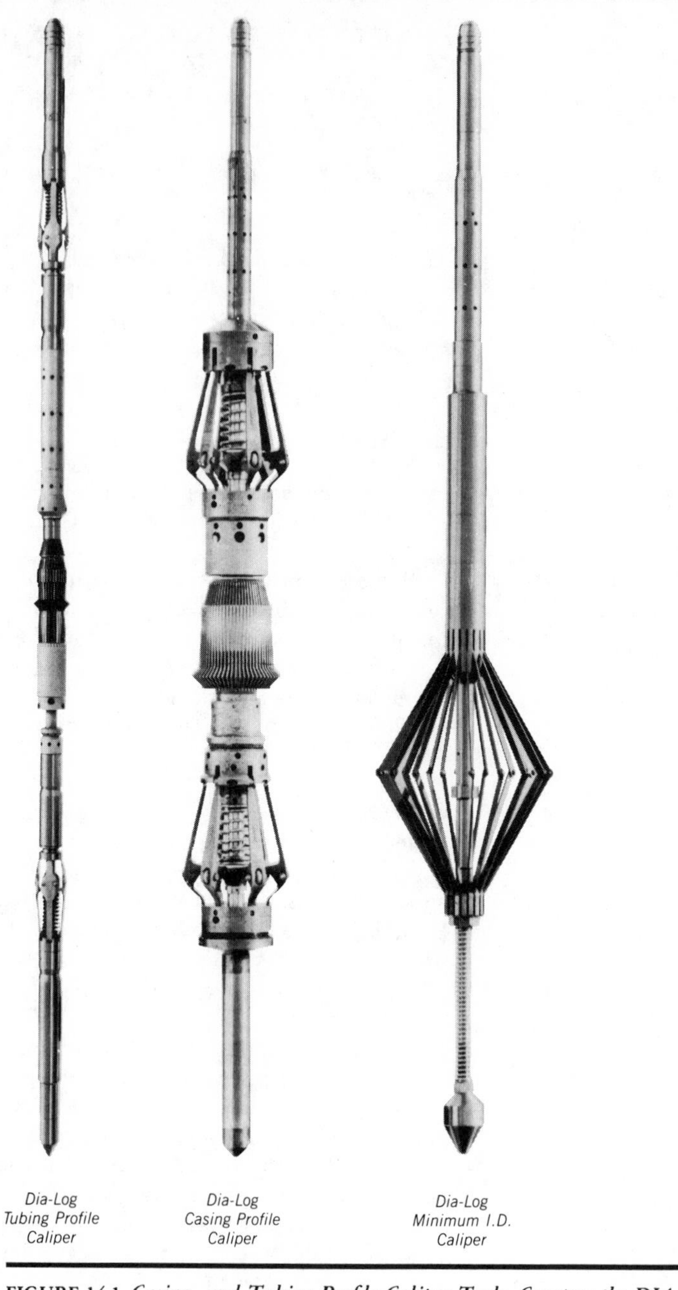

FIGURE 14.1 ***Casing- and Tubing-Profile Caliper Tools. Courtesy the DIA-LOG Company.***

TABLE 14.1 *Caliper Sizes*

Tool Diameter (inches)	Number of Feelers	Casing or Tubing OD (inches)
	CASING PROFILE CALIPERS	
3⅝	40	4½–6
5⅜	64	6⅝–7⅝
7¼	64	8⅝–9
7¾	64	9⅝
8¼	64	10¾
9⁹⁄₁₆	64	11¾
11⁵⁄₁₆	64	13⅜
13⅝	64	16
17⅝	64	20
	TUBING PROFILE CALIPERS	
1½	20	2
1½	20	2¹⁄₁₆
1¾	26	2⅜
2³⁄₁₆	32	2⅞
2¹¹⁄₁₆	44	3½
3¹⁄₃₂	44	4

itor program, or detect damaged tubing joints when "working over" a well.

An accessory tool, which may be run in combination with the tubing-profile caliper, is the Split Detector. This tool, functioning much like a magnetic collar locator, is designed to detect and log vertical splits or hairline cracks in the tubing that might be difficult to locate with the profile caliper. In practice, the split detector is used to log down the tubing, and the profile caliper to log up the tubing. This gives a complete inspection for wall thickness and splits in one run of the cable in the well.

Examples of a tubing profile and a split-detector log are shown in figure 14.2.

Casing Profiles

Casing-profile calipers are available to log 4½- through 20-in. OD casing. The tool is especially valuable where drilling operations have been carried on for an extended period of time through a string of casing. The determination of casing wear is of great importance when deciding if a liner can be safely hung, or if a full production string is required. In producing wells, the casing-profile caliper will locate holes or areas of corrosion that may require remedial work. The tool is also valuable when abandoning wells because it permits

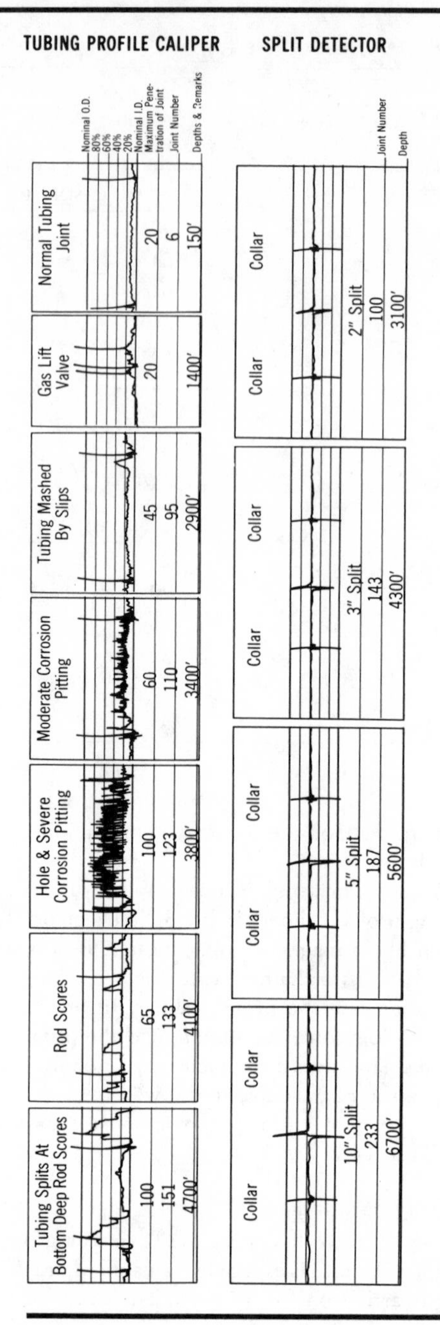

FIGURE 14.2 *Tubing-Profile and Split-Detector Logs. Courtesy the DIA-LOG Company.*

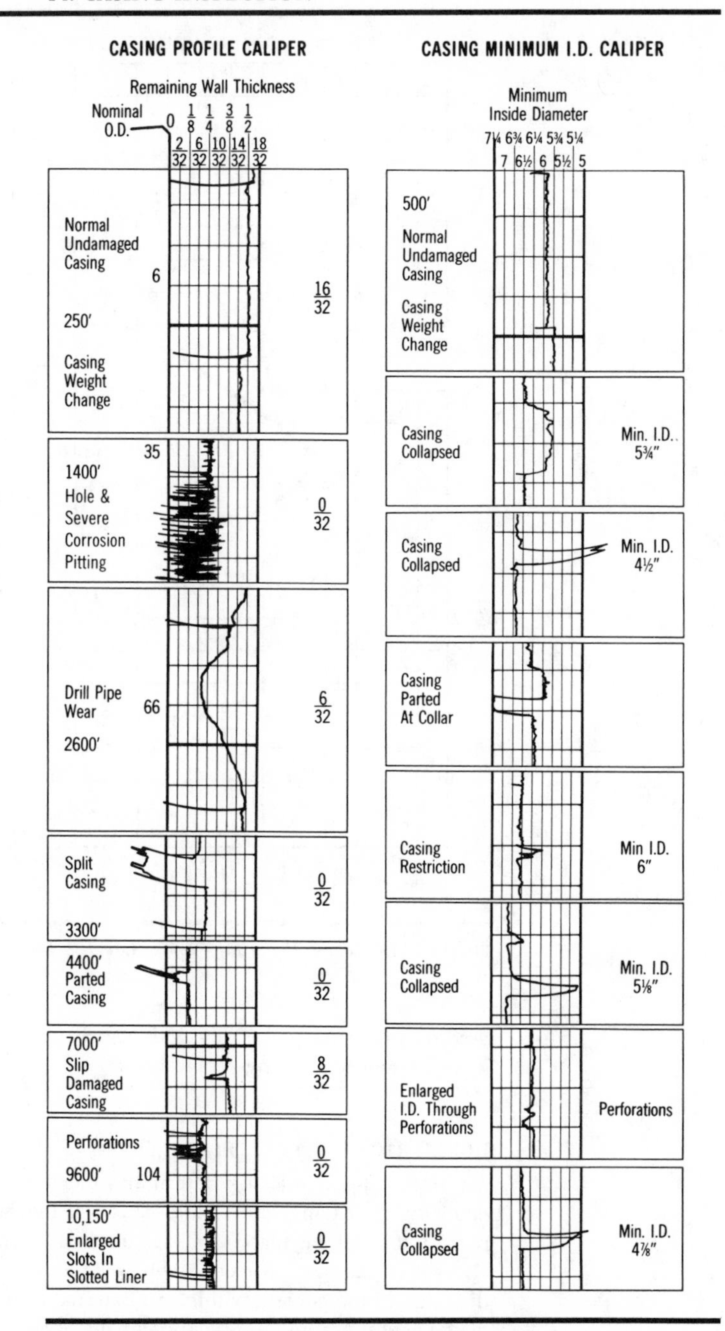

FIGURE 14.3 *Casing-Profile Logs. Courtesy the DIA-LOG Company.*

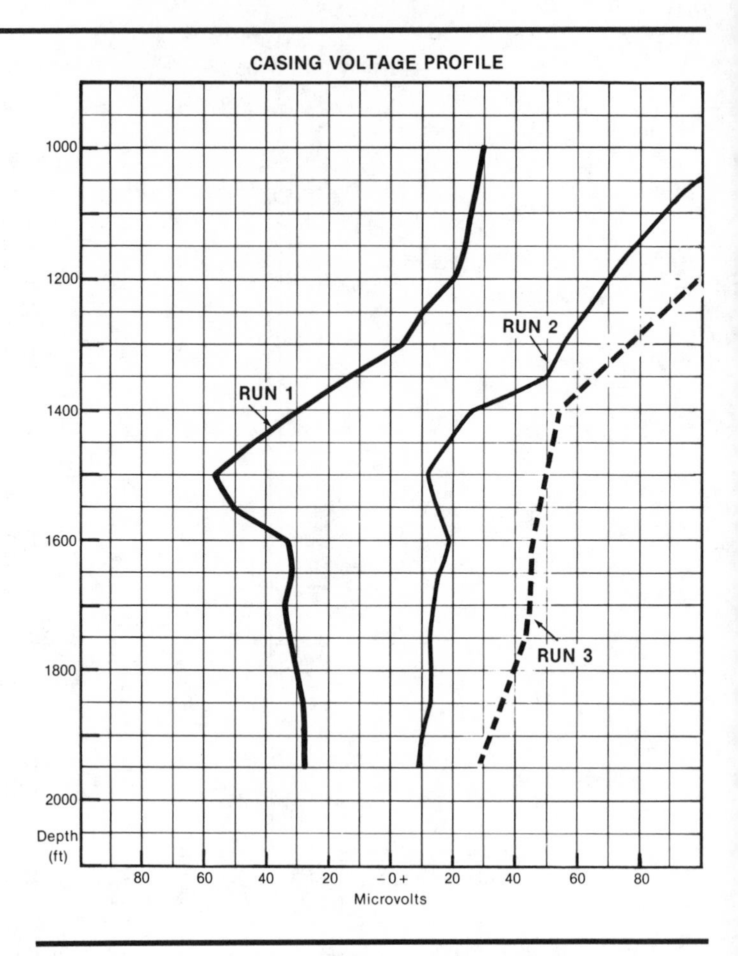

FIGURE 14.4 *Casing-Potential Profile. Courtesy Dresser Atlas.*

grading of casing to be salvaged before it is pulled. Figure 14.3 gives examples of casing-profile logs.

ELECTRICAL-POTENTIAL LOGS

An electrical-potential log determines the flow of galvanic current entering, or leaving, the casing. This will indicate not only where corrosion is taking place, and the amount of iron being lost, but also where cathodic protection will be effective. The magnitude and direction of the current within and external to the casing is derived mathematically from electrical-potential measurements made at fixed intervals throughout the casing string. In order to achieve reliable results from this kind of survey, the borehole fluid must be an electrical insulator; that is, the hole must be either empty or filled with oil or gas. Mud, or other aqueous solutions, will provide a "short"

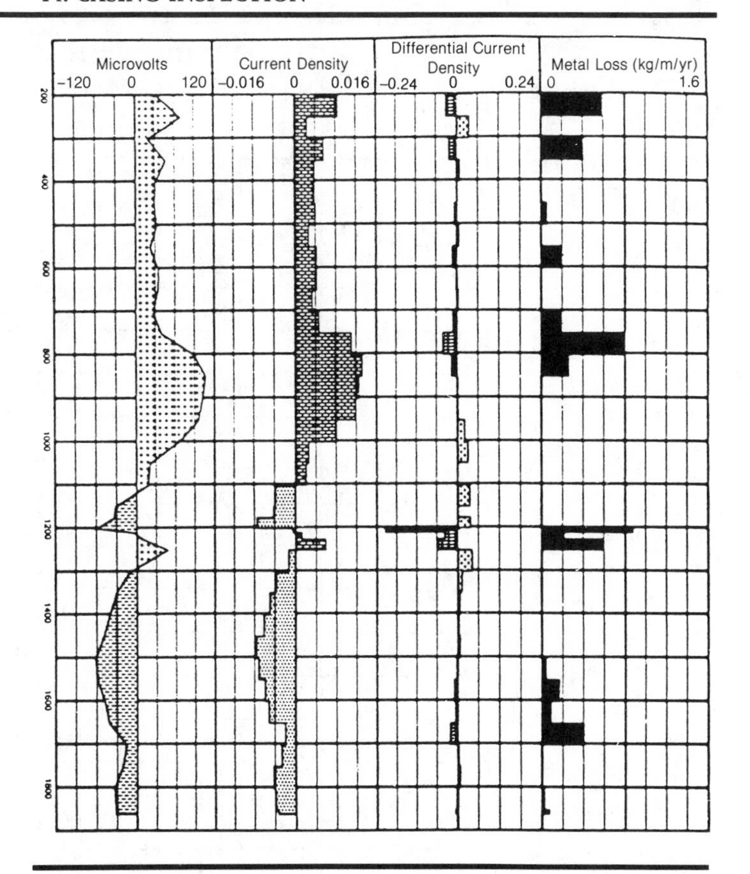

FIGURE 14.5 *Casing-Potential Profile Log Run Before Cathodic Protection. Courtesy Dresser Atlas.*

that invalidates the measurements. The log itself is a recording vs. depth of the small, galvanic voltages detected. Figure 14.4 illustrates such a log.

Figure 14.5 shows an interpretation of casing-potential profile logs run before cathodic protection was installed. Figure 14.6 shows the same after cathodic protection. Note that the metal loss has been reduced to practically zero (fig. 14.6) by the application of appropriate cathodic protection.

ELECTROMAGNETIC DEVICES

The most common inspection tools used to assess casing corrosion are of the electromagnetic type. They come in two versions, those that attempt to measure the remaining metal thickness in a casing string and those that try to detect defects in the inner or outer wall

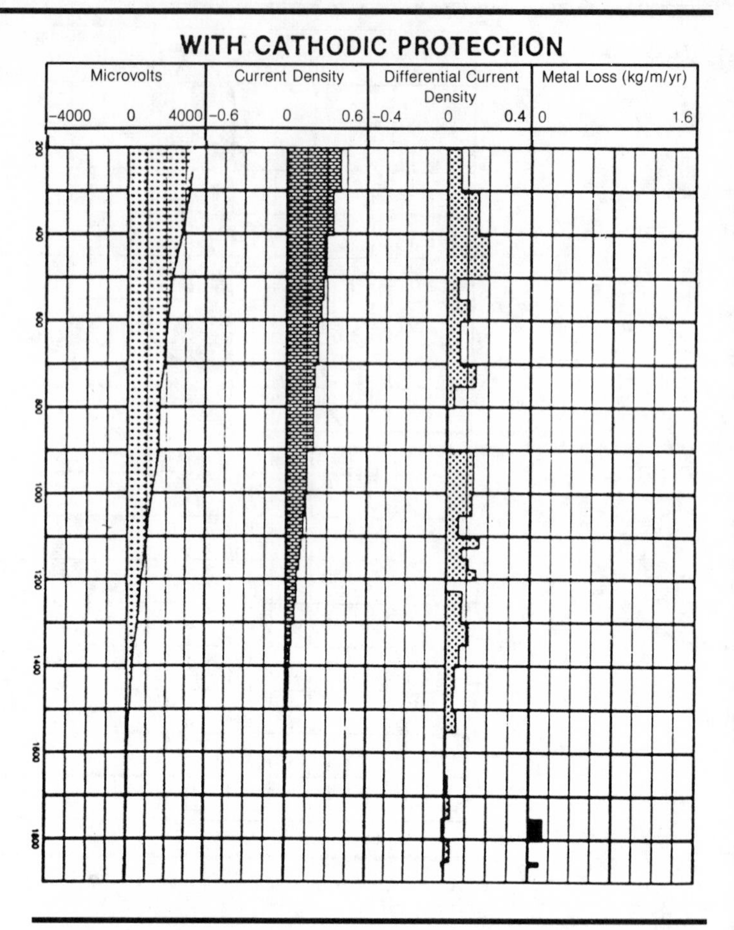

FIGURE 14.6 ***Casing-Potential Profile Log Run with Cathode Protection. Courtesy Dresser Atlas.***

of the casing. Although frequently run together, these tools will be discussed separately.

Electromagnetic Thickness Tool (ETT)

The electromagnetic thickness tools are available under a variety of trade names, such as:

ETT	(Schlumberger)
Magnelog	(Dresser)
Electronic Casing Caliper Log	(McCullough)

They operate in a manner similar to openhole induction tools. Each consists of a transmitter coil and a receiver coil. An alternating current is sent through the transmitter coil. This sets up an alternating magnetic field that interacts with both the casing and the receiver coil (see fig. 14.7). The coils are spaced about three casing-diameters

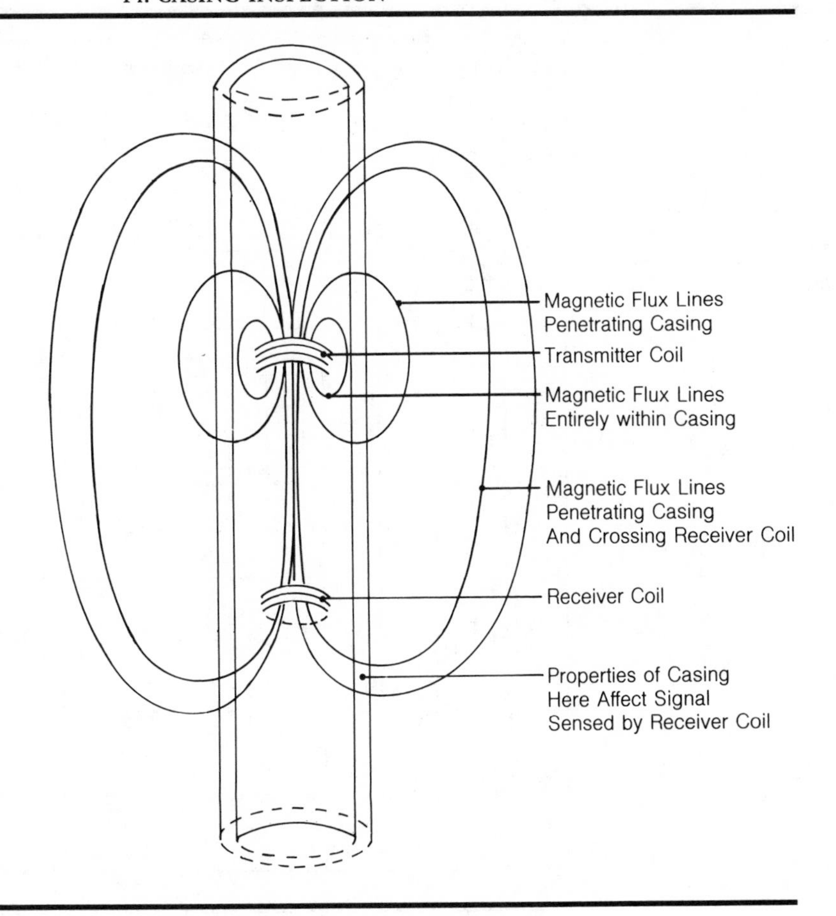

FIGURE 14.7 *Electromagnetic Thickness Tool.*

apart to ensure that the flux lines sensed by the receiver coil are those that have passed through the casing. The signal induced in the receiver coil will be out of phase with the transmitted signal. In general, the phase difference is controlled by the thickness of the casing wall. The raw log measurement is one of phase lag in degrees and the log is scaled in degrees.

Figure 14.8 illustrates an ETT log in severely corroded casing. Note that an increasing thickness corresponds to an increase in the phase-shift angle and vice versa. Some presentations of this log show a rescaling in terms of actual pipe thickness. This rescaling requires that the operator make some calibration readings in a casing of the type present in the well. It is quite common to see quite large differences in thickness between adjacent stands due to a number of variables such as the drift diameter of the pipe, the weight/foot, the relative magnetic permeability of the steel used, etc.

The ETT-type tool is good at finding vertical splits in pipe since

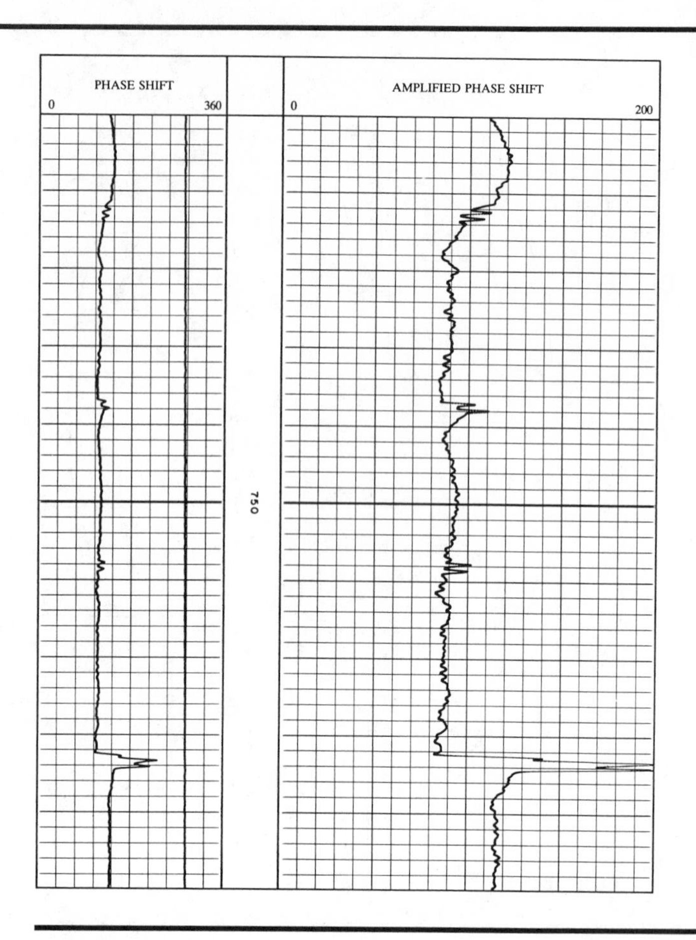

FIGURE 14.8 *Electromagnetic Thickness Log. Courtesy Schlumberger Well Services.*

the magnetic flux lines pass perpendicular to the casing wall. A horizontal circumferential anomaly is less well defined.

Pipe Analysis Log (PAL)

Another closely related measurement uses a slightly different technique and forms the basis of the Pipe Analysis Log (PAL), also known as the Vertilog. Two electromagnetic measurements are of interest in the context of the pipe analysis tool; magnetic-flux leakage and eddy-current distortion.

FLUX LEAKAGE. If the poles of a magnet are positioned near a sheet of steel, magnetic flux will flow through the sheet (figure 14.9). So long as the metal has no flaws, the flux lines will be parallel to the surface. However, at the location of a cavity, either on the surface of the sheet or inside it, the uniform flux pattern will be distorted.

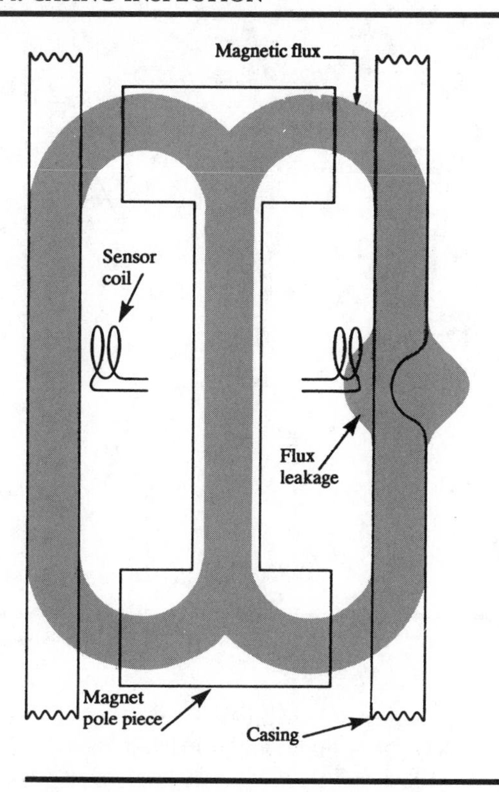

FIGURE 14.9 *Magnetic Flux-Leakage Principle. Courtesy Schlumberger Well Services.*

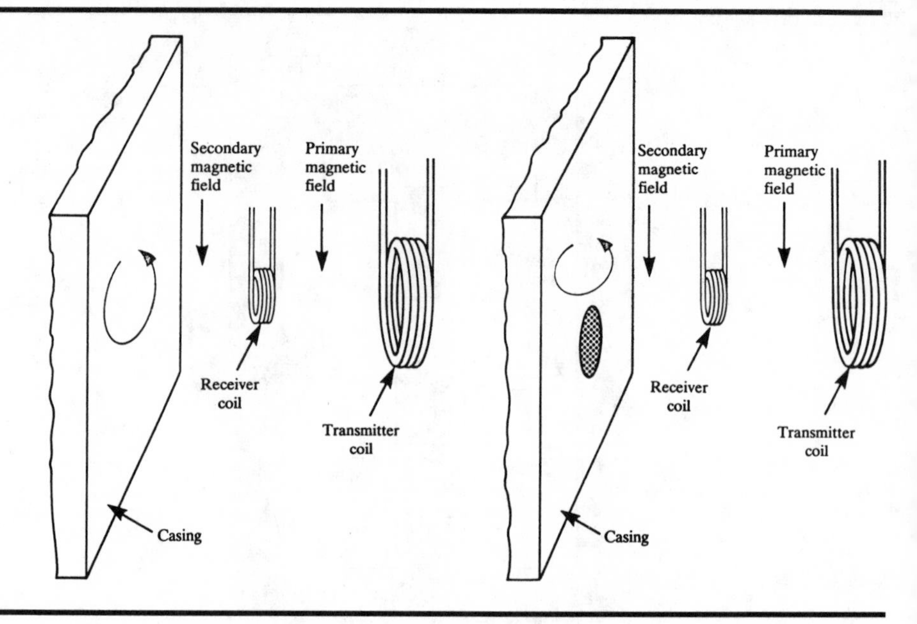

FIGURE 14.10 *Eddy-Current Principle. Courtesy Schlumberger Well Services.*

The flux lines will move away from the surface of the steel at the location of the anomaly, an effect known as flux leakage. The amount of flux distortion will depend upon the size of the defect. If a coil is moved at a constant speed along the direction of magnetic flux parallel to the metal sheet, a voltage will be induced in the coil as it passes through the area of flux leakage. The larger the anomaly, the greater the flux leakage, and therefore the greater the voltage induced. The magnetic flux is distorted on both faces of the sheet, regardless of the location of the defect, and therefore the coil only needs to be moved along one surface to survey the sheet completely. As the coil must be moved through a changing magnetic flux to produce a voltage, no signal is generated when it is moved parallel to the surface of an undamaged sheet of steel.

EDDY CURRENTS. When an alternating current of relatively high frequency is applied to a coil close to a sheet of steel, the resulting magnetic field induces eddy currents in the steel (figure 14.10). These eddy currents in the turn produce a magnetic field that tends to cancel the original field; and the total magnetic field is the vector sum of the two fields. A measure voltage would be induced in a sensor coil situated in the magnetic field. The generation of eddy currents is, at relatively high frequencies, a near-surface effect; so, if the surface of the steel adjacent to the coil is damaged, the magnitude of the eddy currents will be reduced; and, consequently, the total magnetic field will be increased. This will result in a variation in the voltage in the sensor coil. A flaw in the sheet of metal, on

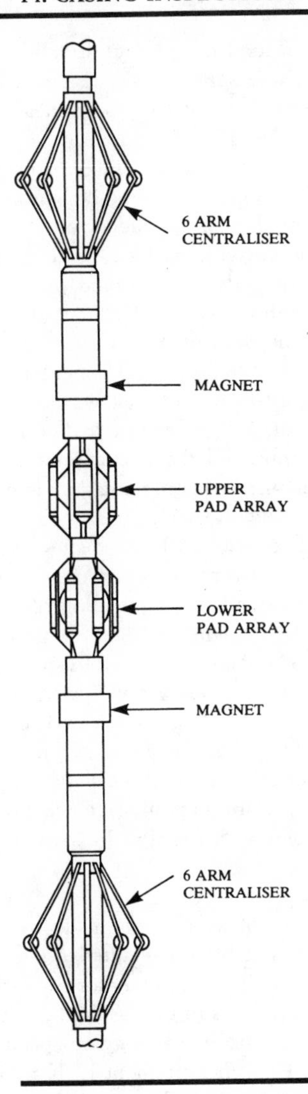

FIGURE 14.11 *The Pipe-Analysis Tool. Courtesy Schlumberger Well Services.*

the surface away from the coils, will not be detected; and, depending upon its distance from the surface, a cavity within the sheet will not influence the eddy currents either.

TOOL PRINCIPLE. The measuring sonde of the pipe-analysis tool (PAT) consists of an iron core with the pole pieces of an electromagnet at each end, and twelve sensor pads in two arrays between the pole pieces (figure 14.11). The two arrays are offset radially to ensure complete coverage of the inner surface of the casing. Each of the pads contains a transmitting coil for eddy-current measurement, and two sensor coils wound in opposite directions for both flux-leakage and eddy-current measurements. The two sensor coils are wound in opposite directions so that for both measurements there is zero voltage so long as no anomaly exists, but a signal will be produced when the quality of the casing is different below the two coils. The same sensor coils can be used for both measurements, as two distinct frequencies are involved. A frequency of 2 kHz is used for measurement of the eddy current, giving a depth of investigation of about 1 mm. The sensor pads are mounted on springs so that they are held in contact with the casing, facilitated through centralization of the sonde. Various sizes of magnet pole pieces are available and are selected according to the inside diameter of the casing, to optimize the signal strength for the flux-leakage measurement.

Six measurements of flux leakage and eddy-current distortion are made on each array, and the maximum signal from each array is sent uphole to the surface instrumentation. Four signals are recorded: the eddy-current and flux-leakage data from the two arrays. The flux-leakage data correspond to anomalies located anywhere in the casing, while eddy-current distortion only occurs at the inside wall of the casing. The standard presentation of the measurements is as shown in figure 14.12, with the data from the two arrays displayed in Tracks 2 and 3. Enhanced data are displayed in Track 1, making any anomalies more obvious. At any particular depth, the larger of the two flux-leakage readings is selected and held for about 0.3 second on the display; the same is done for the eddy-current readings. This enhancement only occurs if the signal amplitudes exceed a certain threshold, to ensure that only significant defects are made more apparent. The holding of the signal allows signal levels to be seen more clearly.

INTERPRETATION. Measurements made by the pipe-analysis tool are generally only suitable for qualitative interpretation. This is because any voltages induced in the sensor coils are not only dependent upon the size of any flaws in the casing, but also upon the magnetic permeability of the casing, the logging speed, and the abruptness of a defect. The PAT measurement is therefore primarily used to locate the presence of small defects in the casing, such as pits and holes; defects such as a gradual decrease of the wall thickness cannot be detected. Since the PAT device will give zero signal in the two extremes of no casing and perfect casing (except at the collars), the electromagnetic thickness tool should also be used to measure the casing-wall thickness, in order to obtain a complete picture of the state of the casing.

Since two sets of data are recorded by the PAL, one set influenced by defects occurring anywhere in the casing and the other by faults on the inner surface, it can be inferred by examination of the log whether the casing is damaged on the inner or outer wall, assuming that there is no defect within the casing. Although the magnetic flux bulges away from both sides of the casing at the location of a

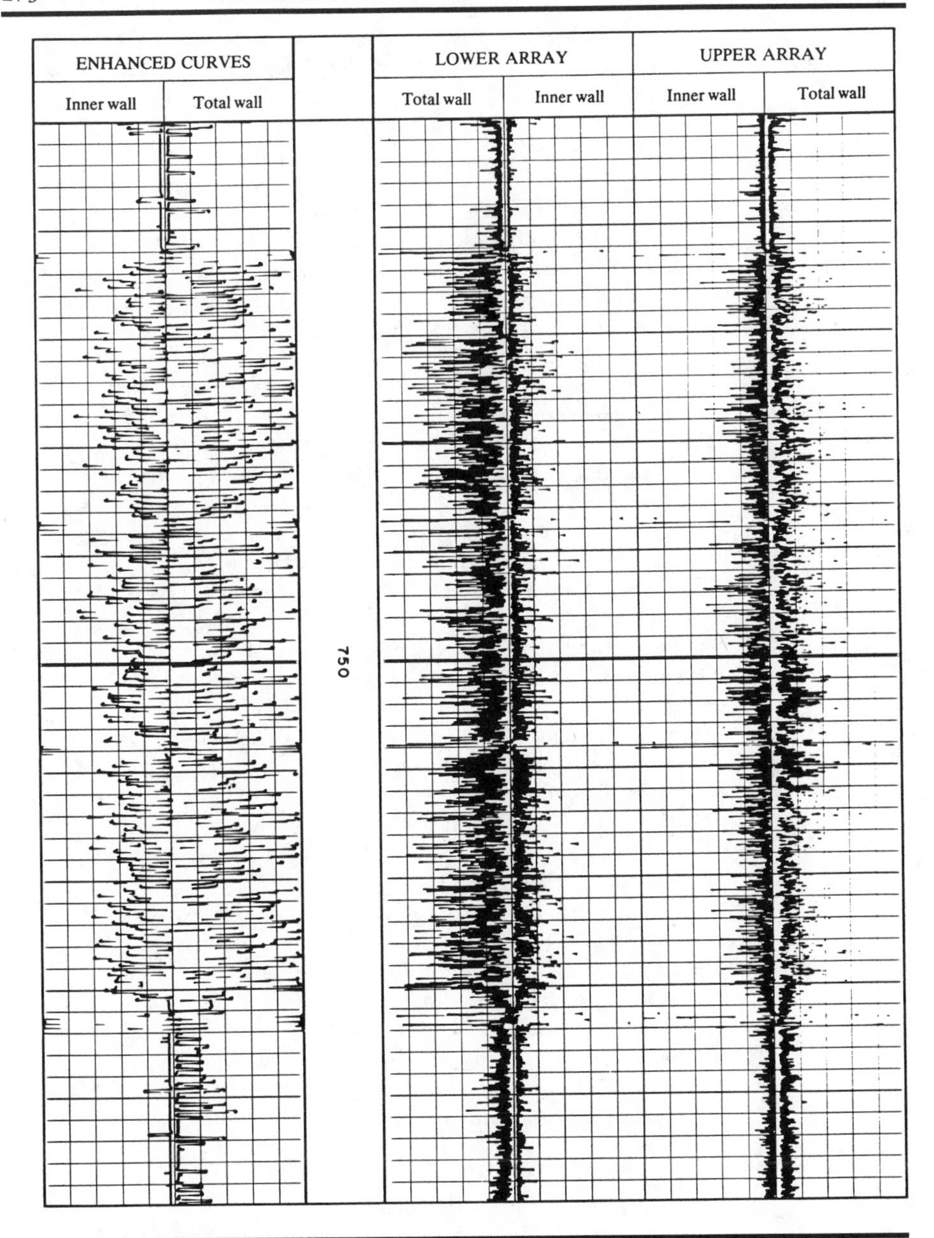

FIGURE 14.12 *The Pipe-Analysis Log in Severely Corroded Casing. Courtesy Schlumberger Well Services.*

defect, the effect is greater on the side of the flaw, hence for the flux-leakage measurement, smaller defects can be detected on the inner surface than on the outer surface. Because of the overlapping configuration of the two pad arrays, all of the inner surface of the casing is surveyed; but there is a casing-diameter-dependent defect size below which the flaw will only be seen by one array, and above which it will be seen by both arrays.

Eddy-current measurements are not able to detect flaws with a diameter smaller than about 0.39 in., while the flux-leakage limit is somewhat lower (0.25 in.). This means that it cannot be determined whether an anomaly less than 0.39 in. in size is on the inner or outer surface. If a deflection is noted on the eddy-current measurement but not on the flux measurement, it is assumed that the defect on the inner wall is less than 1 mm deep, and can usually be ignored. In addition, the flux-leakage readings show events that are not due to casing damage, but rather to the presence of localized magnetization in the casing. This is one reason why a reference PAT survey should be run in new casing, so that a time-lapse technique can be used to determine casing damage at a later time.

The example of figure 14.13 includes sections of perforated casing (798 to 805 m, 807 to 819 m, and 821 to 830 m), and there is a clear indication of damaged and undamaged casing. The flux-leakage (total wall) measurement is responding strongly through the perforated intervals, the eddy-current curve less so. This is probably due to the diameter of the perforations being fairly close to the detection limit of eddy-current measurement. In the upper section of figure 14.13, the tool response is much lower, indicating a certain amount of corrosion on both surfaces of the casing, but probably nothing major. The large defections occurring on all the curves are due to the casing collars.

Concentric casing strings can only be analyzed if (1) both the ETT and PAT are used and (2) there are only two casings. If either casing is corroded, there will be an indication on the ETT readings, although the sensitivity will not be as good as for a single string. The PAT measurement will detect only anomalies on the inner string, but the response will decrease as the magnetic coupling between the two casings increases. Therefore, under suitable conditions, it will be possible to differentiate between inner-string and outer-string damage.

Typical ETT and PAL logs for a variety of defects and anomalies are represented in figures 14.14 through 14.18.

BOREHOLE TELEVIEWERS

Three TV-type tools are available for borehole scanning. The oldest is the Borehole Televiewer (BHTV) which uses a rotating ultrasonic transmitter and receiver to produce an image of the borehole or casing. There is also a borehole television camera that uses a TV camera and an intense light source to transmit a visual image of

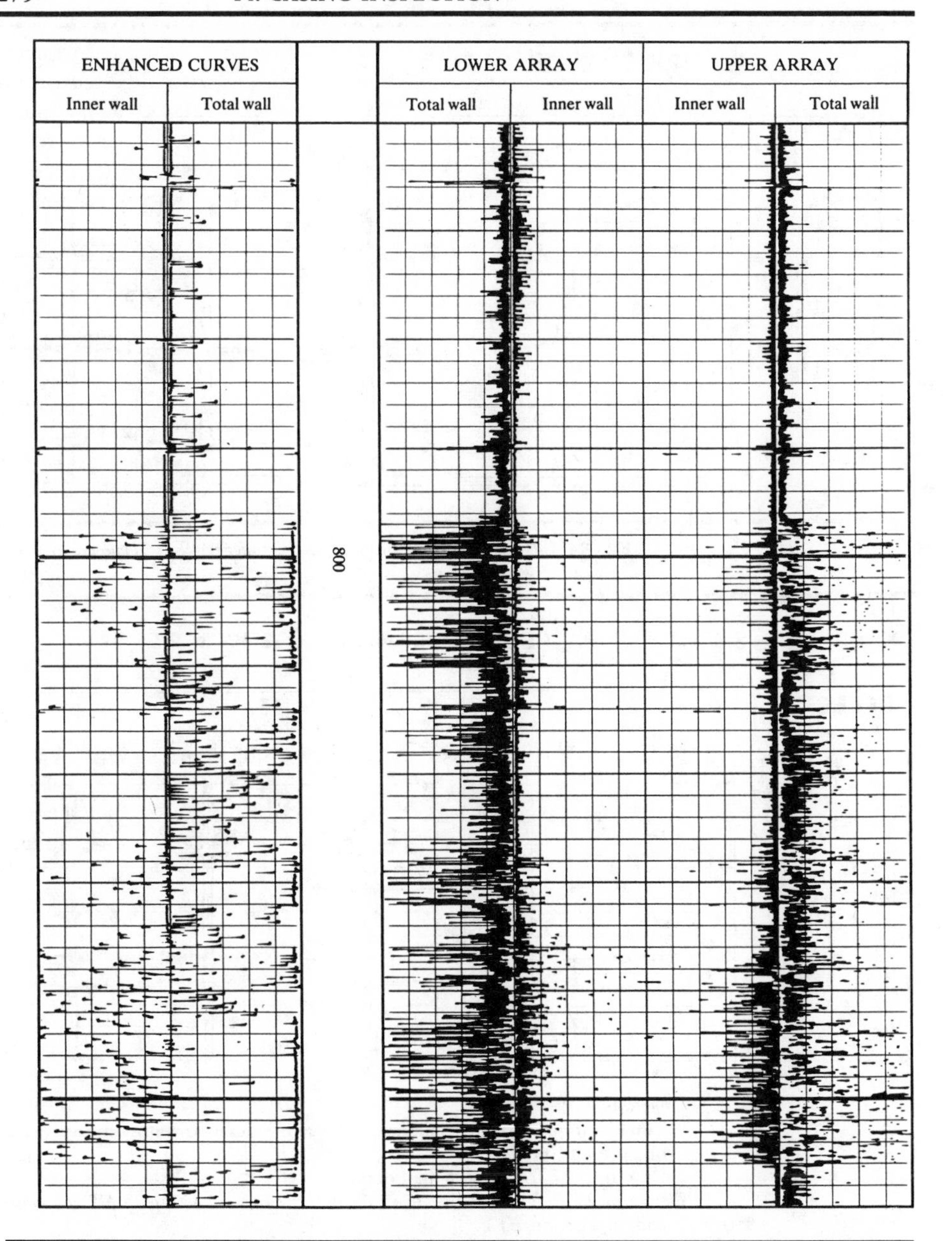

FIGURE 14.13 *Pipe-Analysis Log Over a Perforated Section of Casing. Courtesy Schlumberger Well Services.*

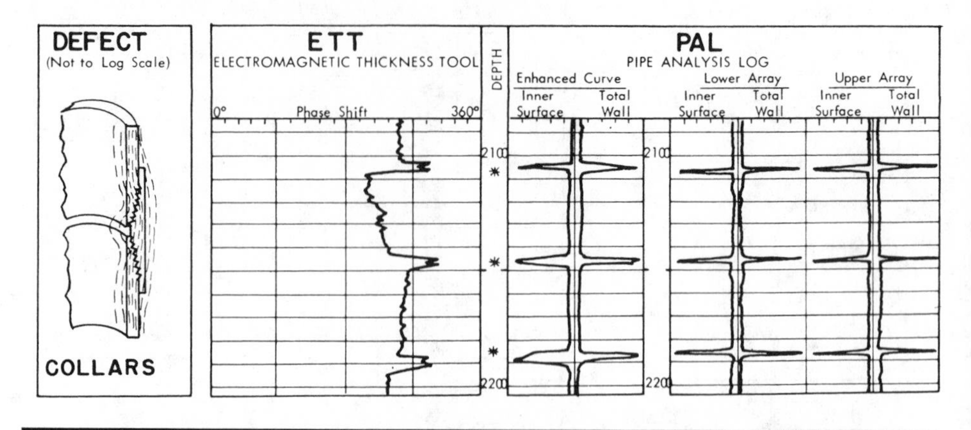

FIGURE 14.14 *Casing Collars. Courtesy Schlumberger Well Services.*

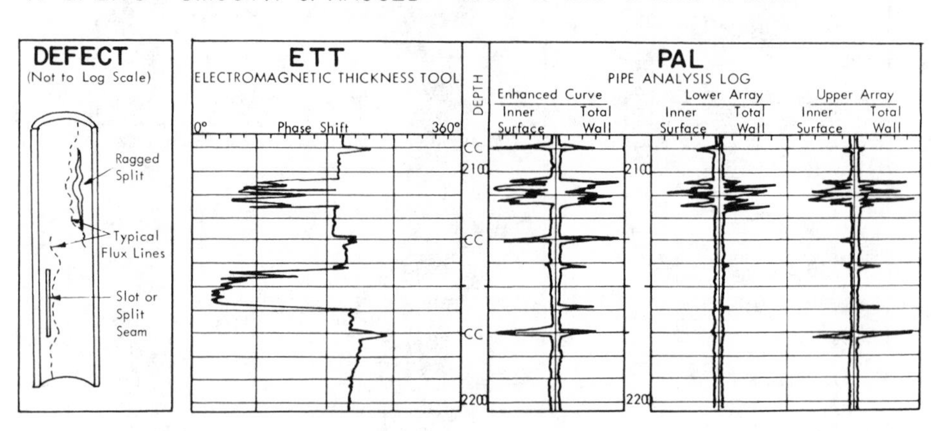

Comments:

No inner-surface anomaly is indicated, therefore defect is neither measureable nor present on ID.

Defect at 2127 ft is larger than defect at 2156 ft.

Defect at 2127 ft may be severe OD defect since it has high PAT response and appears on ETT.

Defect is probably OD pit or internal void due to ETT response at 2127 ft.

OD pits are far more common than any other point defects not on ID.

Magnetic anomalies have been observed in new N-80 casing.

Stress points may occasionally be observed a few feet below collars as a result of clamping joint with tongs.

FIGURE 14.15 *Internal Anomaly and OD Pit. Courtesy Schlumberger Well Services.*

WALL CORROSION - INNER and OUTER

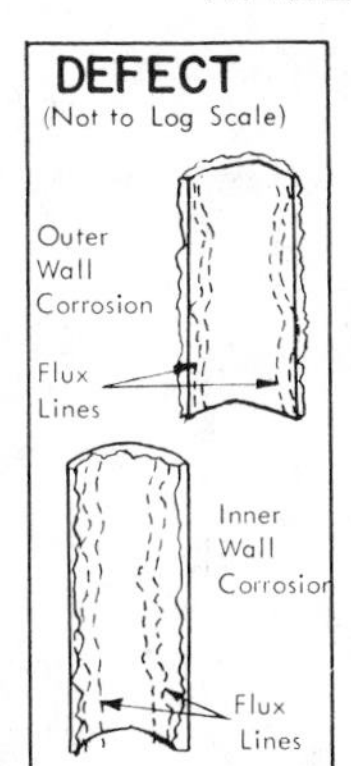

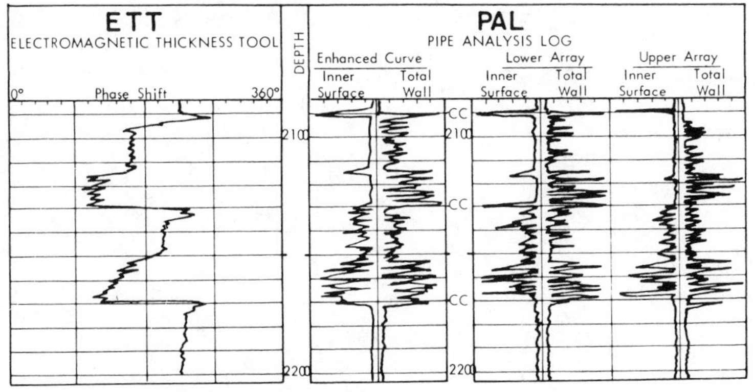

Comments:

If in new casing, defects, if through holes, correspond as follows:

2112 ft ¾ in., through hole (drilled)
2119 ft ½ in., through hole (drilled)
2132 ft ⅜ in., through hole (drilled)
2170 ft 1 through hole (drilled)

Inner-surface eddy-current measurement is not capable of less than ½ in. diameter resolution.

Flux-leakage response larger for deep but small-diameter defects.

Pad overlap effects apparent on both eddy-current and flux-leakage tests.

Defect at 2158 ft does not show on flux-leakage (total wall) test and therefore is minor, $<$ 1 mm deep.

Defect at 2164 ft shows larger on eddy-current than flux-leakage test, therefore shallow ID pit.

FIGURE 14.16 ***Through Hole and ID Pit. Courtesy Schlumberger Well Services.***

✱ POINT DEFECTS NOT ON I.D. - 2127' and 2156'
O.D. PITS, INTERNAL VOIDS, STRESS POINTS, MAGNETIC ANOMALIES

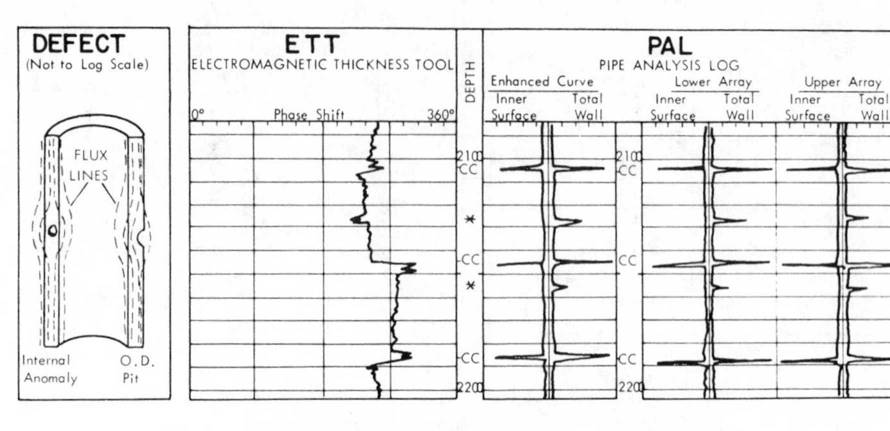

Comments:

Split from 2105 to 2115 ft is ragged. Note dramatic response to eddy-current (inner-surface) and flux-leakage (total wall) tests.

ETT phase shift drops dramatically at split from 2105 to 2115 ft.

Split from 2142 to 2159 ft cannot be determined with certainty using only the PAT. Since flux path leaks into wellbore only at ends of split, PAT indicates only point anomalies at 2142 and 2159 ft.

Split from 2142 to 2159 ft shows up dramatically on ETT. With both ETT and PAT, this anomaly could easily be identified as a split seam in the casing.

FIGURE 14.17 ***Splits. Courtesy Schlumberger Well Services.***

the borehole wall to the surface. Lastly, the volumetric scanning log uses a novel 3-D display technique to give a visual image of the borehole and rotate and tilt the image at will to view any particular anomaly more closely.

An example of the BHTV log is given in figure 14.19. For practical results, this tool requires an unweighted fluid in the hole. With muds above 10 lb/gal, the signal is so attenuated as to give a useless log.

The borehole television camera records on videotape and can be viewed with conventional video playback equipment.

Volumetric scanning tool displays are illustrated in figure 14.20, which shows several wells where the casing shoe was actually gone and the casing was worn and abraded by drilling tools. The vertical scale on these logs is compressed with respect to the horizontal scale.

✳ POINT DEFECTS ON I.D. - 2112', 2119', 2132', 2158', 2164' & 2170' I.D. PITS, THROUGH HOLES

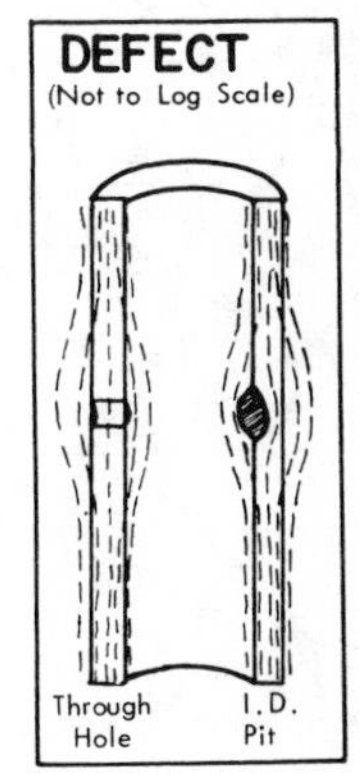

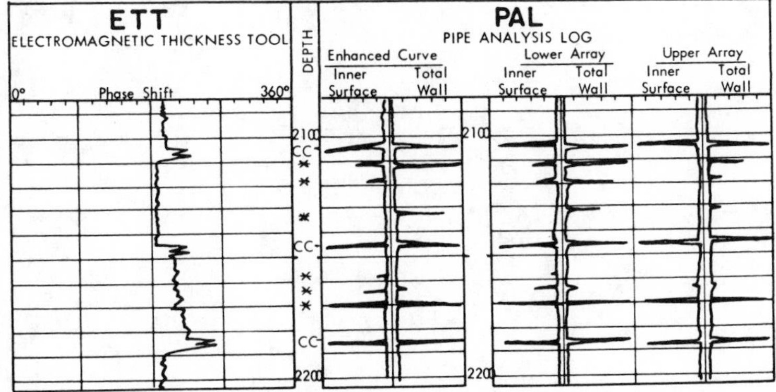

Comments:

Joint from 2090 to 2130 ft has moderate outside-wall corrosion which becomes severe below about 2115 ft.

A through hole appears to exist at 2116 ft.

Note decrease in ETT phase shift from good casing above and along joint, verifying severe corrosion.

Joint from 2130 to 2170 ft has moderate to severe interval-corrosion from 2130 to 2155 ft, and corrosion on inside is increasingly severe for remainder of joint.

ETT verifies increasingly severe corrosion in lower part of joint from about 2130 to 2170 ft.

FIGURE 14.18 ***Inner- and Outer-Wall Corrosion. Courtesy Schlumberger Well Services.***

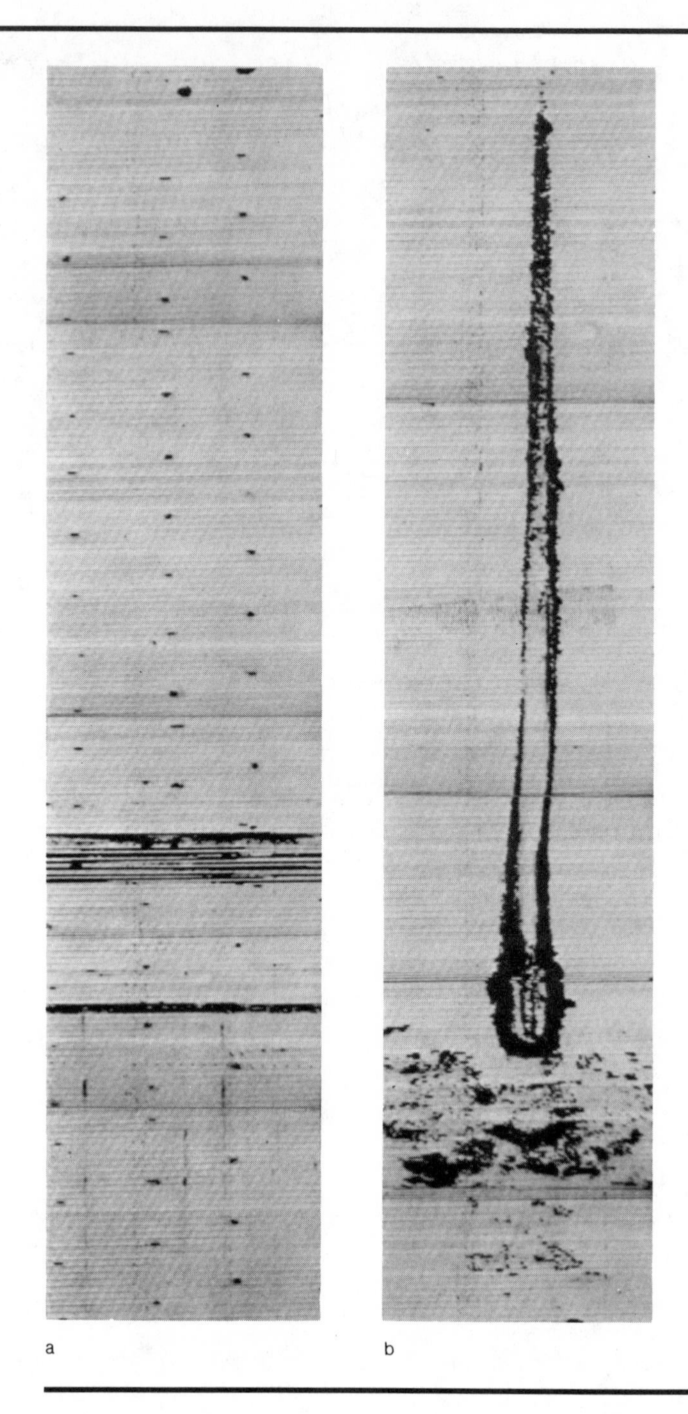

FIGURE 14.19 *BHTV Log.* (a) *Perforations and* (b) *Split Casing.*

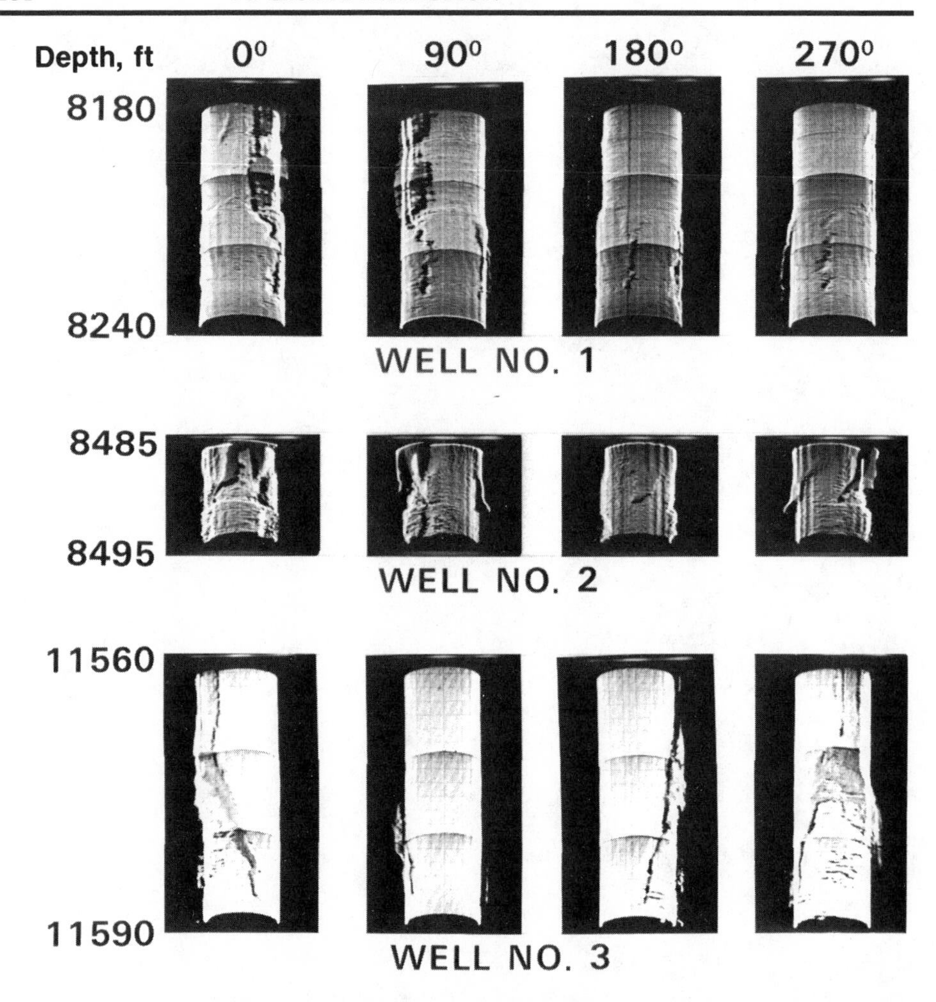

FIGURE 14.20 *Volumetric Scanning Log. Reprinted by permission of World Oil, Gulf Publishing Company from Broding 1982. Copyright © 1982 by Gulf Publishing Company, Houston, Texas. All rights reserved.*

BIBLIOGRAPHY

Broding, R. A.: "Volumetric Scanning Allows 3-D Viewing of the Borehole," *World Oil* (June 1982) 190–196.

"Casing Evaluation Services," Dresser Atlas (1983).

Cuthbert, J. F. and Johnson, W. M., Jr.: "New Casing Inspection Log," presented at the AGA Operating Section Transmission conference, Bel Harbour, Fla., May 19–21, 1975.

"General Catalog," the DIA-LOG Company.

Illiyan, I. S., Cotton, W. J., Jr., and Brown, G. A.: "Test Results of a Corrosion Logging Technique Using Electromagnetic Thickness and Pipe Analysis Logging Tools," *J. Pet. Tech.* (April 1983) 801–808.

APPENDIXES: PRODUCTION LOGGING CHARTS AND TABLES

APPENDIX A: CONVERSION FACTORS BETWEEN METRIC, API, AND U.S. MEASURES

Multiply	by	to find
acres	0.4047	hectares
"	43,560	sq ft
"	4,047	m^2
acre-feet	7,758	bbl[a]
" "	43,560	cu ft
" "	3.259×10^5	gal
atmospheres	76	cm Hg
"	29.92	in. Hg
"	33.93	ft of water
"	1.033	kg/cm^2
"	14.70	psi
barrels (API)[a]	1.289×10^{-4}	acre-ft
" "	158,987	cc[b]
" "	5.615	cu ft
" "	42	gal
" "	9,702	cu in.
" "	1,590	liters
" "	0.1590	m^3
barrels/day[a]	5.615	cu ft/D
" "	0.02917	gal/min
" "	6.625	liters/hr
" "	0.1590	m^3/d
" "	0.006625	m^3/hr
barrels/hour[a]	0.0936	cu ft/min
" "	0.700	gal/min
" "	2.695	cu in./s
bars	0.9869	atm
"	1.020	kg/cm^3
"	14.50	psi
British thermal units	778.57	ft-lb
" " "	0.2520	kcal
" " "	0.2930	W-hr
Btu/minute	0.02357	hp
" "	0.01758	kW
" "	12.97	ft-lb/s
centimeters	3.281×10^{-2}	ft
"	0.3937	in.
"	0.01	m
"	10	mm
cm of mercury	0.01316	atm
" " "	0.4461	ft H_2O
" " "	0.01360	kg/cm^2
cm per second	0.1934	psi
" " "	1.969	ft/min
" " "	0.03281	ft/s
" " "	0.6	m/min

Multiply	by	to find
cubic centimeters	3.531×10^{-3}	cu ft
" "	6.102×10^{-2}	cu in.
" "	10^{-4}	m^3
" "	2.642×10^{-4}	gal
" "	10^{-3}	liters
" "	6.2897×10^{-6}	bbl[a]
cubic feet	0.1781	bbl[a]
" "	2.832×10^4	cc[b]
" "	7.481	gal
" "	1,728	cu in.
" "	0.02832	m^3
" "	28.32	liters
cubic feet/day	1.18	liters/hr
" " "	1.18×10^{-3}	c m/hr
" " "	0.02832	m^3/day
" " "	0.1781	bbl/day[a]
cubic feet/minute	10.686	bbl/hr[a]
" " "	256.5	B/D[a]
" " "	472	cc/s
" " "	7.481	gal/min
" " "	0.472	liters/s
cubic inches	16.39	cc[b]
" "	5.787×10^{-4}	cu ft
" "	1.639×10^{-3}	m^3
" "	4.329×10^{-3}	gal
" "	1.639×10^{-2}	liters
cubic meters	6.2897	bbl[a]
" "	10^6	cc[b]
" "	264.2	gal
" "	6.102×10^4	cu in.
" "	35.31	cu ft
" "	10^3	liters
cubic meters/hour	151.0	B/D[a]
" " "	847.8	cu ft/D
" " "	10^3	liters/hr
" " "	24	m^3
cubic meters/day	0.2621	bbl/hr[a]
" " "	6.2897	B/D[a]
" " "	1.471	cu ft/hr
" " "	35.31	cu ft/D
" " "	41.67	liters/hr
" " "	0.04167	m^3/hr
days	1,440	min
"	86,400	s
feet	30.48	cm
"	12	in.
"	0.3048	m

APPENDIX A (continued)

Multiply	by	to find
feet of water	0.02950	atm
" " "	0.8826	in. Hg
" " "	0.03048	kg/cm^2
" " "	62.43	lb/sq ft
" " "	0.4335	psi
feet/hour	0.008467	cm/s
" "	5.086×10^{-3}	m/min
" "	0.01667	ft/min
feet/minute	0.5080	cm/s
" "	0.01667	ft/s
" "	0.01829	km/hr
" "	0.3048	m/min
feet/second	30.48	cm/s
" "	18.29	m/min
foot-pounds	1.285×10^{-3}	Btu
" "	3.238×10^{-4}	kcal
foot-pounds/minute	3.030×10^{-3}	hp
" " "	2.260×10^{-3}	kW
foot-pounds/second	1.818×10^{-3}	hp
" " "	1.356×10^{-3}	kW
gallons (U.S.)	0.02381	bbl[a]
" "	3,785	cc[b]
" "	0.1337	cu ft
" "	231	cu in.
" "	3.785×10^{-3}	m^3
" "	3.785	liters
gallons (imperial)	1.2009	gal (U.S.)
gallons/minute	1.429	bbl/hr[a]
" "	34.286	B/D
" "	0.1337	cu ft/min
" "	192.5	cu ft/D
" "	3.785	liters/min
" "	90.84	liters/hr
grain (avoir)	0.06480	g
grains/gal	17.12	ppm
" "	142.9	lb/10^4 gal
" "	0.01714	g/liter
grams	15.432	grains
"	10^{-3}	kg
"	0.3215	oz
"	2.205×10^{-3}	lb
grams/cc[b]	62.43	lb/cu ft
" "	8.344	lb/gal
" "	0.03613	lb/cu in.
grams/liter	58.42	grains/gal
hectares	2.471	acres
"	1.076×10^5	sq ft
"	0.010	km^2
horsepower	42.40	Btu/min
"	33,000	ft-lb/min
"	550	ft-lb/s
"	1.014	metric hp
"	10.68	kcal/min
"	0.7457	kW
"	745.7	watts
horsepower-hour	2,544	Btu
" "	641.1	kcal
" "	2.737×10^3	kg-m
" "	0.7455	kW-hr
inches	2.540	cm
"	8.333×10^{-2}	ft
in. of mercury	0.03342	atm
" " "	1.133	ft H_2O
" " "	0.03453	kg/cm^2
" " "	0.4912	psi
in. of water	0.002458	atm
" " "	0.07349	in. Hg
" " "	0.002538	kg/cm^2
" " "	0.03609	psi
kilograms	10^3	g
"	2.205	lb
"	1.102×10^{-3}	tons (short)
kg-calories	3.986	Btu
" "	3,088	ft-lb
" "	1.560×10^{-3}	hp-hr
" "	427	kg-m
" "	1.163×10^{-3}	kW-hrs
kg-calories/min	0.09358	hp
" " "	0.06977	kW
kg/cubic meter	10^{-3}	g/cc[b]
kg/square cm	0.9678	atm
" " "	0.9807	bars
" " "	32.84	ft H_2O
" " "	28.96	in. Hg
" " "	14.22	psi
kilowatts	56.88	Btu/min
"	4.427×10^4	ft-lb/min
"	737.8	ft-lb/s
"	1.341	hp
"	10^3	watts
kilowatt-hours	3,413	Btu
" "	2.656×10^6	ft-lb
" "	1.341	hp-hr

APPENDIX A (continued)

Multiply	by	to find
" "	860	kcal
" "	3.672×10^5	kg-m
liters	10^3	cc[b]
"	6.2897×10^{-3}	bbl[a]
"	0.03531	cu ft
"	0.2642	gal
"	61.02	cu in.
"	10^{-3}	m^3
liters/hour	0.1509	B/D
" "	6.289×10^{-3}	bbl/hr[a]
" "	5.885×10^{-4}	cu ft/min
" "	0.8475	cu ft/d
" "	10^{-3}	m^3/hr
" "	0.02400	m^3/d
meters	3.281	ft
"	39.37	in.
"	10^3	mm
"	6.214×10^{-4}	mile
meters/minute	1.667	cm/s
" "	3.281	ft/min
" "	196.9	ft/hr
" "	0.05468	ft/s
mile	5,280	ft
"	1.609	km
miles/hour	44.70	cm/s
" "	88	ft/min
" "	26.82	m/min
millimeters	0.1	cm
"	3.281×10^{-3}	ft
"	0.03937	in.
minutes	6.944×10^{-4}	days
"	1.667×10^{-3}	hrs
parts/million	0.05835	grains/gal
" "	8.337	lb/10^6 gal
pound	7,000	grains
"	453.6	g
"	0.4536	kg
pounds/cubic ft	0.1337	lb/gal
" " "	0.01602	g/cc[b]
" " "	16.02	kg/m^3
" " "	5.787×10^{-4}	lb/m^3
pounds/square in.	0.06805	atm
" " "	2.309	ft H_2O
" " "	2.036	in. Hg
" " "	51.70	mm Hg
" " "	0.07031	kg/cm^2

Multiply	by	to find
" " "	144	lb/cu ft
seconds	1.157×10^{-3}	days
"	2.778×10^{-4}	hr
"	1.667×10^{-3}	min
square cm	1.076×10^{-3}	sq ft
" "	0.1550	sq in.
" "	10^{-4}	m^2
" "	100	mm^2
square feet	2.296×10^{-5}	acres
" "	929.0	cm^2
" "	144	sq in.
" "	0.09290	m^2
square inches	6.452	cm^2
" "	6.944×10^{-3}	sq ft
" "	645.2	mm^2
square meters	10.76	sq ft
" "	2.471×10^{-4}	acres
" "	1,550	sq in.
°Cent. + 273	1	°K (abs)
°Fahr. + 460	1	°R (abs)
°Cent. + 17.8	1.8	°F
°Fahr. − 32	5/9	°C
°Cent./100 meters	0.5486	°F/100 ft
°F/100 ft	1.823	°C/100 ft
tons (long)	1,016	kg
" "	2,240	lb
tons (metric)	10^3	kg
" "	2,205	lb
tons (short)	2,000	lb
viscosity, lb-s/sq in.	6.895×10^6	viscosity, cp
viscosity, lb-s/sq ft	4.78×10^4	viscosity, cp
viscosity, centistokes	density	viscosity, cp
watts	0.05688	Btu/min
"	44.27	ft-lb/min
"	0.7378	ft-lb/s
"	1.341×10^{-3}	hp
"	0.01433	kcal/min
"	10^{-3}	kW
watt-hours	3.413	Btu
" "	2,656	ft-lb
" "	1.341×10^{-3}	hp-hr
" "	0.860	kcal
" "	367.2	kg-m
" "	10^{-3}	kW-hr

APPENDIX A (continued)

VOLUME CAPACITY OF PIPES[a]

gallons per 1000 ft = 40.3 × (ID in inches)^2
barrels per 1000 ft = 0.9714 × (ID in inches)^2
barrels per 1000 ft = approximately (ID in inches)^2
cubic feet per 1000 ft = 5.454 × (ID in inches)^2
gallons per mile = 215.4240 × (ID in inches)^2
barrels per mile = 5.1291 × (ID in inches)^2

VELOCITY[a]

feet per minute = 0.127324 (cubic feet per day) ÷ (ID in inches)^2
feet per minute = 1,029.42 (barrels per minute) ÷ (ID in inches)^2
feet per second = 0.4085 (gallons per minute) ÷ (ID in inches)^2

TANK VOLUMES[a]

barrels per foot in round tank = (diameter, in feet)^2 ÷ 7.14
barrels per inch in round tank = (diameter, in feet)^2 ÷ 85.7
barrels per inch in square tank = 0.0143 (length, in feet) × (width, in feet)
cubic feet per inch in square tank = 0.0833 (length, in feet) × (width, in feet)

OIL GRAVITY

$$\text{sp. gr. @ 60°F} = \frac{141.5}{131.5 + \text{API gravity}}$$

$$\text{API gravity} = \frac{141.5}{\text{sp. gr. @ 60°F}} - 131.5$$

GAS GRAVITY[c]

$$\text{gas specific gravity} = \frac{\text{density of gas at sc (g/cc)}}{0.00122} = \frac{\text{density of gas at sc (lb/cu ft)}}{0.0762}$$

$$\begin{array}{c}\text{gas specific gravity}\\ \text{(ideal gas-law conditions)}\end{array} = \frac{\text{density of gas}}{\text{density of air at same temp. and press.}} = \frac{\text{molecular weight of gas}}{28.966}$$

[b] The metric symbol, cc, for cubic centimeters has been replaced by the SI symbol cm^3, but it is still widely used.
[c] sc (standard conditions) = 60°F (15.56°C) and 14.7 psia (one atmosphere).

APPENDIX B: AVERAGE FLUID VELOCITY VS. TUBING SIZE, API NONUPSET TUBING

Description				Fluid Velocity for Flow Rate of:											
Nom. OD	Wt.	Int. Diameter		1000 B/D			10 m^3/hr			100 m^3/d			1000 cu ft/D		
in. (mm)	lb/ft	in.	mm	m/min	cm/s	ft/min	m/min	cm/s	ft/min	m/min	cm/s	ft/min	m/min	cm/s	ft/min
1.9 (48.3)	2.75	1.610	40.89	84.54	140.9	276	127.11	211.8	417.0	52.9	88.11	173.5	15.0	25.0	49.12
2⅜ (60.3)	4.00	2.041	51.84	52.56	87.6	172	79.00	131.7	259.2	32.8	54.79	107.8	9.31	15.5	30.56
	4.60	1.995	50.67	55.02	91.7	180	82.70	137.8	271.3	34.4	57.32	112.8	9.75	16.3	31.99
	5.80	1.867	47.42	62.82	104.7	205	94.42	157.4	309.8	39.3	65.47	128.9	11.1	18.6	36.53
2⅞ (73.0)	6.40	2.441	62.00	36.78	61.3	120	55.28	92.1	181.4	22.9	38.31	75.5	6.51	10.9	21.37
	8.60	2.259	57.38	42.90	71.5	140	64.48	107.5	211.5	26.8	44.72	87.9	7.60	12.7	24.95
3½ (88.9)	7.70	3.068	77.93	23.28	38.8	76.0	34.99	58.3	114.8	14.5	24.25	47.7	4.12	6.87	13.53
	9.20	2.992	76.00	24.48	40.8	79.8	36.79	61.3	120.7	15.3	25.50	50.2	4.33	7.22	14.22
	10.20	2.922	74.22	25.68	42.8	83.7	38.60	64.3	126.6	16.1	26.75	52.7	4.54	7.57	14.91
	12.70	2.750	69.85	28.98	48.3	94.5	43.56	72.6	142.9	18.1	30.20	59.4	5.13	8.55	16.84
4 (101.6)	9.50	3.548	90.12	17.40	29.0	56.7	26.15	43.6	85.8	10.9	18.14	35.7	3.08	5.14	10.11
4½ (114.3)	12.60	3.958	100.53	13.98	23.3	45.6	21.01	35.0	68.9	8.7	14.56	28.7	2.48	4.13	8.128

AVERAGE FLUID VELOCITY VS. TUBING SIZE, API EXTERNAL UPSET TUBING

1.05 (26.67)	1.20	0.824	20.93	322.62	537.7	1053	484.90	808.2	1590.9	201.71	336.21	661.81	57.2	95.3	187.5
1.315 (33.40)	1.80	1.049	26.64	199.14	331.9	650	299.31	498.8	982.0	124.51	207.50	408.51	35.3	58.8	115.7
1.660 (42.16)	2.40	1.380	35.05	115.02	191.7	375	172.88	288.1	567.2	71.92	119.85	235.95	20.4	34.0	66.86
1.9 (48.26)	2.90	1.610	40.89	84.54	140.9	276	127.11	211.8	417.0	52.88	88.11	173.47	15.0	25.0	49.12
2⅛	4.70	1.995	50.67	55.02	91.7	180	82.70	137.8	271.3	34.40	57.32	112.86	9.75	16.3	31.99
(60.32)	5.95	1.867	47.42	62.82	104.7	205	94.42	157.4	309.8	39.28	65.48	128.88		18.6	36.53
2⅞	6.50	2.441	62.00	36.78	61.3	120	55.28	92.1	181.4	22.99	38.31	75.46	11.1	10.9	21.37
(73.02)	8.70	2.259	57.38	42.90	71.5	140	64.48	107.5	211.6	26.82	44.72	88.02	7.60	12.7	24.95
3½	9.30	2.992	76.00	24.48	40.8	79.8	36.79	61.3	120.7	15.30	25.50	50.21	4.33	7.22	14.22
(88.9)	12.95	2.750	69.85	28.98	48.3	94.5	43.56	72.6	142.9	18.12	30.20	58.45	5.13	8.55	16.84
4 (101.6)	11.00	3.476	88.29	18.12	30.2	59.2	27.23	45.4	89.3	11.32	18.89	38.15	3.21	5.35	10.54
4½ (114.3)	12.75	3.958	100.53	13.98	23.3	45.6	21.01	35.0	68.9	8.74	14.56	28.66	2.48	4.13	8.128

APPENDIX C: AVERAGE FLUID VELOCITY VS. CASING SIZE

Description				Fluid Velocity for Flow Rate of:											
Nom. OD	Wt.	Int. Diameter		1000 B/D			10 m³/hr			100 m³/d			1000 cu ft/D		
in. (mm)	lb/ft	in.	mm	m/min	cm/s	ft/min	m/min	cm/s	ft/min	m/min	cm/s	ft/min	m/min	cm/s	ft/min
	9.50	4.090	103.9	13.08	21.8	42.7	19.74	32.9	64.8	8.21	13.7	26.9	2.320	3.867	7.613
4½	11.60	4.000	101.6	13.44	22.4	44.7	20.40	34.0	66.9	8.49	14.1	27.8	2.426	4.043	7.958
(114.3)	13.50	3.920	99.6	14.22	23.7	46.6	21.48	35.8	70.5	8.94	14.9	29.3	2.530	4.216	8.299
	15.10	3.826	97.2	15.00	25.0	48.8	22.50	37.5	73.8	9.36	15.6	30.7	2.651	4.419	8.698
	11.50	4.560	115.8	10.50	17.5	34.4	15.84	26.4	52.0	6.59	11.0	21.6	1.866	3.110	6.123
5	13.00	4.494	114.2	10.80	18.0	35.4	16.26	27.1	53.4	6.76	11.3	22.2	1.921	3.202	6.304
(127.0)	15.00	4.408	112.0	11.28	18.8	36.8	16.98	28.3	55.7	7.06	11.8	23.2	1.997	3.329	6.553
	18.00	4.276	108.6	11.88	19.8	39.1	17.94	29.9	58.9	7.46	12.4	24.5	2.123	3.538	6.964
	13.00	5.044	128.1	8.64	14.4	28.1	12.96	21.6	42.5	5.39	8.9	17.7	1.525	2.542	5.004
	14.00	5.012	127.3	8.70	14.5	28.5	13.08	21.8	42.9	5.44	9.1	17.8	1.545	2.575	5.069
5½	15.50	4.950	125.7	8.94	14.9	29.2	13.44	22.4	44.1	5.59	9.3	18.3	1.584	2.640	5.196
(139.7)	17.00	4.892	124.3	9.12	15.2	29.9	13.74	22.9	45.1	5.72	9.5	18.7	1.622	2.703	5.320
	20.00	4.778	121.4	9.60	16.0	31.3	14.40	24.0	47.3	5.99	9.9	19.7	1.700	2.833	5.577
	23.00	4.670	118.6	10.02	16.7	32.8	15.12	25.2	49.6	6.29	10.5	20.6	1.780	2.966	5.838
	17.00	6.135	155.8	5.82	9.7	19.0	8.76	14.6	28.7	3.64	6.1	11.9	1.031	1.719	3.383
	20.00	6.049	153.6	5.94	9.9	19.5	9.00	15.0	29.5	3.74	6.2	12.3	1.061	1.768	3.480
6⅝	24.00	5.921	150.5	6.24	10.4	20.4	9.42	15.7	30.9	3.92	6.5	12.8	1.107	1.845	3.632
(168.3)	28.00	5.791	147.1	6.54	10.9	21.3	9.90	16.5	32.5	4.12	6.8	13.5	1.157	1.929	3.797
	32.00	5.675	144.1	6.78	11.3	22.2	10.20	17.0	33.5	4.24	7.1	13.9	1.205	2.008	3.953
	17.00	6.538	166.1	5.10	8.5	16.7	7.68	12.8	25.2	3.19	5.3	10.5	.908	1.513	2.979
	20.00	6.456	164.0	5.22	8.7	17.2	7.86	13.1	25.8	3.27	5.4	10.7	.931	1.552	3.055
	23.00	6.366	161.7	5.40	9.0	17.6	8.16	13.6	26.8	3.39	5.6	11.1	.958	1.596	3.142
7	26.00	6.276	159.4	5.52	9.2	18.2	8.34	13.9	27.4	3.47	5.8	11.4	.985	1.642	3.233
(177.8)	29.00	6.184	157.1	5.70	9.5	18.7	8.64	14.4	28.4	3.51	6.0	11.8	1.015	1.691	3.329
	32.00	6.094	154.8	5.88	9.8	19.3	8.88	14.8	29.1	3.69	6.2	12.1	1.045	1.745	3.429
	35.00	6.004	152.5	6.06	10.1	19.8	9.12	15.2	29.9	3.79	6.3	12.4	1.077	1.794	3.532
	38.00	5.920	150.4	6.24	10.4	20.4	9.42	15.7	30.9	3.92	6.5	12.8	1.107	1.846	3.633

7⅝ (193.7)	20.00	7.125	181.0	4.32	7.2	14.1	6.48	10.8	21.3	2.69	4.4	8.86	.764	1.274	2.508
	24.00	7.052	178.4	4.44	7.4	14.5	6.54	10.9	21.5	2.72	4.5	8.94	.786	1.310	2.580
	26.40	6.969	177.0	4.50	7.5	14.7	6.78	11.3	22.2	2.82	4.7	9.23	.799	1.332	2.622
	29.70	6.875	174.6	4.62	7.7	15.1	6.96	11.6	22.8	2.90	4.8	9.48	.821	1.369	2.694
	33.70	6.765	171.8	4.80	8.0	15.6	7.20	12.0	23.6	2.99	4.9	9.81	.848	1.413	2.782
	39.00	6.625	168.3	4.92	8.2	16.3	7.44	12.4	24.4	3.09	5.2	10.2	.884	1.474	2.901
8⅝ (219.1)	24.00	8.097	205.7	3.33	5.55	10.9	5.02	8.38	16.5	2.09	3.5	6.86	.592	.9865	1.942
	28.00	8.017	203.6	3.39	5.66	11.1	5.13	8.55	16.8	2.13	3.6	6.98	.604	1.006	1.981
	32.00	7.921	201.2	3.48	5.81	11.4	5.25	8.75	17.2	2.18	3.6	7.15	.618	1.031	2.029
	36.00	7.825	198.8	3.55	5.92	11.7	5.35	8.93	17.6	2.23	3.7	7.32	.634	1.056	2.079
	40.00	7.725	196.2	3.63	6.05	12.0	5.46	9.10	17.9	2.27	3.8	7.44	.650	1.084	2.134
	44.00	7.625	193.7	3.75	6.25	12.3	5.64	9.40	18.5	2.35	3.9	7.69	.668	1.113	2.190
	49.00	7.511	190.8	3.87	6.45	12.7	5.82	9.70	19.1	2.42	4.0	7.94	.688	1.147	2.257
9⅝ (244.5)	29.30	9.063	230.2	2.66	4.44	8.70	4.00	6.68	13.1	1.66	2.77	5.45	.472	.787	1.550
	32.30	9.001	228.6	2.69	4.49	8.83	4.05	6.75	13.3	1.68	2.80	5.53	.479	.799	1.572
	36.00	8.921	226.6	2.74	4.58	8.98	4.17	6.95	13.7	1.73	2.89	5.69	.488	.813	1.600
	40.00	8.835	224.4	2.80	4.67	9.16	4.23	7.05	13.9	1.76	2.93	5.78	.497	.829	1.631
	43.50	8.755	222.4	2.85	4.75	9.33	4.29	7.15	14.1	1.78	2.97	5.86	.506	.844	1.661
	47.00	8.681	220.5	2.88	4.81	9.49	4.35	7.25	14.3	1.81	3.02	5.95	.515	.859	1.690
	53.50	8.535	216.8	3.00	5.00	9.81	4.51	7.53	14.8	1.88	3.13	6.16	.533	.888	1.748
10¾ (273.0)	32.75	10.192	258.9	2.10	3.50	6.88	3.17	5.28	10.4	1.32	2.19	4.33	.374	.623	1.226
	40.50	10.050	255.3	2.16	3.60	7.08	3.25	5.42	10.7	1.35	2.25	4.45	.384	.641	1.261
	45.50	9.950	252.7	2.20	3.68	7.22	3.33	5.55	10.9	1.39	2.31	4.53	.392	.653	1.286
	51.00	9.850	250.2	2.25	3.75	7.37	3.39	5.65	11.1	1.41	2.35	4.61	.400	.666	1.312
	55.50	9.760	247.9	2.29	3.82	7.51	3.45	5.75	11.3	1.44	2.39	4.70	.408	.679	1.337
	60.70	9.660	245.4	2.34	3.91	7.66	3.52	5.88	11.5	1.46	2.44	4.78	.416	.693	1.364
	65.70	9.560	242.8	2.40	4.00	7.82	3.60	6.00	11.8	1.50	2.49	4.91	.425	.708	1.393
11¾ (298.5)	38.00	11.150	283.2	1.76	2.94	5.75	2.64	4.41	8.66	1.10	1.83	3.60	.312	.520	1.024
	42.00	11.084	281.5	1.77	2.96	5.82	2.67	4.45	8.76	1.11	1.85	3.64	.316	.526	1.036
	47.00	11.000	279.4	1.81	3.02	5.91	2.73	4.55	8.96	1.14	1.89	3.73	.321	.534	1.052
	54.00	10.880	276.4	1.84	3.08	6.04	2.79	4.65	9.15	1.16	1.93	3.81	.328	.547	1.076
	60.00	10.772	273.6	1.88	3.14	6.16	2.83	4.73	9.29	1.18	1.97	3.86	.334	.557	1.097

	48.00	12.715	323.0	1.35	2.25	4.42	2.04	3.40	6.69	0.85	1.41	2.78	.240	.400	.7875
13⅜	54.50	12.615	320.4	1.37	2.29	4.49	2.07	3.46	6.79	0.86	1.44	2.82	.244	.406	.8001
(339.7)	61.00	12.515	317.9	1.39	2.33	4.56	2.11	3.52	6.92	0.88	1.46	2.87	.248	.413	.8129
	68.00	12.415	315.3	1.41	2.36	4.64	2.13	3.56	6.99	0.89	1.48	2.91	.252	.420	.8261
	72.00	12.347	313.6	1.43	2.39	4.69	2.16	3.60	7.09	0.90	1.49	2.95	.255	.424	.8352
	55.00	15.376	390.6	0.92	1.54	3.02	1.39	2.32	4.56	0.58	0.86	1.90	.164	.274	.5385
16	65.00	15.250	387.4	0.93	1.56	3.07	1.41	2.36	4.63	0.59	0.91	1.92	.167	.278	.5475
(406.4)	75.00	15.124	384.2	0.95	1.59	3.13	1.44	2.40	4.72	0.60	0.99	1.96	.170	.283	.5566
	84.00	15.010	381.3	0.97	1.62	3.17	1.46	2.44	4.79	0.61	1.02	1.99	.172	.287	.5651
20 (508.0)	94.00	19.124	485.8	0.60	1.00	1.95	0.90	1.50	2.96	0.37	0.62	1.23	.106	.177	.3481

APPENDIX D: AVERAGE FLUID VELOCITY

When a production logging tool is present in the casing or tubing, the average fluid velocity in the tool/pipe annulus may be determined from the following charts.

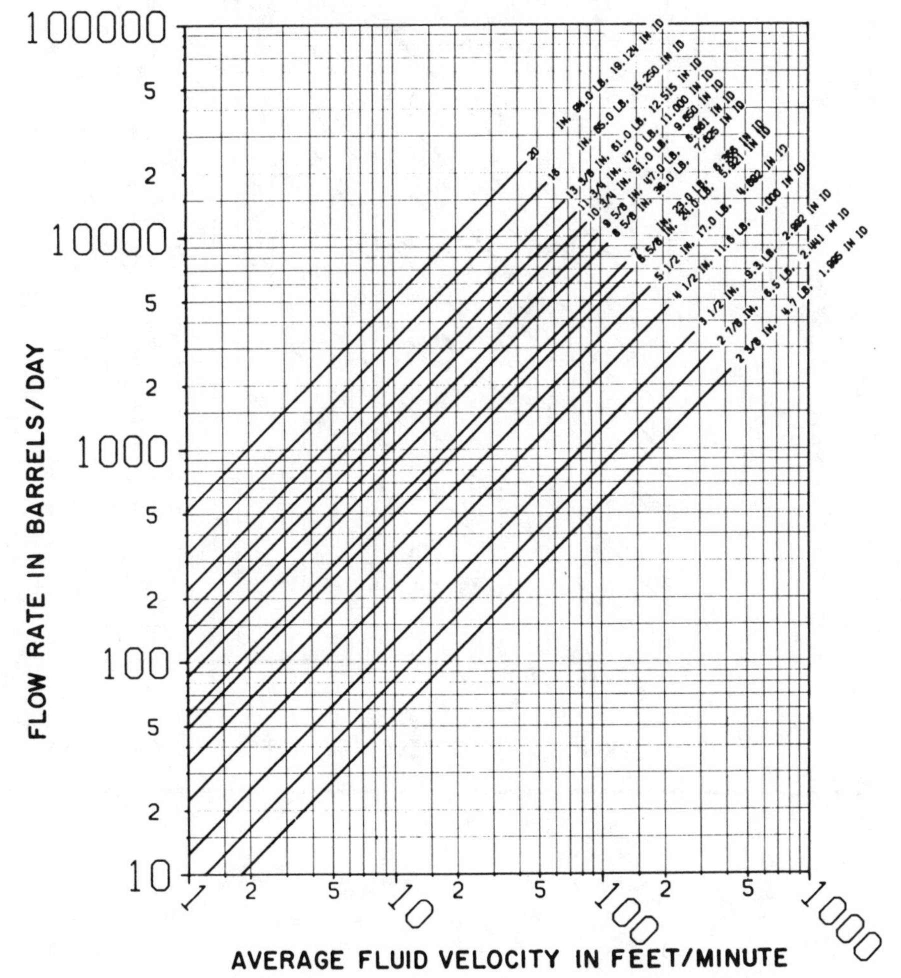

Courtesy Schlumberger Well Services, from "Fluid Conversions in Production Log Interpretation" (1974) 54–56.

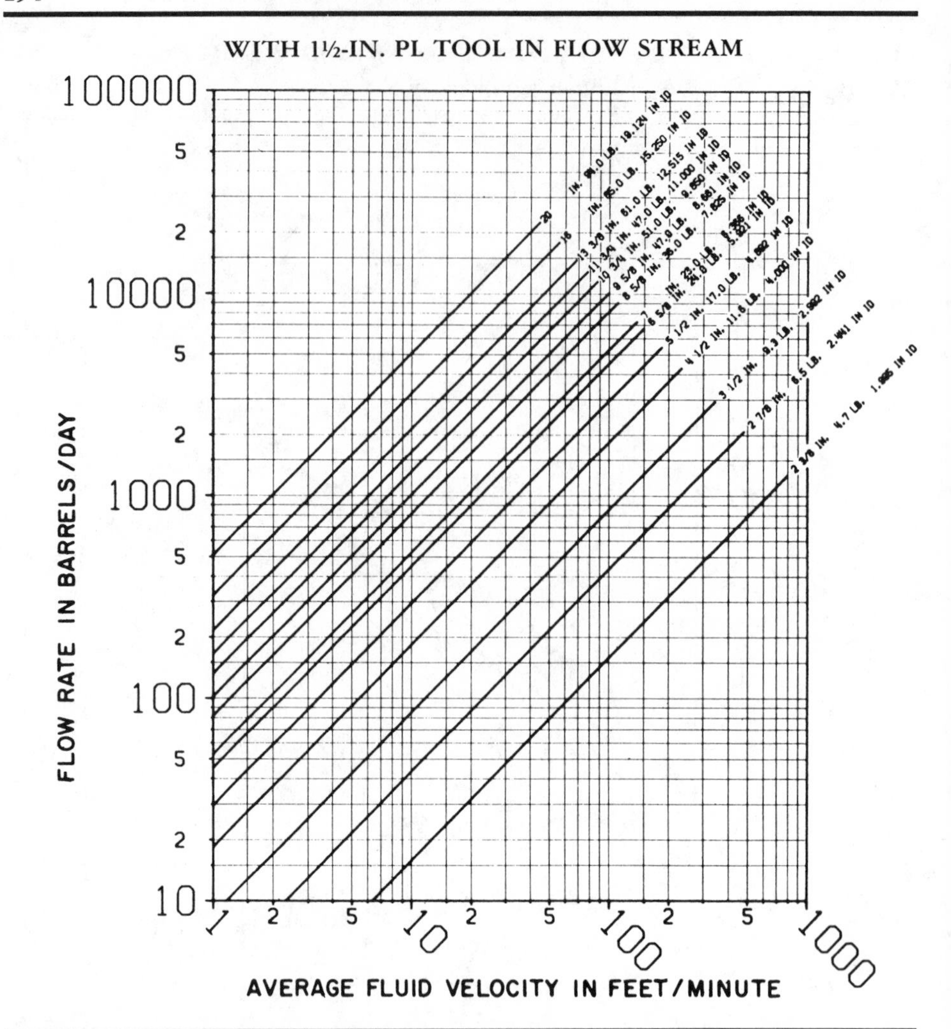
WITH 1½-IN. PL TOOL IN FLOW STREAM
FLOW RATE IN BARRELS/DAY
AVERAGE FLUID VELOCITY IN FEET/MINUTE
100000
5
2
10000
5
2
1000
5
2
100
5
2
10
1
2
5
10
2
5
100
2
5
1000
20 IN. 94.0 LB. 19.124 IN ID
16 IN. 65.0 LB. 15.250 IN ID
13 3/8 IN. 61.0 LB. 12.515 IN ID
11 3/4 IN. 47.0 LB. 11.000 IN ID
10 3/4 IN. 51.0 LB. 9.850 IN ID
9 5/8 IN. 47.0 LB. 8.681 IN ID
8 5/8 IN. 36.0 LB. 7.825 IN ID
7 IN. 23.0 LB. 6.366 IN ID
6 5/8 IN. 24.0 LB. 5.921 IN ID
5 1/2 IN. 17.0 LB. 4.892 IN ID
4 1/2 IN. 11.6 LB. 4.000 IN ID
3 1/2 IN. 9.3 LB. 2.992 IN ID
2 7/8 IN. 6.5 LB. 2.441 IN ID
2 3/8 IN. 4.7 LB. 1.995 IN ID

WITH 2⅛-IN. PL TOOL IN FLOW STREAM

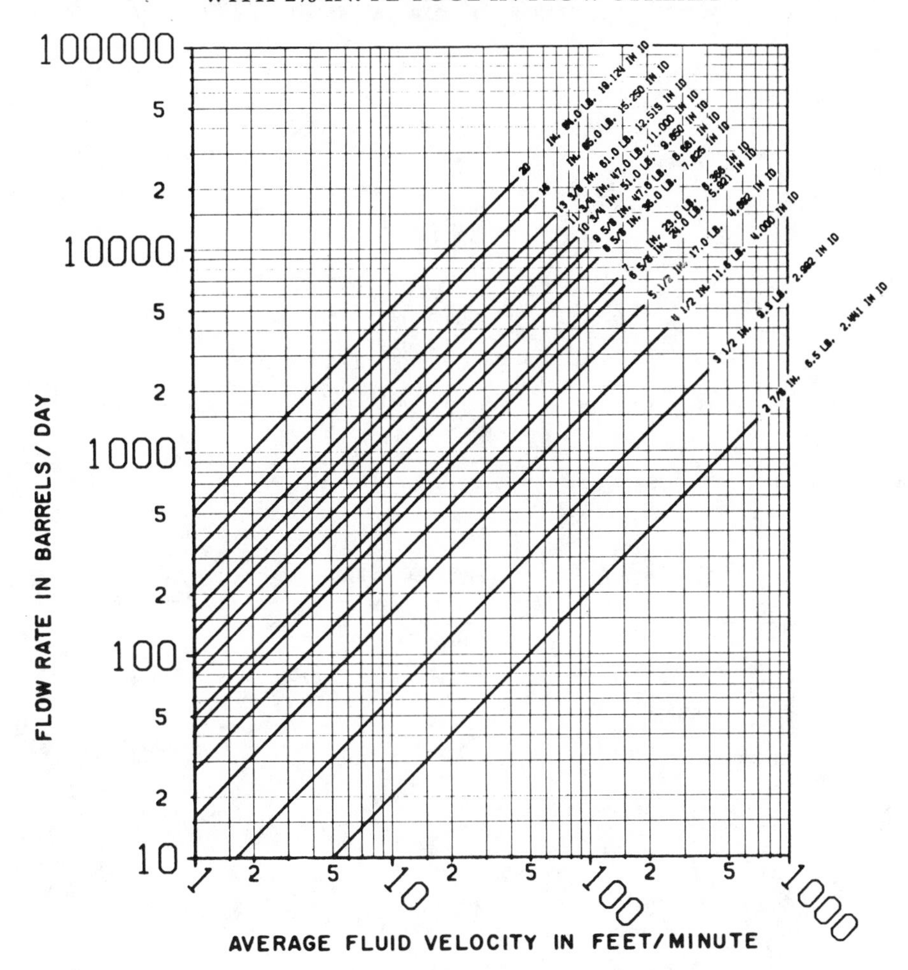

APPENDIX E: QUICK GUIDE TO BIPHASIC FLOW INTERPRETATION

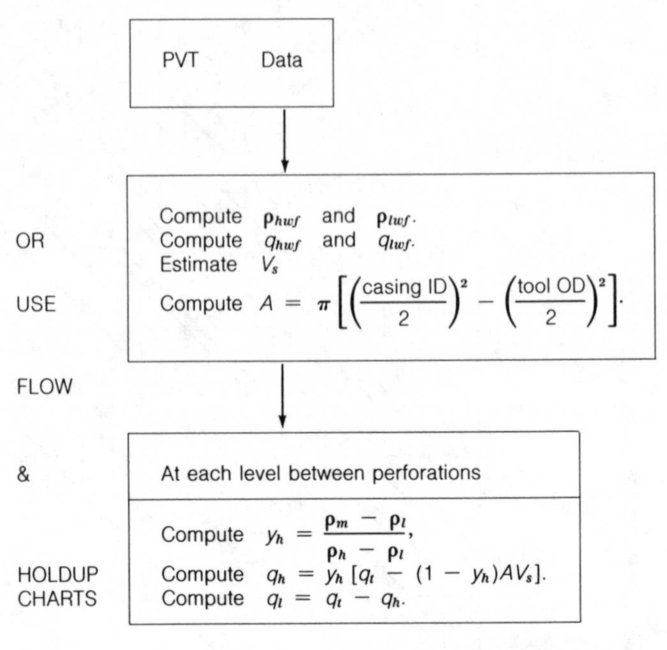

Where: h is for heavy phase.
l is for light phase.
m is for mixture.

APPENDIX F: HOLDUP AND FLOW RATE CHARTS

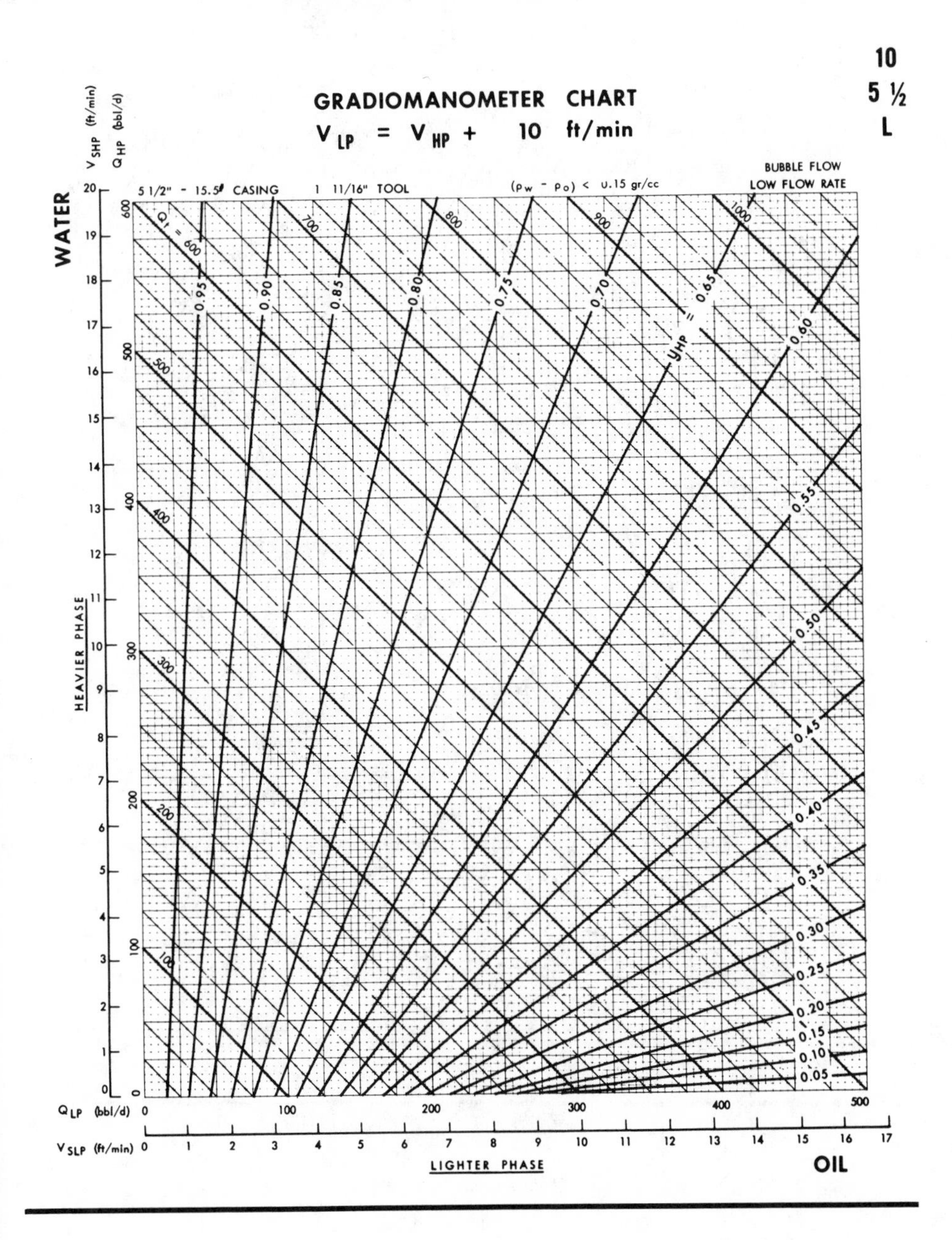

Courtesy Schlumberger Well Services, from "Production Log Interpretation" (1970)8C-11–8C-26.

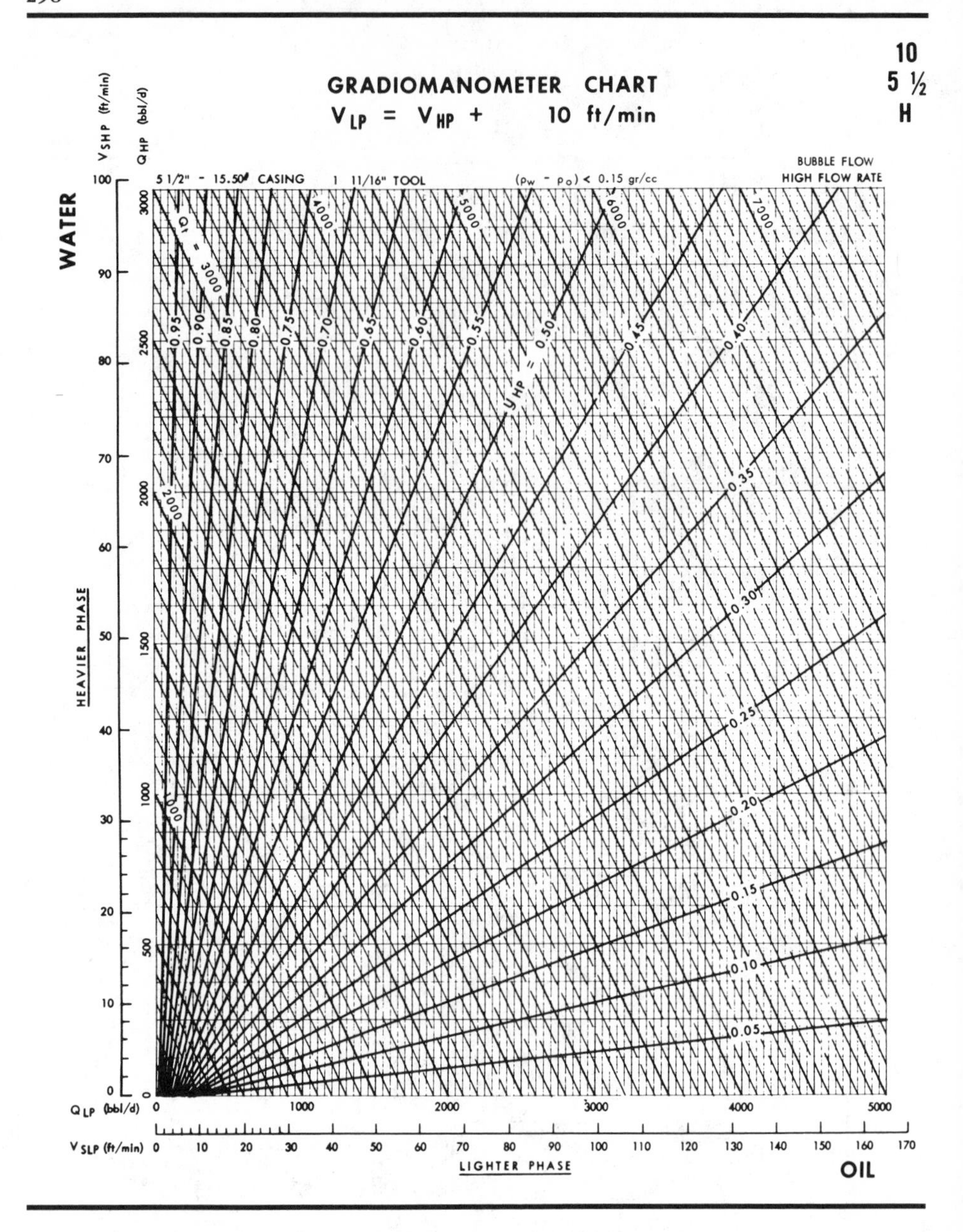

10
5 ½
H
GRADIOMANOMETER CHART
V_{LP} = V_{HP} + 10 ft/min
5 1/2" - 15.50# CASING
1 11/16" TOOL
(ρ_w - ρ_o) < 0.15 gr/cc
BUBBLE FLOW
HIGH FLOW RATE
V_{SHP} (ft/min)
Q_{HP} (bbl/d)
WATER
HEAVIER PHASE
Q_t = 3000
4000
5000
6000
7000
2000
1000
y_{HP} = 0.50
0.95
0.90
0.85
0.80
0.75
0.70
0.65
0.60
0.55
0.45
0.40
0.35
0.30
0.25
0.20
0.15
0.10
0.05
Q_{LP} (bbl/d)
V_{SLP} (ft/min)
LIGHTER PHASE
OIL

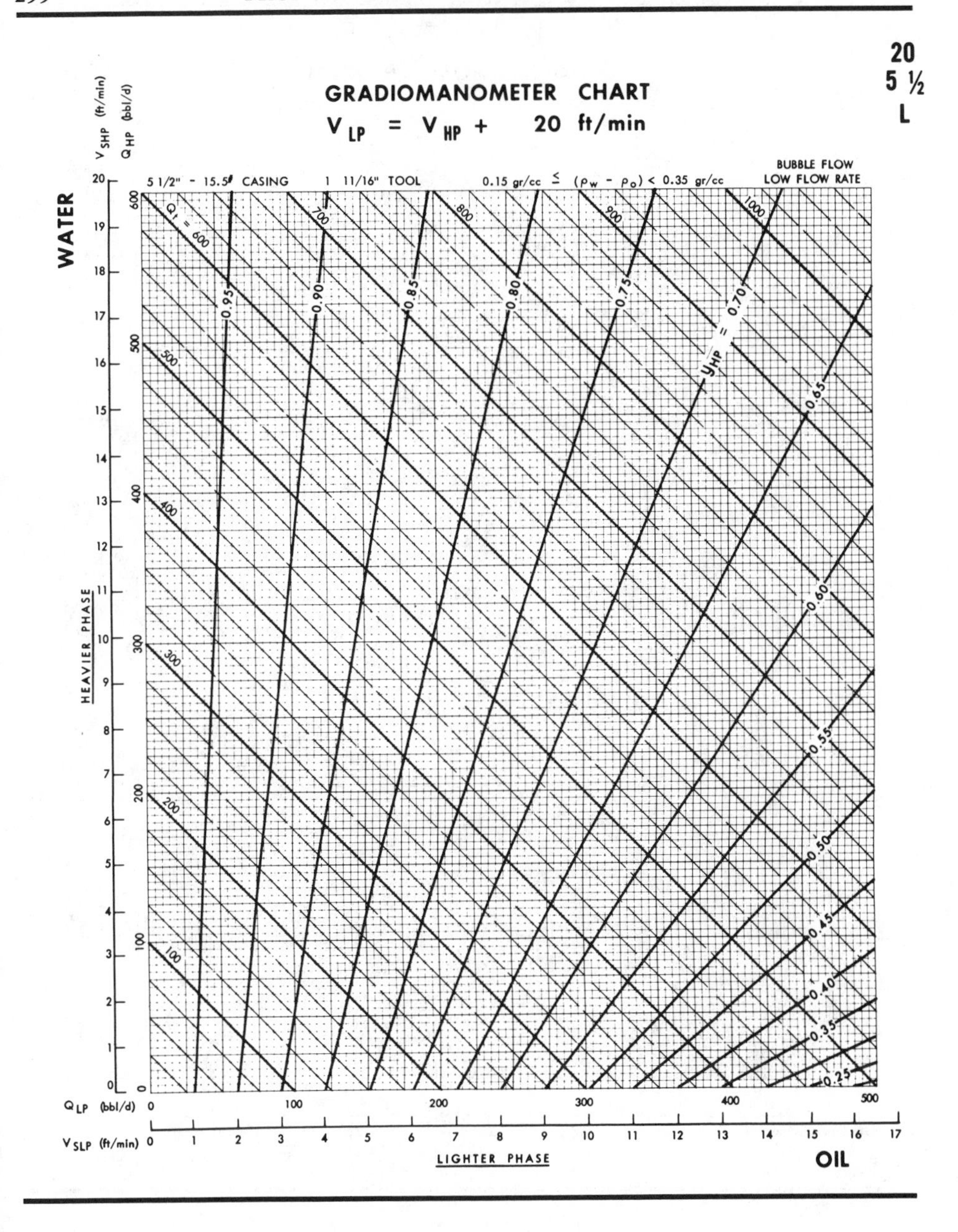
20
5 ½
L
GRADIOMANOMETER CHART
V LP = V HP + 20 ft/min
5 1/2" - 15.5# CASING
1 11/16" TOOL
0.15 gr/cc ≤ (ρw - ρo) < 0.35 gr/cc
BUBBLE FLOW
LOW FLOW RATE
WATER
V SHP (ft/min)
Q HP (bbl/d)
HEAVIER PHASE
Qt = 600
yHP = 0.70
Q LP (bbl/d)
V SLP (ft/min)
LIGHTER PHASE
OIL

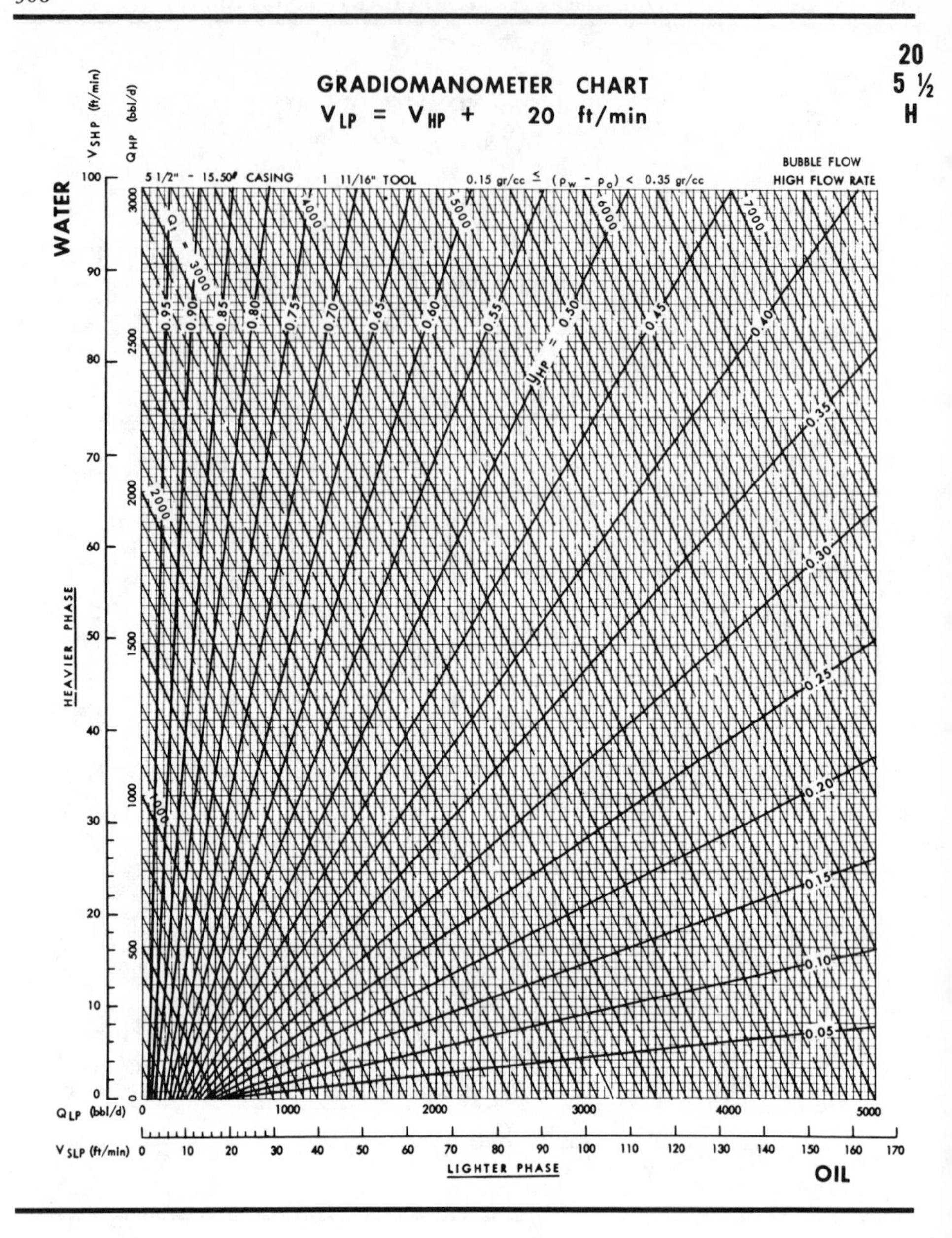
20
5 ½
H
GRADIOMANOMETER CHART
VLP = VHP + 20 ft/min
BUBBLE FLOW
HIGH FLOW RATE
5 1/2" - 15.50# CASING
1 11/16" TOOL
0.15 gr/cc ≤ (ρw - ρo) < 0.35 gr/cc
WATER
HEAVIER PHASE
VSHP (ft/min)
QHP (bbl/d)
Qt = 3000
1000
2000
4000
5000
6000
7000
yHP = 0.50
0.95
0.90
0.85
0.80
0.75
0.70
0.65
0.60
0.55
0.45
0.40
0.35
0.30
0.25
0.20
0.15
0.10
0.05
QLP (bbl/d)
VSLP (ft/min)
LIGHTER PHASE
OIL

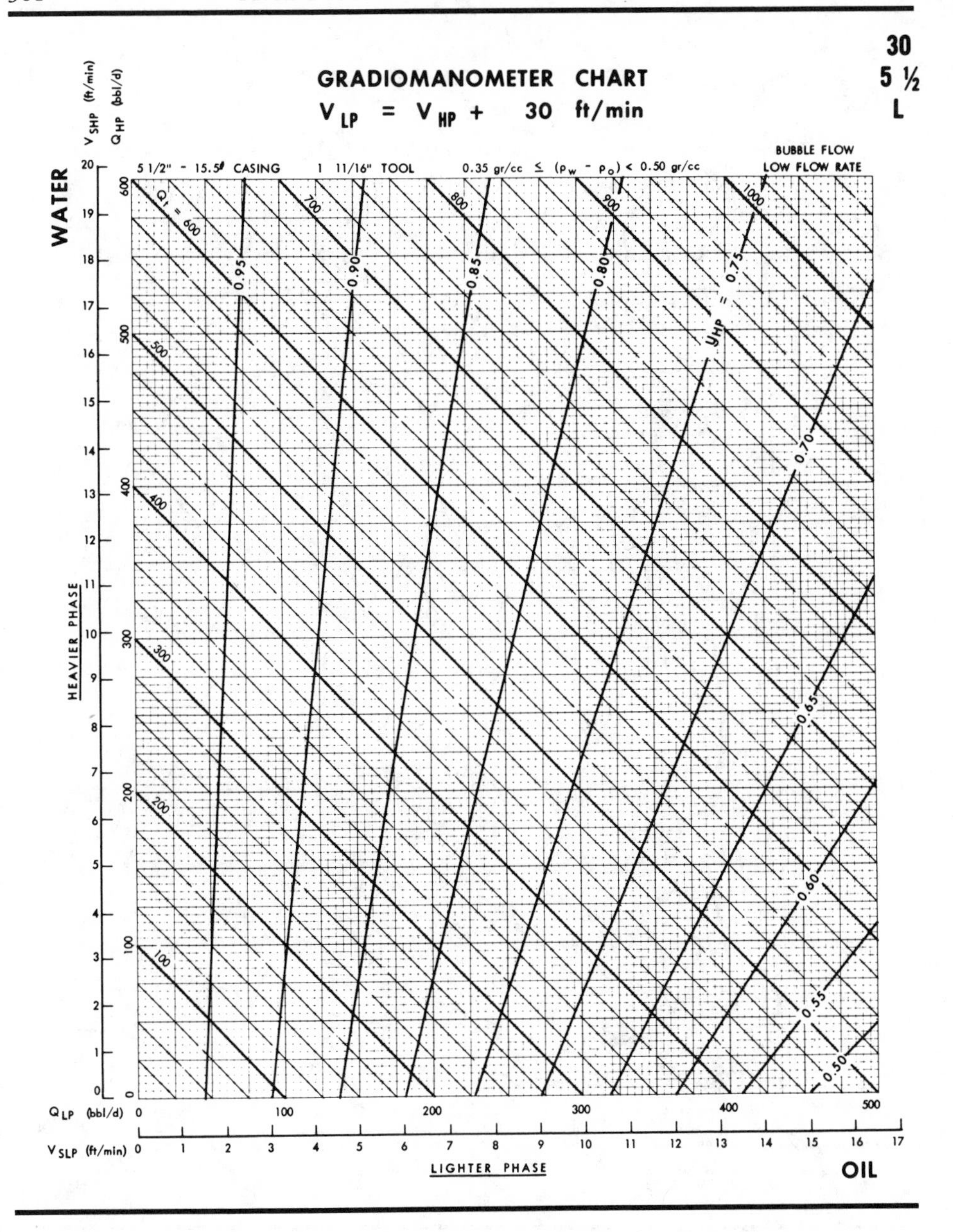
30
5 ½
L
GRADIOMANOMETER CHART
$V_{LP} = V_{HP} + 30$ ft/min
5 1/2" - 15.5# CASING
1 11/16" TOOL
0.35 gr/cc ≤ ($\rho_w - \rho_o$) < 0.50 gr/cc
BUBBLE FLOW
LOW FLOW RATE
WATER
HEAVIER PHASE
V_{SHP} (ft/min)
Q_{HP} (bbl/d)
Q_t = 600
y_{HP} = 0.75
Q_{LP} (bbl/d)
V_{SLP} (ft/min)
LIGHTER PHASE
OIL

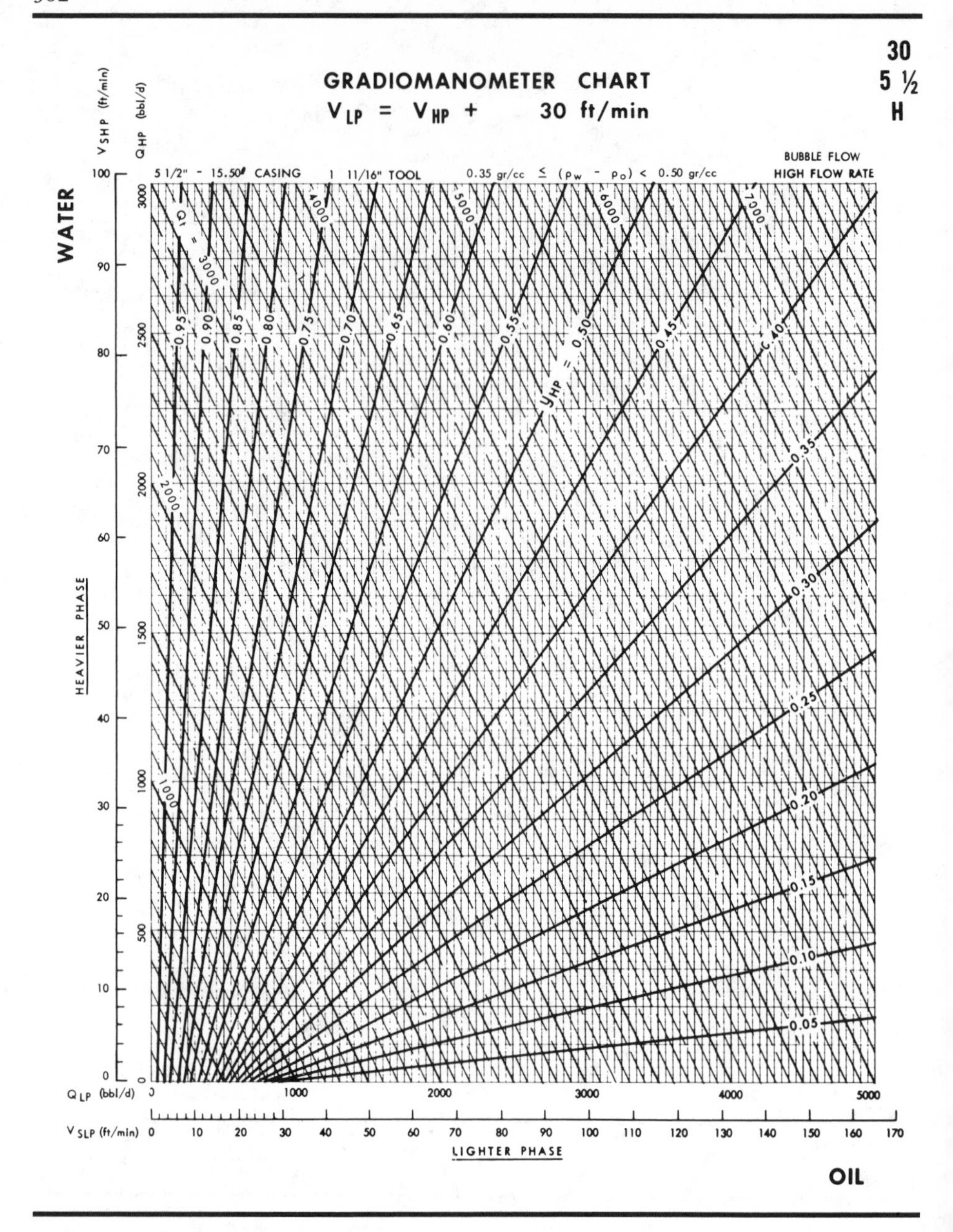

30
5 ½
H
GRADIOMANOMETER CHART
V LP = V HP + 30 ft/min
BUBBLE FLOW
HIGH FLOW RATE
5 1/2" - 15.50# CASING
1 11/16" TOOL
0.35 gr/cc ≤ (ρw - ρo) < 0.50 gr/cc
WATER
HEAVIER PHASE
V SHP (ft/min)
Q HP (bbl/d)
Qt = 3000
4000
5000
6000
7000
2000
1000
0.95
0.90
0.85
0.80
0.75
0.70
0.65
0.60
0.55
y HP = 0.50
0.45
0.40
0.35
0.30
0.25
0.20
0.15
0.10
0.05
Q LP (bbl/d)
V SLP (ft/min)
LIGHTER PHASE
OIL

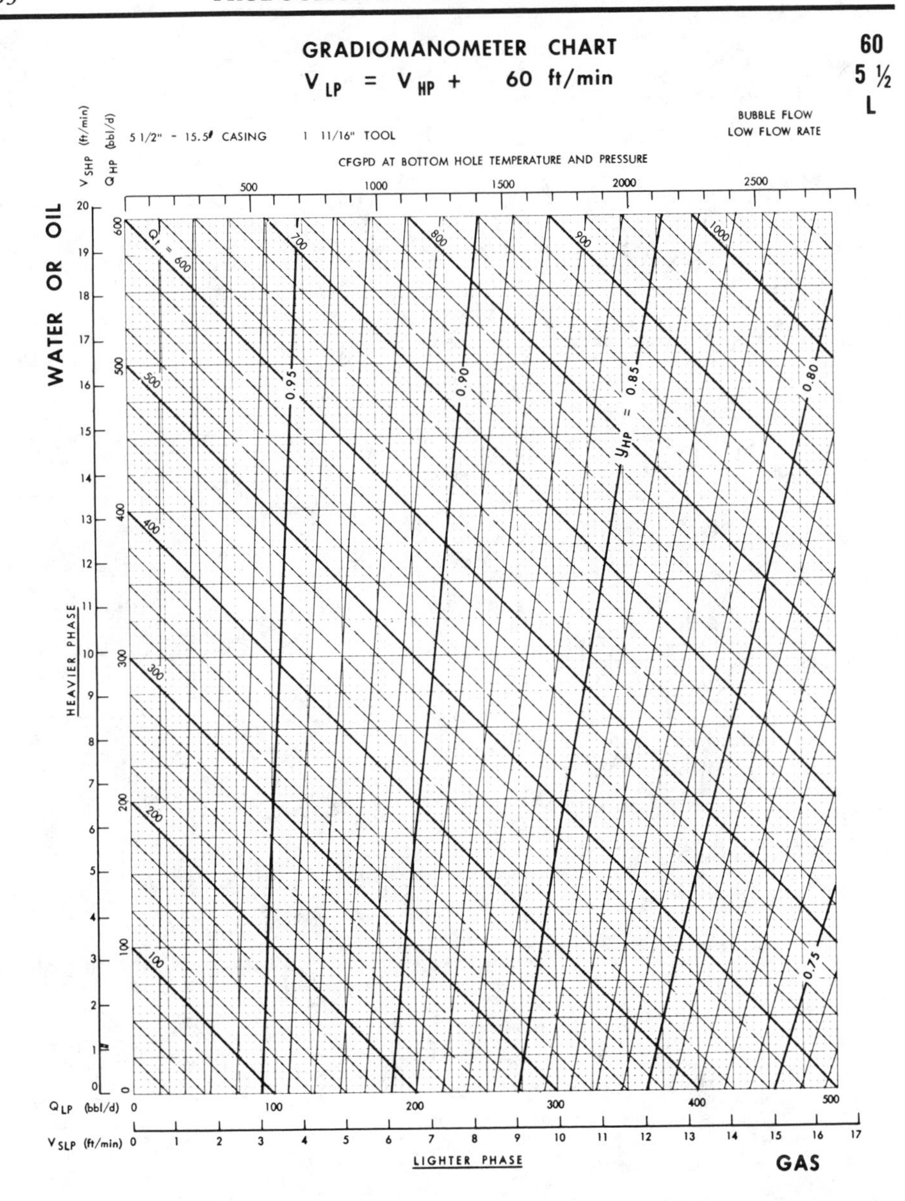
GRADIOMANOMETER CHART
V LP = V HP + 60 ft/min
60
5 ½
L
BUBBLE FLOW
LOW FLOW RATE
5 1/2" - 15.5# CASING
1 11/16" TOOL
CFGPD AT BOTTOM HOLE TEMPERATURE AND PRESSURE
500
1000
1500
2000
2500
V SHP (ft/min)
Q HP (bbl/d)
WATER OR OIL
HEAVIER PHASE
Qt = 600
700
800
900
1000
500
400
300
200
100
0.95
0.90
yHP = 0.85
0.80
0.75
Q LP (bbl/d)
V SLP (ft/min)
LIGHTER PHASE
GAS

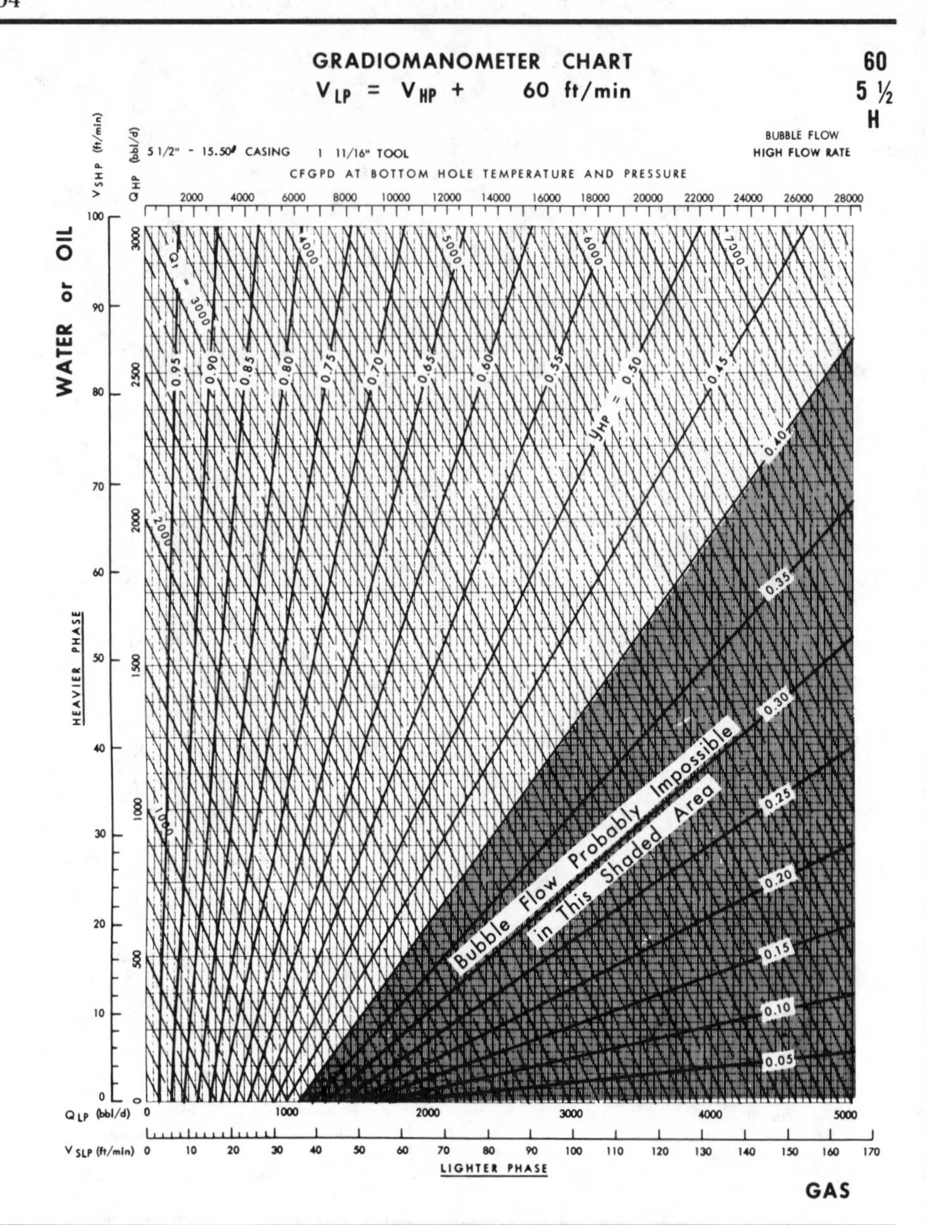
GRADIOMANOMETER CHART
V LP = V HP + 60 ft/min
60
5 ½
H
5 1/2" - 15.50# CASING
1 11/16" TOOL
BUBBLE FLOW
HIGH FLOW RATE
CFGPD AT BOTTOM HOLE TEMPERATURE AND PRESSURE
V SHP (ft/min)
Q HP (bbl/d)
WATER or OIL
HEAVIER PHASE
Q l = 3000
yHP = 0.50
Bubble Flow Probably Impossible
in This Shaded Area
Q LP (bbl/d)
V SLP (ft/min)
LIGHTER PHASE
GAS

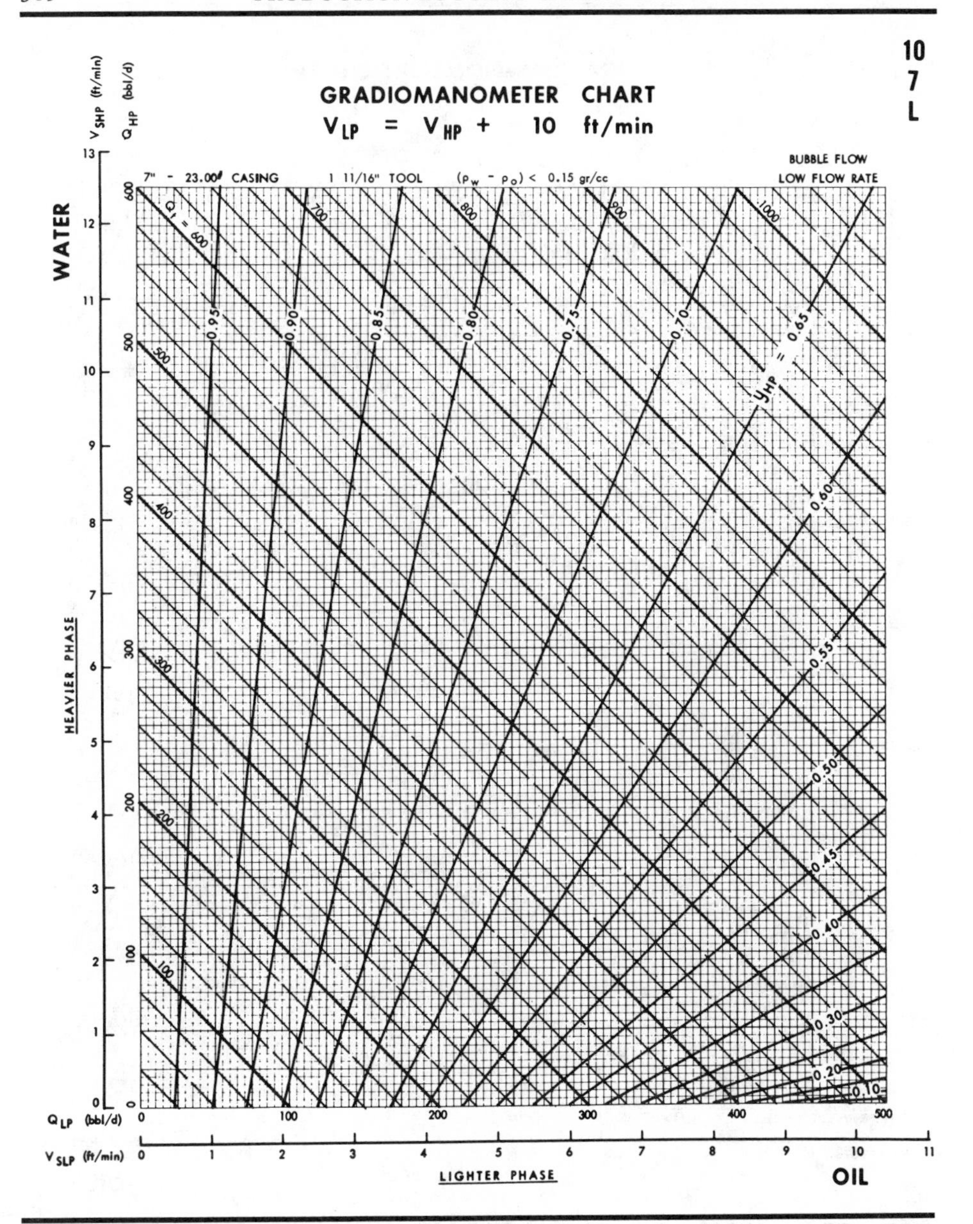
10
7
L
GRADIOMANOMETER CHART
V LP = V HP + 10 ft/min
7" - 23.00# CASING
1 11/16" TOOL
(ρw - ρo) < 0.15 gr/cc
BUBBLE FLOW
LOW FLOW RATE
WATER
HEAVIER PHASE
V SHP (ft/min)
Q HP (bbl/d)
Q LP (bbl/d)
V SLP (ft/min)
LIGHTER PHASE
OIL
Qt = 600
yHP = 0.65

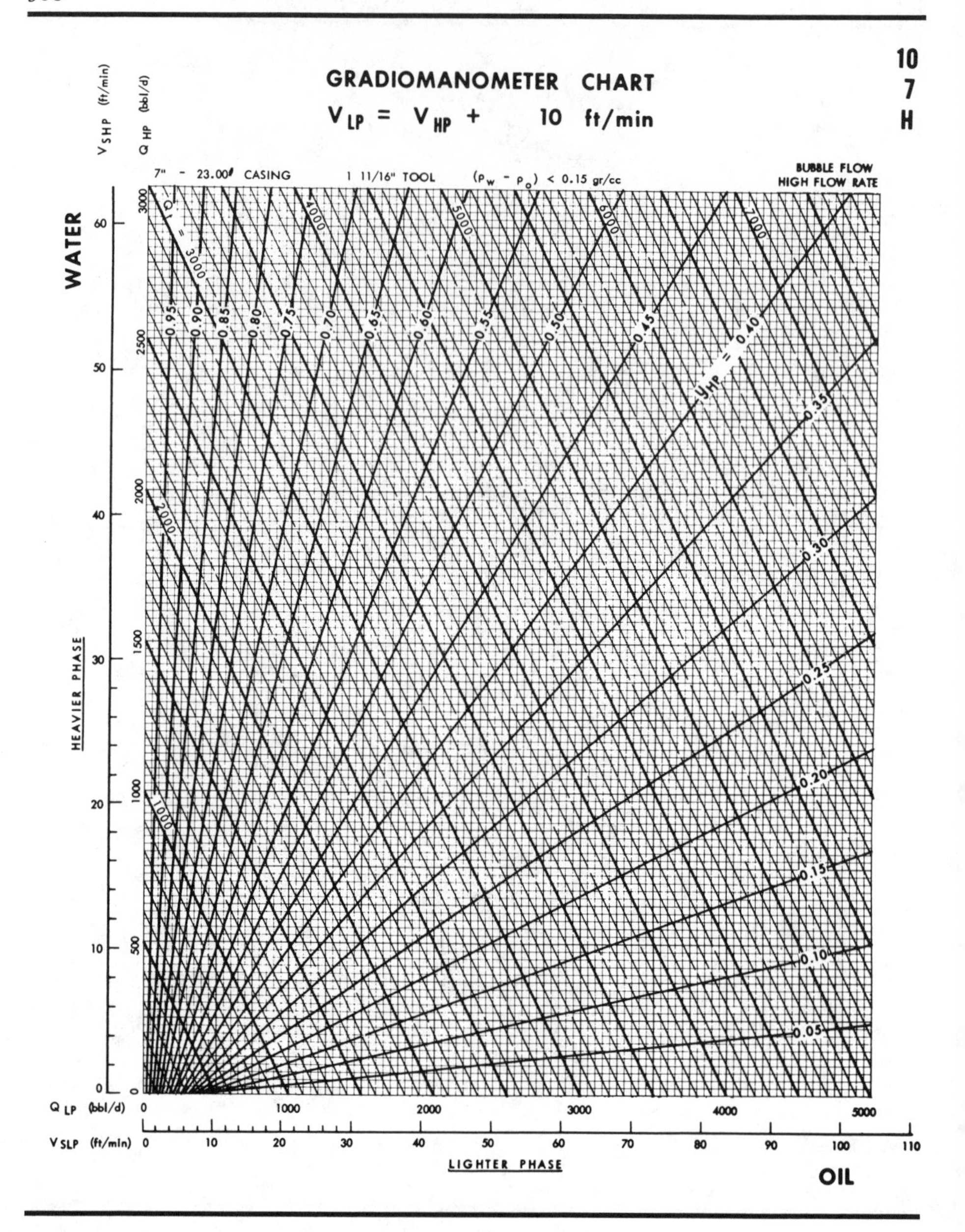

GRADIOMANOMETER CHART
$V_{LP} = V_{HP} + 10$ ft/min
10
7
H
7" - 23.00# CASING
1 11/16" TOOL
$(\rho_w - \rho_o) < 0.15$ gr/cc
BUBBLE FLOW
HIGH FLOW RATE
V_{SHP} (ft/min)
Q HP (bbl/d)
WATER
HEAVIER PHASE
Q_T = 3000
4000
5000
6000
7000
2000
1000
0.95
0.90
0.85
0.80
0.75
0.70
0.65
0.60
0.55
0.50
0.45
y_{HP} = 0.40
0.35
0.30
0.25
0.20
0.15
0.10
0.05
Q LP (bbl/d)
V_{SLP} (ft/min)
LIGHTER PHASE
OIL

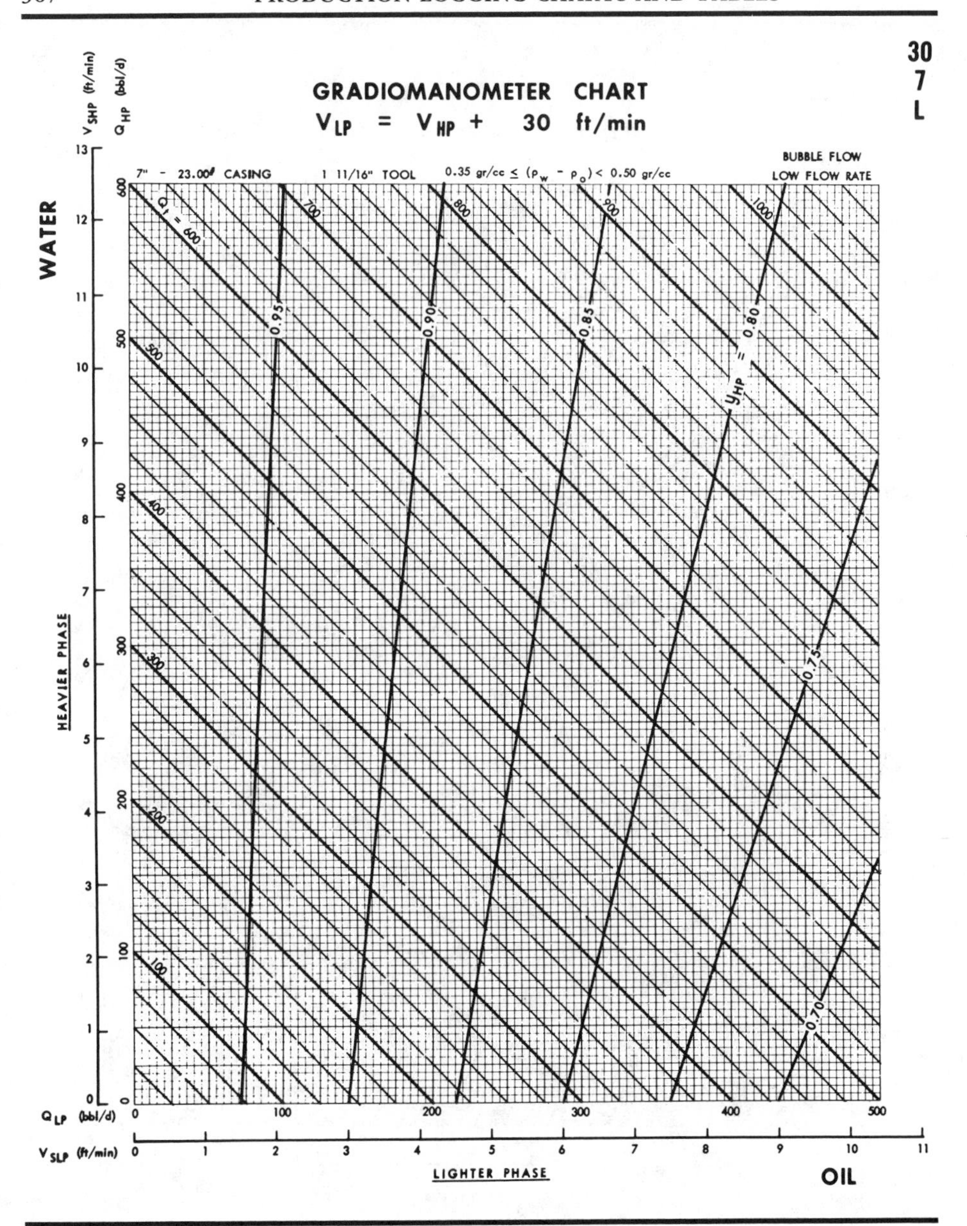
GRADIOMANOMETER CHART
V LP = V HP + 30 ft/min
7" - 23.00# CASING
1 11/16" TOOL
0.35 gr/cc ≤ (ρw - ρo) < 0.50 gr/cc
BUBBLE FLOW
LOW FLOW RATE
WATER
HEAVIER PHASE
V SHP (ft/min)
Q HP (bbl/d)
Qt = 600
700
800
900
1000
500
400
300
200
100
0.95
0.90
0.85
yHP = 0.80
0.75
0.70
Q LP (bbl/d)
V SLP (ft/min)
LIGHTER PHASE
OIL

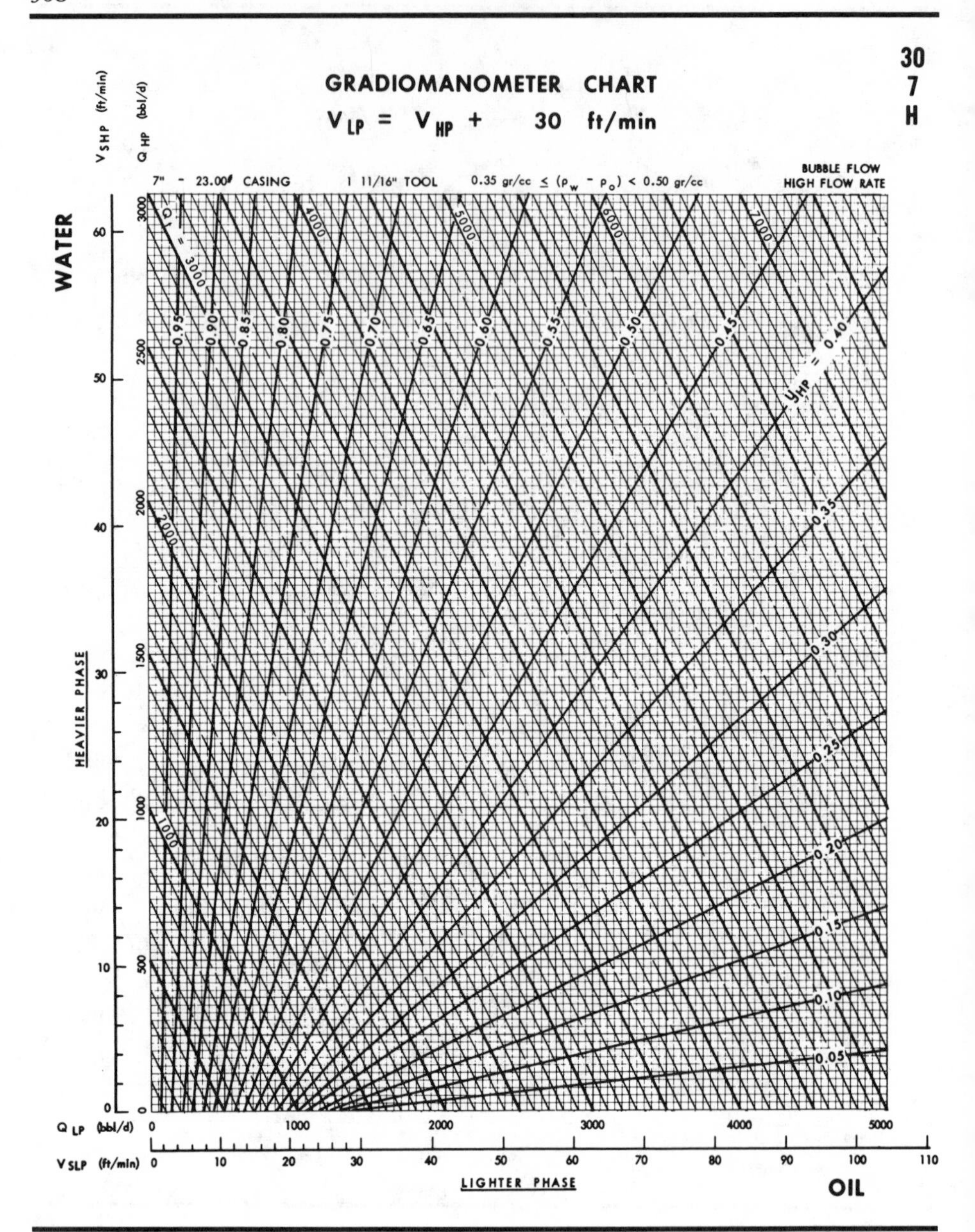
30
7
H
GRADIOMANOMETER CHART
V LP = V HP + 30 ft/min
V SHP (ft/min)
Q HP (bbl/d)
7" - 23.00# CASING
1 11/16" TOOL
0.35 gr/cc ≤ (ρw - ρo) < 0.50 gr/cc
BUBBLE FLOW
HIGH FLOW RATE
WATER
HEAVIER PHASE
Qt = 3000
4000
5000
6000
7000
2000
1000
0.95
0.90
0.85
0.80
0.75
0.70
0.65
0.60
0.55
0.50
0.45
yHP = 0.40
0.35
0.30
0.25
0.20
0.15
0.10
0.05
Q LP (bbl/d)
V SLP (ft/min)
LIGHTER PHASE
OIL

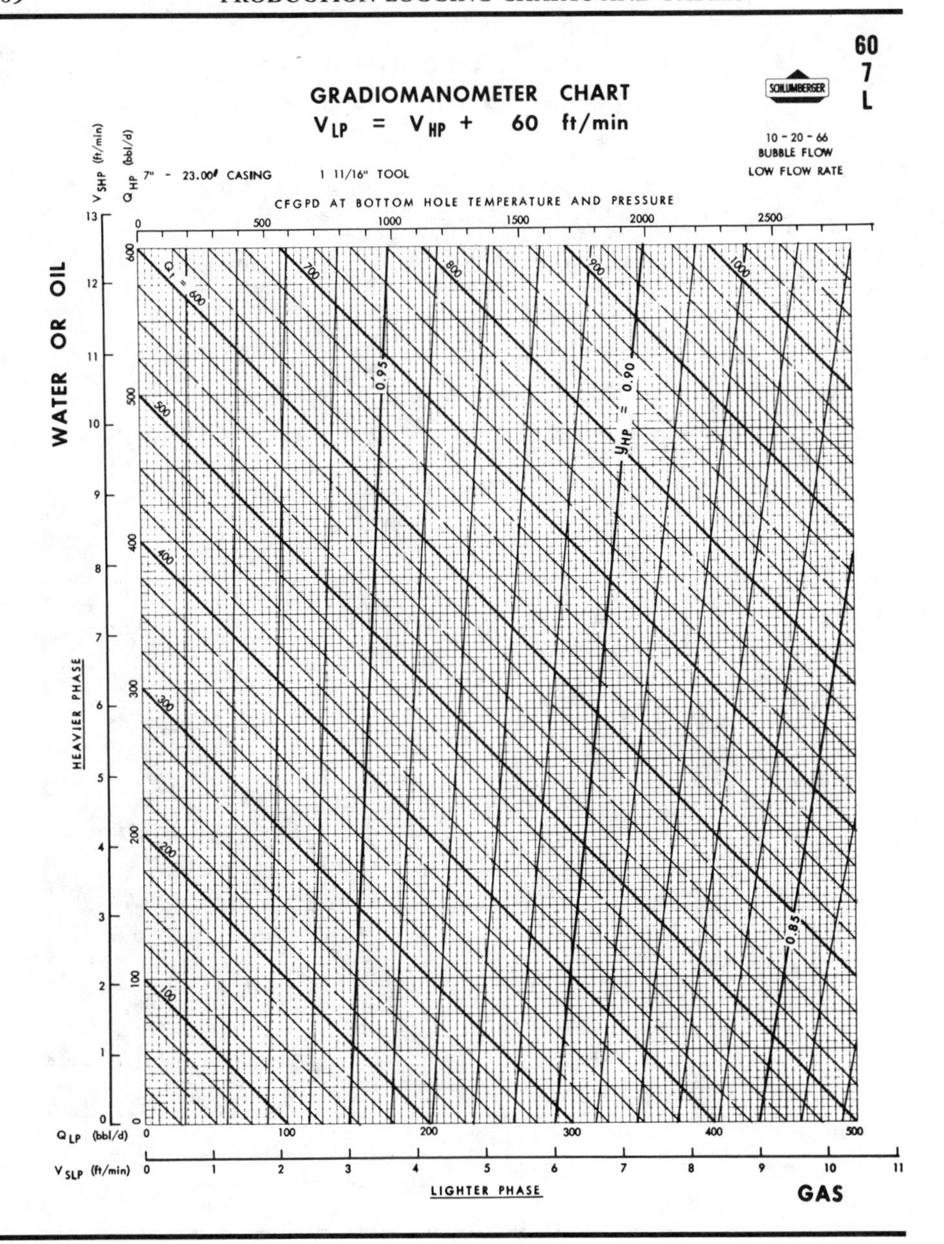
60
7
L
GRADIOMANOMETER CHART
V_{LP} = V_{HP} + 60 ft/min
SCHLUMBERGER
10 - 20 - 66
BUBBLE FLOW
LOW FLOW RATE
7" - 23.00# CASING
1 11/16" TOOL
CFGPD AT BOTTOM HOLE TEMPERATURE AND PRESSURE
0 500 1000 1500 2000 2500
V_{SHP} (ft/min)
Q_{HP} (bbl/d)
WATER OR OIL
HEAVIER PHASE
Q_t = 600
700
800
900
1000
500
400
300
200
100
0.95
y_{HP} = 0.90
0.85
Q_{LP} (bbl/d) 0 100 200 300 400 500
V_{SLP} (ft/min) 0 1 2 3 4 5 6 7 8 9 10 11
LIGHTER PHASE
GAS

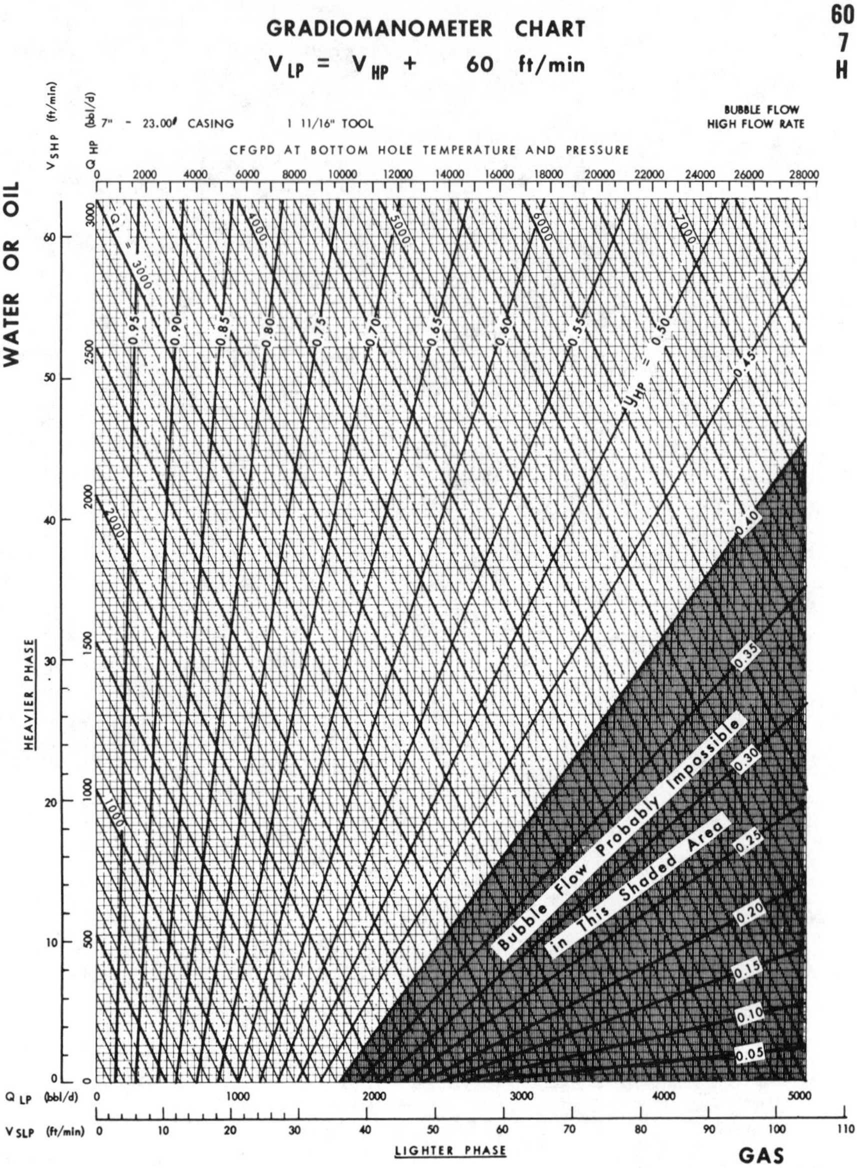

60
7
H

GRADIOMANOMETER CHART

$V_{LP} = V_{HP} + 60$ ft/min

7" - 23.00# CASING

1 11/16" TOOL

BUBBLE FLOW
HIGH FLOW RATE

CFGPD AT BOTTOM HOLE TEMPERATURE AND PRESSURE

0 2000 4000 6000 8000 10000 12000 14000 16000 18000 20000 22000 24000 26000 28000

WATER OR OIL

HEAVIER PHASE

V_{SHP} (ft/min) 0 10 20 30 40 50 60

Q_{HP} (bbl/d) 0 500 1000 1500 2000 2500 3000

Q_T = 1000 2000 3000 4000 5000 6000 7000

0.95 0.90 0.85 0.80 0.75 0.70 0.65 0.60 0.55

y_{HP} = 0.50

0.45 0.40 0.35 0.30 0.25 0.20 0.15 0.10 0.05

Bubble Flow Probably Impossible
in This Shaded Area

Q_{LP} (bbl/d) 0 1000 2000 3000 4000 5000

V_{SLP} (ft/min) 0 10 20 30 40 50 60 70 80 90 100 110

LIGHTER PHASE

GAS

BIBLIOGRAPHY FOR APPENDIXES

"Fluid Conversions in Production Logging," Schlumberger (1974).
"Production Log Interpretation," Schlumberger (1973).

GENERAL BIBLIOGRAPHY

Dresser Atlas

"Carbon/Oxygen Log" (1981).
"Casing Evaluation Services" (1983).
"Interpretative Methods for Production Well Logs" second edition (1982).
"Production Services" (1981).
"Spectralog" (1981).
"Wireline Services Catalog" (1982).

DIA-LOG

"Borehole Sound Survey."
"General Catalog."

N.L. McCullough

"Noise Logging Service" (1976).

Petroleum Extension Service, The University of Texas at Austin

"Lesson 6—Well Cementing" (1983).

Schlumberger Well Services

"The Essentials of Cement Evaluation" (1976).
"Fluid Conversions in Production Logging" (1974).
"Production Log Interpretation" (1970).
"Production Services" (March 1975).
"Reservoir and Production Fundamentals" (1980).
"Services Catalog" (1970).
"Well Evaluation Developments—Continental Europe" (1982).

INDEX

Activation logs, 173
Albite, 190
Alkali olivine basalt, 169
Allanite, 168, 169
Aluminum, 188
Amplitude measurements, 230, 231
Amplitude-frequency analysis, 129
Andesite, 169
Anhydrite, 190
Anorthite, 190
Apatite, 169
API gamma ray standard, 152
API units, 153
Attenuation measurement, 231
Attenuation rate, 228, 229
Audible spectrum, 129
Audio Log, 100, 129
Autunite, 168

Baltwoodite, 168
Bandpass filters, 131
Basalt, 169
Basket flowmeter, 77
Bastnaesite, 168
BATS, 17, 100
Bauxite, 169
Bentonite, 169, 227
Betafite, 168
BHTV, 274, 280
Biotite, 169
Biphasic flow, 137–138, 169, 296
Bond index, 242
Borax, 190
Borehole audio tracer survey (BATS), 129
Borehole environment, 27
Borehole fluid sampler, 100
Borehole sound survey, 129
Borehole televiewers, 259, 274
Borehole TV, 259
Boron, 188
Bottomhole gas density, 49
Brannerite, 168
Bubble flow, 68, 102
Bubble-point pressure, 51, 53

Cadmium, 188
Calcite, 169, 190
Calcium, 188
Calibration of continuous flowmeter, 83
Caliper logs, 259
Calipers, 6, 22
Capacitance (dielectric), 99
Capacitance (dielectric), tools, 105
Capacitance watercut meter, 105
Capture cross sections, 174, 188–189
Capture units, 175
Carbon, 188
Carbon/oxygen logs, 4, 6, 29, 173
Carbonates, 169
Carnalite, 190
Carnotite, 168
Casing inspection, 259–281
Casing OD weight and thickness, 244
Casing potential profile log, 264–266
Casing-collar location, 6
Casing-collar log, 5, 19
CAT, 21
CBL, 5, 18, 223
CBL, amplitude interpretation, 243
CBL, tool, 230
CBL-VDL, 18
CCL, 5, 19
Cement
  API classifications of, 226
  compressive strength of, 226, 242
  effect of accelerator on, 227
  physical properties of, 227
Cement bond log, 5–6, 18, 29, 223–257
Cement bond logging, 223
  principles of, 226
Cement evaluation tool, 252
Cement top, 113, 117
CET, 5, 18, 230, 252–257
CET presentation, 255
Chalk, 169
Cheralite, 164
Chlorine, 188
Chlorite, 190
Christmas-tree, 28
Cinnabar, 190
Clay minerals, 169
Clay typing, 171
C/O, 4, 10
Combination tools, 80, 108, 110

Completion problems, 1, 91
Compressive strength, 228
Condensate reservoirs, 34
Continuous flowmeter, 12, 29, 75–76, 78–80, 88
Conversion factors, 284
Cycle skipping, 234, 237–239

Delta-t stretching, 233, 234, 236
Densimeter, 29, 99
Densities of NaCl solutions, 41
Density, 227
Departure curves, 207, 215
Depositional environment, 171
Depth of investigation, 215
Deviated holes, 240
Diabase, 169
Diagenetic changes, 171
Diatomaceous earth, 227
Dielectric constant, 105
Differential-temperature surveys, 121–122, 124
Diorite, 169
DNLL, 173, 176, 178
Dolomite, 169, 190–191
Dowle and Cobb method, 113
Dry-gas reservoir, 36
Dual detector neutron lifetime log, 173
Dual spacing thermal decay time log, 173
Dual water method, 216
Dunite, 169

Eddy current principle, 270
Electrical potential logs, 259, 264
Electromagnetic devices, 265
Electromagnetic inspection devices, 259
Electromagnetic thickness tool, 6, 20, 29, 266–267
Electronic casing caliper log, 266
Epidote, 169
Eschynite, 168
ETT, 5, 20, 266–267, 274
Euxenite, 168

Far counts, 184
Fergusonite, 168
Filter response, 132
Fixed gate, 238–239
Floating gate, 238–239
Flow
  laminar, 65
  rate of, 68, 91
  turbulent, 65
Flow profiles, 91
Flow regimes, 65–73
Flowmeters, 4, 6, 75–88, 141
  basket, 75
  correction, 84
  packer, 75
Fluid, properties of, 37
Fluid density, 4, 6
Fluid density log, 104
Fluid density logging, 15
Fluid density tool, 99, 103–104
Fluid identification, 99–110
Fluid sampler, 99, 106
Fluid sampler tool, 106
Fluid velocity, 288
Formation volume factor, 40
Fractures, 159
Free pipe, 239, 248
  in deviated hole, 248
Frequency analysis, 129
Friction gradients, 102
Friction reducer, 227
Froth flow, 68
Full waveform display, 233, 235
Full-bore flowmeter, 29, 70, 75–79, 88

Gabbro, 169
Gamma absorption, 99
Gamma ray, 6, 147
  calibration of, 152
Gamma ray corrections, 156
Gamma ray detector, 147
Gamma ray logs, 4, 8, 147–148
  perturbing effects on, 155
Gamma ray spectra, interpretation of, 159, 161
Gamma ray spectral log, 9, 162
Gamma ray spectrometry, 161
Gamma ray spectroscopy, 159
Gas, solubility in water, 42
Gas channel, 127
Gas density, 51
Gas formation volume factor, 42, 49
Gas injection, 121
Gas production, 120
Gas tracers, 96
Gas viscosity, 50, 52
Gas/liquid flow regimes, 69
Gating systems, 238, 240
Geothermal gradient, 113, 115, 119
Geothite, 190
Glauconite, 169, 190
GR, 4, 8
GR spectra, 6
Gradio, 15
Gradiomanometer, 6, 29, 94–100, 141
Gradiomanometer log, 103
Granite, 170

Granodiorite, 170
Graywackes, 171
GST, 4
Gypsum, 190

Half-wave display, 233, 236
Halite, 190
Hermatite, 190
High-resolution thermometer, 114
Holdup, 65–72, 102, 137–143
and flow rate charts, 140, 297
Horner plot, 108
HRT, 14
Huttonite, 168
Hydraulic seal, 223
Hydrogen, 188
Hydrolog, 100

IGT, 4, 10
Illite, 149, 169
Inelastic gamma log, 4, 6, 10
Iron, 188

Kaolinite, 149, 169
Kermite, 190

Limestone, 169, 191
Limonite, 190
Liquid production, 119
Lithium, 188
Log inject log, 207, 214
Logging speed, 153
Lost-circulation materials, 227
Lost-circulation zone, 113, 117–118
Low-water-loss materials, 227

Magnelog, 20, 266
Magnesium, 188
Magnetic flux leakage, 269
Magnetite, 190
Manganese, 188
Manganite, 190
Manometer, 99–100, 106
Marine black shales, 171
Marine deposits, 171
Mass flow rate, 119
Mercury, 188
Mica, 169
Mica/biotite, 190
Microannulus/channeling, 250, 252
Microcline, 169
Mineral identification, 162
Mist flow, 68
Monazite, 168, 169
Monitoring reservoir performance, 173
Montmorillonite, 149, 169
Multicomponent hydrocarbon system, 33
Multifingered calipers, 29
Muscovite, 169

Natural fracture systems, 171
Natural gamma rays, origin of, 147
Natural gamma spectra, interpretation of, 159, 161
Natural gamma spectra logs, 4
Natural gamma spectroscopy, 147
Natural gases, composition of, 36, 38
Natural oils, composition of, 36
Natural radioactive deposits, 96
Natural-gas deviation factor, 48
Naturally occurring radionuclides, 167
Near counts, 184
Near/far count-rate display, 187
Neutron absorbers, 174
Neutron generator, 173
NGT, 4, 9
Nitrogen, 188
NLL, 4, 7
Noise amplitude, 129
Noise logs, 4–6, 17, 29, 129–232
Noise spectrum, 130–131
Nuclear flolog, 91

Oil density, 57, 58
Oil formation volume factor, 54
Oil reservoirs, 33
Oil shales, 170
Oil viscosity, 57, 59
Oilwell cementing, 223
Orthoclase, 169, 190
Oxidizing environment, 171
Oxygen, 188

Packer/diverter flowmeter, 13
Packer flowmeter, 29, 75–76, 88
PAL, 5, 21, 268
Partial cementation, 242
PAT, 271, 274
Peak noise, 131
Periodite, 170
Permafrost, 113
Petroleum reservoirs, 2
Phosphates, 170, 171
Phosphorus, 188
Phyrite, 190
Pilbarite, 168
Pipe analysis log, 6, 21, 268, 271–275
Plagioclase, 169

Planning, 25
Plateau basalt, 169
Potassium, 147, 159, 169, 188
  distribution of, in rocks, 169
Pozzolan, 227
Pressure control equipment, 25
Primary cementing, 224
Production combination tool, 82
Production log, choice of, 28–29
Production logging charts, 283
Production logging tools, 3
Pseudo-critical natural-gas parameters, 48
Pseudo-critical pressure, 47
Pseudo-critical temperature, 47
Pseudo-reduced pressure, 47
Pseudo-reduced temperature, 47
Pulsed neutron log, 4, 6, 29, 173
Pulsed neutron logging, 173, 221
Pulsed neutron tool, 174
Pyrolusite, 190

Quartz, 170, 190
Quartz-crystal pressure gauge, 100, 108
Quartzite, 170

Radial differential temperature logs, 121
Radial differential thermometer, 29
Radial differential-temperature tool, 125
Radioactive elements, minerals and rocks, 166
Radioactive salts, deposit of, 91
Radioactive trace materials, choice of, 95
Radioactive tracer logs, 4, 6, 16, 29, 75–76, 91
Ratio curve, 184
RDT, 14
RDT tool, 126
Reservoir fluid properties, 31–63
Reservoir monitoring, 204
Reservoir oils, 39
Reservoir performance, 1
  monitoring of, 173
Resonators (vibrators), 99, 105
Retarder, 227
Reynolds number, 66
Rhyolite, 170
Riser requirements, 25

Samarkite, 168
Sand, 191, 227
Sandstones, 170
Saturated oil, 55
Schist, 170
Scintillation counter, 152
Scintillation detectors, 149
Shale, 149, 170, 190
Shale content, estimating, 155
Shut-in temperature survey, 122, 123
Sibilation, 17
Siderite, 190
Sigma, 175, 184
Sigma curve, 184
Sigma gas, 199
Sigma oil, 197
Sigma water, 196
Sigma-ratio crossplot, 200
Silica, 170
Silicon, 188
Single-component hydrocarbon system, 32
Slip velocity, 69–70, 138–139
Slug flow, 68
Sodium, 188
Sodium chloride, 227
Solubility of gas in water, 43
Sonan log, 16, 100, 129
Source rock potential, 171
Spectral gamma ray log, 9, 162
Spectralog, 9
Sphene, 169
Spinner response, 82
Split detector, 261
Stage cementing, 225
Standard pressures and temperatures, 61
Static formation temperature, 116
Stratigraphic correlations, 171
Sulfur, 188
Superficial velocity, 65
Surface temperature, 113
Syenite, 170
Sylvite, 190

Tau, 175, 184
TDT, 4, 7
TDT-K, 173, 175, 184
TDT-K gating system, 176
TDT-K log presentation, 185
TDT-M, 173, 177–178
Temperature, 4, 6
Temperature log, 29
Temperature logging, 14, 113–128
Temperature profiles, 118
  gas-injection, 122
  gas-production, 120
  liquid-production, 119
  water-injection, 121, 123
Temperature sensors, 113
Thermal conductivity, 119
Thermal multigate decay log, 173

Thermal neutron, 174
Thermal neutron decay curves, 175
Thermometers, 75
Thickening time, 227
Thocholite, 168
Tholeiites, 169
Thorianite, 168
Thorite, 168
Thorium, 159
  distribution in rocks, 169
  group, 147
  series, 167
Thorium-bearing minerals, 168
Thorium/potassium crossplot, 163
Thorium/potassium ratios, 163
Thorogummite, 168
Through-tubing caliper, 86
Time constants, 153
Time-lapse logging, 213
Time-lapse technique, 204
Timed-run analysis, 94
TMD, 173, 179–180
Tracer ejector tool, 91, 93
Tracer logs, 91–98
Tracers
  gas, 96
  oil-soluble, 96
  water-soluble, 96
Travel-time measurement, 232
Treatment effectiveness, 2, 91
Tuff, 170
Tyuyamunite, 168

Undersaturated oil, 55
Uranite, 168
Uranium, 159
  distribution in rocks, 169
Uranium group, 147
Uranium minerals, 168

Variable density display, 223, 228, 232
VDL, 223, 228, 232
Velocity profiles, 66
Velocity shot, 91
Velocity shot log, 94
Vertilog, 21, 268
Viscosity, 227
Vugs, 171

Water holdup, 71
Water injection, 120
Water viscosity, 42
Watercutmeter, 100
Wave train, 231
Wave train display, 232
Wave train recording, 228
Wave train signatures, 242
Weathered soils, 171
Weeksite, 168
Well-cemented pipe, 249, 251

Xenotime, 169

Yttrocrasite, 168

Zircon, 168, 169